FALCO TINNUNCULUS, Linn.
Thurmfalke.

Altes Weibchen.   Altes Männchen.

The Common Kestrel

The Common Kestrel

Richard Sale

SNOWFINCH PUBLISHING

Snowfinch Publishing
The Beeches
Coberley Road
Coberley
Gloucestershire GL53 9QY
UK

© in the text and figures: Richard Sale, 2020.
© in the Chapter head drawings: Kerry Jane, 2020.
© in all unattributed photographs: Richard Sale, Graham Anderson and Keith Burgoyne.

ISBN 978-0-9571732-4-8

All rights reserved. No parts of this publication may be reproduced, stored in a retrieval system, or transmitted in any form or by any means, electronic, mechanical, photocopying, recording or otherwise, without the prior written permission of the Publisher.

The left-hand front and rear endpapers are from a book on the birds of Sweden, with text by Professor Einar Lönnberg and drawings by Marcus and Wilhelm von Wright, hand-coloured by the same artists, published in 1830.

The right-hand from endpaper is of an engraving by Georg Michael Kirn (1810-1872) an artist/engraver working in the Stuttgart area. Kirn produced engravings for several books, this image appearing in *Das Buch der Welt*, first published in 1853 by Hoffmannsche Verlag, Stuttgart.

The right-hand rear endpaper is of an engraving by Otto von Riesenthal for his folio of images of Birds of Prey (raptors and owls) sold in a slip case entitled *Die Raubvögel Deutschlands und des angrenzenden Mittel-Europas* published by Cassel Th. Fischer in 1876.

Acknowledgements

This book would not have been possible without the assistance of many people – for their knowledge and their willingness to share it, for their photographs, and for their help in the field. The limitation of space means I cannot name them all, but I thank each one of them.

Over many years my work on Kestrels in Scotland has depended on the friendship and field-craft of Graham Anderson and Keith Burgoyne, and I thank them for their continuing help and for allowing me to use data they have collected on the falcons and for the use of their photographs.

Torsten Prohl has provided many photos for use in the book, a gift for which mere thanks does not do justice to his generosity and friendship. Gordon Riddle has been equally kind in offering photos and information. I also thank Dave Anderson, Dave Hetherington, Gordon Kirk, Ant Messenger, Mike Price, Mark Rafferty and Ralf Wassmann for the use of photographs, and Stefan Kupko, Sonja Kübler and Marco Lodder for both information and photographs.

In England my thanks go to Philippa Page, and Ann and Scott Parry for allowing me to use the nestboxes on their land for photography, and especially for their kindness in allowing me to install additional perches to aid the work. As my own attempts at building work invariably end in failure and minor injury, I also thank Chris Perry, Barry Guy and Dave Hetherington for actually carrying out the work.

Chapter 8 is devoted to an extensive study over several years of Kestrels breeding in a nestbox in a barn in Hampshire. I am heavily indebted to Simon Newberry, the farm manager, for not only agreeing to the study, but for his continuing enthusiasm for the project. The cameras and recording equipment were installed by Adam Cook and Nathan Sale, to both of whom I offer my thanks. Particular thanks go to Nathan Sale who was involved throughout the project and was always willing to offer his time and expertise to solve problems and make modifications. I also thank Lydia Newberry, not only for her enthusiasm for the project but also for training *Kai*, the male Kestrel who was used to provide flight data. Particular thanks are also due to Judith Jeffery, Lidar and Meteorological Instruments Manager, Chilbolton Observatory for providing the meteorological data which underpins the study for 2017-2020, and to Yvonne Jeffrey of the Middle Wallop Meteorological Station for data for 2017.

I am also indebted to Dave Hughes for allowing me to fly his Kestrel *Kevin* to provide flight data on 'hovering'. Both *Kevin* and *Kai* carried Inertial Measuring Units (IMUs) to provide dynamic data. I offer sincere thanks for assistance with the construction of these units to Waldo Cervantes, Natan Moran and, most particularly, Seb Madgwick. I also thank Adam Cook and Nathan Sale for operating the high-speed cameras which filmed *Kevin*.

I also thank the library services of Oxford University, and in particular Sophie Wilcox of the Alexander Library of Ornithology, for help and assistance, especially with obtaining some of the more obscure reference material.

Finally, I thank my wife for her willingness to put up with my absences, and my bad humours when the weather, other circumstances and most especially the 2020 Covid-19 crisis threatened the project. This book would not have been possible without her unfailing support and enthusiasm.

A female Kestrel brings a mouse to her brood in a nest box in southern Scotland.

Contents

Introduction

1 The Falcons	10
2 The Common or Eurasian Kestrel	24
3 Diet	54
4 Hunting *Flight Characteristics, Techniques and Strategies*	90
5 Food Consumption and Energy Balance	150
6 Breeding Part 1 *Pair Formation to Nest Sites*	166
7 Breeding Part 2 *Eggs to Fledglings*	206
8 Four Seasons in a Barn	268
9 Movements and Winter Grounds	320
10 Friends and Foes	330
11 Population *Survival and Population Numbers*	340
References	374
Index	390

Introduction

It is not so many years since any reasonable journey along the UK's motorway system was enlivened by the sight of a bird hovering above the verges which lined the route. For many young people that vision of a bird apparently hanging almost motionless was the well-spring for a life of interest in all things wild, birds in particular. The bird was, of course, the Common Kestrel.

Over the last few years the sight of a Kestrel above a motorway verge has been become much less common. There are many reasons for the decline – changes in the way land is managed and farmed in Britain, the increase in population of other raptors, particularly Buzzards and Goshawks to name but two – the net effect of which has been a steep drop in the Kestrel population. The last chapter of this book explores the reasons in more detail. Ironically, the publication of this book was delayed by the Covid-19 crisis which engulfed Britain and the rest of the World, in 2020. During the crisis there were numerous media reports noting that the absence of people in the countryside, a product of the lockdown rules, would be beneficial for birds as their breeding season would be subjected to much less disturbance than usual. But at the same time there were other reports that the slaughter of birds of prey by illegal shooting and poisoning had increased, primarily, but not exclusively, over moorland controlled for grouse shooting. Those with an interest in destroying raptors were taking advantage of the lack of prying eyes and a police force dealing with more important issues. While most of the recovered dead birds were those associated with moorland, history records that Kestrels straying into upland areas are also targeted. As is usually the case, when it comes to population numbers, humans hold the key.

My own interest in birds began with my father who was an ardent birdwatcher. Family holidays were not to the seaside, but to inland areas, moors, forests and so on, and were planned not for the height of summer, but for the breeding season. My father taught me to watch birds – to really watch them, not just to learn their names. And I am grateful for his having done so. He hoped I might study zoology, but physics interested me more. But birds remained my passion and after life as a working physicist I retired early to study them in greater depth, firstly in the Arctic and latterly in the UK where my interest is mainly in the flight of falcons. We are lucky in this country in having four breeding falcons which specialise in different hunting techniques – Peregrines which stoop for, and Merlins which chase, birds; Hobbies which use agile flight to hunt insects and birds; and Kestrels which hunt mammals. These are generalisations, but largely true.

Each technique fascinated me, but the Kestrel's hovering flight was the most intriguing. That, and a continuing interest in the breeding biology of raptors, has led to the writing of this book. The book attempts to draw together what we know about Kestrels and why they are declining in Britain.

This male Kestrel has just delivered prey to a nestbox in southern England.

1   The Falcons

Early studies of the way various bird species were related used fossils and current morphology to piece together an overall family tree. In recent years genetic analysis has been added to the tools scientists may use to investigate ancestry. Studies of molecular phylogenetics across 169 species of bird (Hackett *et al.*, 2008) suggested that the Passeriformes (perching birds, species which include more than half of all avian species) were all related. The work suggested that the Falconiformes had branched from an ancestral line which also included corvids, finches and thrushes among others. But while it might be assumed – indeed for many years it had been assumed – that the falcons and the hawks were related, forming a large order, the Falconiformes (which also includes the vultures, the osprey and the Secretary Bird (*Sagittarius serpentarius*), the latter a seemingly unlikely cousin, but one which has often been referred to as the 'pedestrian eagle'), Hackett and co-workers found the two were distinct, confirming the suggestion of those who had previously noted differences in terms of morphology and behaviour.

More surprisingly, the researchers found that the closest ancestral link of the falcons was to the parrots[1]. This finding caused considerable debate, but was confirmed by the further genetic analysis of Suh *et al.* (2011). Confirmation also came from Pyle (2013) who examined 4500 museum specimens of 65 falcon species and 375 species of parrot to investigate moult sequences and noted similarities in the two families (Fig. 1), similarities which are not shared by other avian groups. The exception was the Kakapo (*Strigops habroptila*) whose moult pattern resembles that of most other birds.

---

[1] The suggestion of an ancestral link had been previously mooted by Ericson *et al.* (2005), but many had been unconvinced by the suggestion, believing it was an artefact of the study.

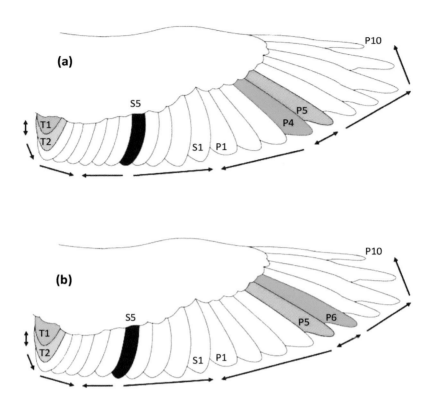

Figure 1. Moult sequences of the primaries (P), secondaries (S) and tertials (T) of falcons (a) and parrots (b). Falcons have 10 primaries and 12 or 13 secondaries (including tertials): 13 are shown here. Parrots have 10 primaries and 12 secondaries/tertials. Moulting starts from nodal feathers which are illustrated in colour. In falcons the nodal primary is P4 or P5, in parrots it is P5 or P6. S5 is the nodal secondary in both. The arrows indicate the direction of moult. Redrawn from Pyle (2013).

The link between falcons and parrots was further strengthened by results from the work of the Avian Phylogenomics Consortium, a group of some 200 scientists from 20 countries. Late in 2014 the Consortium produced around two-dozen papers across the entire spectrum of ornithological scientific journals suggesting an ancient link between falcons, parrots, the seriemas of South America and all Passeriformes. The Consortium suggest these four groups form the 'Australaves' clade, one of two clades which comprise core landbirds, the other being the 'Afroaves' which include, among others, the remaining diurnal birds of prey, owls and woodpeckers. For more information on the Consortium's work see, for instance, Jarvis *et al.* (2014). The evolutionary connection between parrots and falcons, at first glance improbable, received an expert seal of approval with a decision by the American Ornithologists' Union to re-order falcons in the 53rd supplement to their North American bird check-list (Chesser *et al.*, 2012).

In their genetic studies, Wright *et al.* (2008), established that the split between the ancestral parrot/falcon occurred on Gondwanaland in the Cretaceous era, *i.e.* sometime between 65 million and 145 million years ago. However, while there is now a consensus on this evolution, it is worth recalling that the science of genetics is a rapidly-changing one, and further advances may shed doubt, rather than light, on this assumed history.

The family Falconidae of the Order Falconiformes is now most often divided into two sub-families, the Polyborinae, which are largely confined to neotropical South America and include the caracaras and the Forest-falcons, and the Falconinae, which are more widespread and include the falconets and pygmy falcons, and the 'True Falcons' (species of the genus *Falco*)[2].

Further genetic and morphological studies suggest that the Polyborinae probably evolved in neotropical South America. The Falconinae are more widespread, with some family members in the New World, though the majority are Old World species. In their genetic analysis of the Falconinae, Fuchs *et al.* (2015) considered that species divergence probably began about 16 million years ago, but that the rapid divergence of the 'True Falcons' began much later, around the time of transition from the Miocene to the Pliocene, coinciding with the development of, and rapid expansion of, the C4 grasslands (so-called because the plants had evolved a new method of fixing carbon from the atmosphere) primarily in Africa. That the divergence of the 'True Falcons' and the expansion of the C4 grasslands are linked explains why the falcons are grassland/savannah rather than woodland species.[3]

The 'True Falcons' do not build nests, using nests constructed by other birds in which to lay their eggs, making a scrape on the ledge of a cliff or on the ground, or utilising convenient holes in trees or cliffs. They are now considered to comprise four main groups, but with several outliers, some of which may be associated with those groups, though opinions as to which vary.

The first of the four groups are the Hierofalcons, the largest of all falcons, comprising Gyrfalcon (*Falco rusticolus*), Lanner (*Falco biarmicus*), Saker (*Falco cherrug*) and Laggar (*Falco jugger*). The Gyrfalcon has a circumpolar distribution, but is restricted to the Arctic; the Lanner breeds in southern Europe and north Africa; the remaining two species are Asian.

The second group comprises the many sub-species of Peregrine, (*Falco peregrinus*), together with the Barbary Falcon (*Falco pelegrinoides*), which inhabits North Africa, the Middle East and west central Asia, and the Prairie

---

[2] Some experts define three families, adding the Herpetotherinae (usually including the Laughing Falcon (*Herpetotheres cachinnans*) alone, but sometimes also including the *Microstur* Forest-falcons).

[3] In his classic book on falcons, the late Tom Cade (Cade, 1982) noted that the expansion of open grasslands had stimulated not only the development of falcons but of mankind, as early fossil hominids are also found around the transition of Miocene to Pliocene, and wondered whether those early men, gazing at the speeding birds in the sky, were the start of the association between man and falcon.

A Kestrel chick, aged about 28 days, peers down from a corvid nest in southern Scotland.

Female Kestrel (*above*) and male Kestrel (*below*) delivery prey to their broods.

Falcon (*Falco mexicanus*) of North America. Many experts consider the Barbary to be a Peregrine sub-species rather than full species, but there is much more debate over the position of the Prairie Falcon. Once considered to be a Nearctic hierofalcon, an idea dismissed by genetic analysis, and then a Peregrine sub-species, it is now accepted by many to be a full species. The distribution of the Peregrine is extraordinary. Almost all the 'True Falcons' are birds of the tropics or of temperate zones. The exceptions are the Gyrfalcon and the Merlin, which are cold climate specialists (though the latter has a distinctly more southerly range). The Peregrine breeds further north than the Merlin, and perhaps even as far north as the Gyrfalcon, though this northward extension of its more usual southerly range appears to be less a specialism than a natural by-product of the falcon's astonishing success as a predator. Peregrines breed on all continents except Antarctica, a feature shared by only five other terrestrial species – the Barn Owl (*Tyto alba*), Cattle Egret (*Bubulcus ibis*), Glossy Ibis (*Plegadis falcinellus*), Great Egret (*Ardea alba*) and Osprey (*Pandion haliaetus*). But while breeding on six continents, the Peregrine's range is curiously asymmetric – it is largely absent from South America and equatorial Africa, and from some islands: it does not, for instance, breed on Iceland, despite breeding on Greenland to the west and the British Isles/Scandinavia to the east, whereas the Gyrfalcon breeds across all the northern land masses of the Atlantic. (The Merlin (*Falco columbarius*), which breeds across northern North America and Eurasia, also breeds on Iceland, but is absent from Greenland.)

A third *Falco* group comprises the Hobbies – the Eurasian Hobby (*Falco subbuteo*), African Hobby (*Falco cuvierii*), Oriental Hobby (*Falco severus*) and Australian Hobby (*Falco longipennis*). Many experts also include the Sooty Falcon (*Falco concolor*) and Eleonora's Falcon (*Falco eleonorae*) in the Hobby group. The geographical spread of these six species, from Europe eastwards to Indo-China and Australia, suggests a tightly-knit group with strong evolutionary links, but the morphological and behavioural similarities of the Bat Falcon (*Falco rufigularis*) of Central and South America, and the Aplomado Falcon (*Falco femoralis*) of South and Central America, and the southern USA, ensure an ongoing debate about the Hobby group's composition.

The final group is the Kestrels:

The Common, or Eurasian, Kestrel (*Falco tinnunculus*), of Eurasia and Africa;
Lesser Kestrel (*Falco naumanni*), of southern Europe and central Asia;
American Kestrel (*Falco sparverius*), of North and South America;
Fox Kestrel (*Falco alopex*), which ranges across central Africa;
Greater or White-eyed Kestrel (*Falco rupicoloides*), of southern and eastern Africa;
Rock Kestrel (*Falco rupicolus*) of southern Africa;

Female Kestrel with prey south-west England.

Seychelles Kestrel (*Falco araea*), which is endemic to the Seychelles;
Mauritius Kestrel (*Falco punctatus*), which is endemic to Mauritius;
Madagascar Kestrel (*Falco newtoni*), of Madagascar, Mayotte and the Comores, together with an Aldabran sub-species;
Moluccan or Spotted Kestrel (*Falco moluccensis*), of Indonesia;
Australian or Nankeen Kestrel (*Falco cenchroides*), of Australia and New Guinea;
Grey Kestrel (*Falco ardosiaceus*), of central Africa;
Dickinson's Kestrel (*Falco dickinsoni*), of central Africa;
Madagascan Banded Kestrel (*Falco zoniventris*), which is endemic to Madagascar.

The Mauritius Kestrel had the unenviable claim to being one of the world's most endangered species when, in 1974, the population was believed to comprise just four birds. A conservation programme boosted the population to several hundred, but the carrying capacity of the island is probably only about 1000 birds so the falcon will likely remain vulnerable: recently the population has begun to decline, and the species is listed as Endangered by the IUCN. The population of the Seychelles Kestrel is also limited by the carrying capacity of the island group. It is believed the population is about 400 breeding pairs, the majority of which are on the main island of Mahé. The species is IUCN listed as Vulnerable.

But while the above 14 species share the name 'kestrel', the genetic analyses of Wink and Sauer-Gürth (2000) and Wink *et al.* (2007) suggest the group to be rather more loosely associated. Wink *et al.* considered that while the bulk of the Kestrel group branched from a single ancestral form, the American Kestrel was more closely related to the Aplomado Falcon, and that the two share an ancestral link to the red-footed falcons. Furthermore, Wink and co-workers consider Dickinson's Kestrel to be far-removed from the Kestrel group, being genetically closer to the New Zealand Falcon (*Falco novaeseelandiae*).

To further complicate the story, many experts consider the Eastern Red-footed or Amur Falcon (*Falco amurensis*), of China, the Western Red-footed Falcon (*Falco vespertinus*), of eastern Europe and west central Asia, together with the Red-headed (or Red-necked) Falcon (*Falco chicquera*) of India, belong to the Kestrel group. However, this view is at odds with the analysis of Wink *et al.* (2007), not only on the position of the red-footed falcons, but that of the Red-headed Falcon, Wink *et al.* considering the latter to be more closely related to the larger, Peregrine/Hiero, falcons. It is to be hoped that further genetic analysis may soon clarify the Kestrel family tree.

This four-group system leaves several species as outliers – the Orange-breasted Falcon (*Falco deiroleucus*) of central and South America, the Taita Falcon (*Falco fasciinucha*), a rare species of east and southern Africa, the Black Falcon (*Falco*

The Common Kestrel

Fledgling Kestrels, south-west England.

subniger), Brown Falcon (*Falco berigora*) and Grey Falcon (*Falco hypoleucus*), all of which are endemic to Australia, the New Zealand Falcon and the Merlin a northern species that will be known to most readers of this book.

The kestrels are a complex group. Many species are found in Africa, which Groombridge *et al.* (2002) considered the evolutionary home of the group. Using genetic analysis to study not only kestrel evolution, but the interdependence of various African/near African sub-species, Groombridge and co-workers found the analysis supported the idea of an African origin, with an ancient move to the Nearctic and a more recent move from Africa towards Madagascar, and on to Mauritius and the Seychelles. The timing of the movement to Australia is ambiguous. However, while the idea that kestrels evolved in Africa fits with the distribution of the various species, it is not the only possible interpretation, complicating factors being the position of the red-footed falcons which may be related to the American Kestrel and the fact that the Lesser Kestrel is so distinct from the other Palearctic kestrels that it seems basal to the entire family.

In his excellent book on the Common Kestrel, Village (1990) attempts to enumerate the characteristics which define a member of the kestrel group. He notes that the obvious one, hovering[4], which would be chosen as diagnostic by most observers in Britain, fails as almost half the kestrel species use this form of flight rarely if at all: while a 'hovering' falcon will probably be a kestrel, a kestrel does not have to hover. Plumage colour, again so diagnostic for observers in Britain, also fails, for while most of the kestrel group have upperparts which are chestnut/rufous or mainly so, a contrast to most other Falconiformes, and are more definitely patterned than other *Falcos*, three forms (Grey, Dickinson's and the Madagascan Banded) are grey. Grey and Dickinson's both breed in central and eastern Africa and were long thought to be closely related: they were even considered allopatric (*i.e.* they could interbreed), but there was no range overlap. However, as noted above, the work Wink *et al.* (2007) suggests they may not be closely related at all. Another aspect of coloration might also be considered to define a kestrel, for sexual dimorphism in plumage is uncommon among 'True Falcons' other than kestrels. But only half of the kestrel group species exhibit the characteristic. As Village (1990) concludes, it is easier to define what a kestrel is not than what it is. The map overleaf, illustrates the distribution of the 14 species of the kestrel group.

The map also indicates the ranges of the 10 presently recognised sub-species of the Common (or Eurasian) Kestrel, of which the nominate (the form found in Britain) has the most extensive range. But the idea of 10 sub-species should not be seen as fixed: it is still debated, with some experts considering the number too high as most are minimally different. For instance, many believe

---

[4] While 'hovering' is the popular term used for the particular Kestrel hunting mode of hanging apparently motionless in the air, the scientific community prefer 'flight-hunting' as the Kestrel does not 'hover' in the way that, for instance, hummingbirds do, those species being able to remain stationary in still air. The Kestrel is utilising the wind to remain stationary and cannot 'hover' in still air for more than 1-2s.

there is only one sub-species on the Cape Verde archipelago and question whether the Canary islands sub-species (*F. t. canariensis*) really has recently evolved into a separate sub-species (*F. t. dacotiae*) on the eastern archipelago (Fuerteventura, Lanzarote and the Chinijo islets) as the German ornithologist Ernst J. O. Hartert (1859-1933) maintained in 1913. However, recent mitochondrial analysis of Kestrels on Madeira, the Canaries, the Balearic islands and nearby mainland sites in Spain and North Africa (Kangas *et al.*, 2018) found that the two Canary sub-species have diverged both from each other, though only to a small degree, and from mainland birds. Kestrels from Madeira (*F. t. canariensis*) suggested an affinity both with the Canary islands and mainland birds implying hybridisation at some stage. Overall, gene flow with mainland birds has occurred with all the insular sub-species, the degree dependent on the distance to the island.

Among the currently recognised sub-species, the Cape Verde Kestrel is the smallest, *F. t. neglectus* being marginally larger than *F. t. alexandrei* if the latter is distinct. The nominate Kestrel is the largest of the sub-species. This could be seen as conforming to Bergmann's Rule (that body size is larger in cooler climates) were it not for the fact that the east Asian *F. t. interstinctus* is about the same size.

For many years, most experts had considered that the kestrel of South Africa (*F. t. rupicolus*), known locally as the Rock Kestrel, deserved recognition as a full species. One possible indicator of its having species status was the fact that while it is sexually dimorphic, the size differential between female and male is significantly less than other sub-species. Recently, the majority view has been accepted, the Rock Kestrel now being listed as a full species on the Integrated Taxonomic Information System.

Gray (1958) records two presumed natural hybrids between male Merlin and female Common Kestrel each dating to the 1890s, but no verified hybridisation records have been identified. However, there are several known instances of hybridisation between Common Kestrels and Lesser Kestrels.

Of the 14 kestrels, the Common Kestrel is the most widespread, breeding on Atlantic coasts in Europe, north of the Arctic Circle in Fennoscandia, across Asia to Japan, in Arabia, in northern and north-eastern Africa and in a belt across central Africa. The Common Kestrel is also the most numerous Old World falcon, though perhaps not the most numerous worldwide, that distinction likely belonging to the American Kestrel. Until the early years of the new century Kestrels were probably the most numerous of British raptors, but a decline in population – which we will return to later in the book – coupled with an increase in Common Buzzard (*Buteo buteo*) numbers, means the species is perhaps no longer the nation's most populous diurnal raptor, and may also be less populous than the Tawny Owl (*Strix aluco*). When the Kestrel was the most numerous, it was also, arguably, the best known, its enthusiasm for 'hovering' above motorway and railway verges meaning it was more often

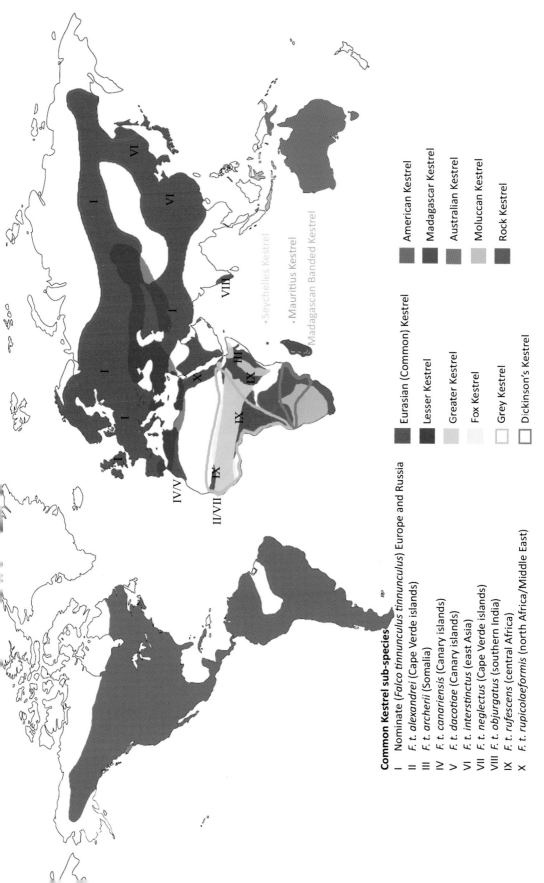

The Common Kestrel

seen. The recent upsurge in interest in Peregrines breeding within cities, the use of video cameras at nest sites and live feeds to websites allowing many to follow the hatching and growth of chicks, means that the Kestrel is no longer either the most numerous or the best known. But it remains one of the most fascinating.

The Falcons

Opposite Female Kestrel.

Above Male Kestrel.

I caught this morning morning's minion, king-
    dom of daylight's dauphin, dapple-dawn-drawn Falcon, in his riding
Of the rolling level underneath him steady air, and striding
High there, how he rung upon the rein of a wimpling wing
In his ecstasy! then off, off forth on swing,
As a skate's heel sweeps smooth on a bow-bend the hurl and gliding
    Rebuffed the big wind. My heart in hiding
Stirred for a bird – the achieve of, the mastery of the thing!

Brute beauty and valour and act, oh, air, pride, plume, here
    Buckle! AND the fire that breaks from thee then, a billion
Times told lovelier, more dangerous, O my chevalier!

No wonder of it: shéer plód makes plough down sillion
Shine, and blue-bleak embers, ah my dear,
    Fall, gall themselves, and gash gold-vermilion.

The Windhover, Gerard Manley Hopkins (1844-1889).

The illustration is from *A History of British Birds* by the Reverend Francis Orpen Morris, published in six volumes during 1850-1857. The books were financed by Benjamin Fawcett who employed Alexander Francis Lydon to produce the drawings and the wood-block engravings for colour printing.

2 THE COMMON OR EURASIAN KESTREL

The species' name almost certainly derives from the French, either from *quercerelle*, a ratchet, or *crécelle*, a rattle, each an onomatopoeic form of the falcon's main call, a persistent *kee-kee-kee*. In France the bird is *Faucon crécerelle*, a name which seems to be close enough to either suggested origin to confirm the derivation, and even closer to a combination of the two. The French name probably crossed the Channel into Britain along with the Norman conquerors, transposing into the medieval English *castrell* which became *kestrel* when emphasis moved from the second to the first syllable and spellings were regularised. Lockwood (1984) notes that in the earliest reference to the name, dating from 1544, the spelling was given as both *kestrel* and *kastrel*, the latter surviving into the 19th century in some localities[1]. By the late 17th century most written works were using *kestrel*, though *kestril* was occasionally seen as late as the mid-19th century. Lockwood also notes the use of other local names for the falcon. Most prominent was not, as might be expected, *windhover*, but *stanniel*[2] (in various forms such as *stannel* and *stanyel*) which likely derived from the Old English *stangella*, meaning *stone yeller*, with *yeller* being from the persistent and piercing call, *stone* being added for emphasis, as with 'stone deaf'. *Stanniel* seems to have been a northern England name, the more familiar *windhover* being common in the south and west. Its derivation is obvious – a diminutive of *wind hoverer* – and had local variants such as *windbibber*, *windfanner* and *windfucker*, the latter apparently deriving from an original (16th century) use of a now vulgar (but ubiquitous) word meaning 'beat' or 'strike'.

---

[1] Shakespeare, writing after 1544, uses *coystrill* (*Twelfth Night*, Act 1, Scene 3) and *coistrel* (*Pericles*, Act 4, Scene 6) in each case to denote a scoundrel or knave which has led to the suggestion that *castrell*, and therefore *kestrel*, derive from knave because of 'a kestrel for a knave' usually, but not correctly, associated with the *Book of St Albans* mentioned later in this Chapter.

[2] Shakespeare most definitely uses *stanniel* for a Kestrel in *Twelfth Night* (Act 2, Scene 5) when Sir Toby Belch compares the falcon's hovering flight with Malvolio's reading of a letter deliberately dropped in front of him by Maria.

The second part of the species' scientific name, *Falco tinnunculus*, derives from the Latin *tinnulus*, ringing or tinkling (which is also the root of *tinnitus*), again referencing the shrill nature of the bird's call.

In Germany the Kestrel is the *turmfalke*, 'tower falcon', referencing the species' enthusiasm for nesting on buildings[3]. *Tärnfalk*, is the Scandinavian name, for the same reason. In Spanish the falcon is *cernícalo*, a word which almost certainly derives from *cernerse*, to hover. Curiously, the name doubles as a vulgar term for a brute, idiot or lout: in Spanish *coger un cernical* is the familiar expression for being drunk.

If the origin of the Kestrel's name seems straightforward, the derivation of *falcon*, the family name, is less so. Majority opinion favours an origin in the Latin *falx* (the genitive of *falcis*) a curved blade or sickle, probably from the shape of the bird's talons or beak, perhaps even the shape of the wing. That fits with the idea that falconry as a sport was known to the Romans who acquired the skills from the east. However, there is very little evidence that the Romans practiced falconry: the most often quoted reference is to the 1st century AD poet Marcus Valerius Martialis – commonly known as Martial – who wrote:

> *He used to prey upon birds, now he is the servant of the bird-catcher, and deceives birds, repining that they are not caught for himself.*[4]

During the Roman era, writers documented every aspect of life and there are no other references to falconry, so while some see Martial's quote as an indication of Roman falconry, most authorities believe the poet was referring to the opportunistic use of raptors to scare birds into nets, a method known to have been practiced by the Egyptians and Greeks. As an example, Aristotle, writing in the 4th century BC, notes the inhabitants of Thrace working with hawks in marshland. Birds would be flushed, hawks chasing them so that the birds panicked and came back to earth, allowing the hunters to easily gather them up. As a reward the hunters would throw some captured birds into the air for the hawks to take. A similar technique, known as 'daring' was practiced in medieval times, using a falconry bird to scare, for instance, larks into nets.

The sport of falconry probably began in western Asia several thousand years ago, spreading east to China, Korea and Japan, and westward a little later. Although it was probably practiced in Mesopotamia (what is now south-east Turkey, eastern Syria and Iraq) before the rise of the Greek and Roman civilisations it is likely it reached Europe no earlier than 500AD, mosaics in the 'Villa of the Falconer' at Argos in Greece, which clearly shows falconry, dating from around that time. That later arrival is also supported by considerable

---

[3] Annually in Berlin at least 300 pairs of Kestrels breed, some in nestboxes placed, in part, by Stefan Kupko who has been studying the falcons there for over 40 years. At present the city is also home to a half-dozen or so of pairs of Peregrines (Stefan Kupko pers. comm.).

[4] From Martial's *Epigrams* Book 14, Number 216. The translation is from Bohn's Classical Library of 1897, and includes the now rarely heard *repining*, meaning *upset* or *miserable*.

archaeological evidence from Germany and Scandinavia. Recent studies of finds in eastern England (Wallis, 2017; Wallis, 2020) present persuasive evidence that the Anglo-Saxons brought falconry to Britain in the late 6th or early 7th centuries.

But if the Saxons brought the sport to Britain, it was the Norman conquest which brought the language of falconry. That does not, of course, mean that *falcon* is not Latin in origin. The old French *faucon* was the word for a diurnal raptor and may well have derived from the Latin: *raptor* itself derives from the Latin *rapere*, to snatch or carry off. As French was the language of the English nobility after 1066, *faucon* became the old English form for the birds until about the 15th century when scholarly Englanders added an 'l' to conform to the assumed Latin origins. Other terms for falcons which also derive from the sport of falconry, and which remain in (relatively) common use, include *tiercel* (occasionally *tercel*) for a male falcon. This probably derives from the Latin *tertius*, meaning one-third, through the old French *terçuel* (occasionally *tiercelet*) because male birds are, in general, one-third smaller than females. Another term is *eyas* (occasionally *eyass*) for a nest-bound chick which is assumed to derive from the Latin *nidus*, nest, through the old French *niais*, with *un niais*, being modified over time. Chicks are also occasionally referred to as *pulli*, though this is often used to refer to the chicks of any bird rather than exclusively those of a falcon or raptor. Here the derivation is more certain, the Latin *pullus* meaning a young animal, though as it was also the Latin for chicken (the base of the French *poulet*, which became Middle English *polet* and hence *pullet* for a hen), there is room for negotiation, particularly as *pullulare* was the Latin verb 'to sprout' which could also be used to imply young growth.

Medieval literature abounds with books on falconry, arguably the most remarkable being one of the first, *De Arte Venandi cum Avibus* (*The Art of Hunting with Birds*), a scientific treatise as well as a practical guide, based on observation and practice, written by Frederick II (1194-1250), the Holy Roman Emperor. But if Frederick's book is one of the more impressive treatises, perhaps the most famous is the *Book of St Albans*. Published in 1486 the book was the last of only eight from the St Albans Press which had been established in 1479 in the Benedictine Monastery of St Albans (in the English county of Hertfordshire). The book was a compilation of information of interest to gentlemen of the day covering the topics of falconry, hunting and heraldry. The falconry section is the supposed work of Dame Juliana Barnes or Berners, 'Dame' in this context probably meaning Prioress. Though no reference to the existence of a prioress of that name at Sopwell Priory, which lies close to St Albans, has been discovered, Sopwell's records are not complete so it is possible that Juliana was indeed a real woman, perhaps becoming the priory's head as such positions were often taken by the daughters of nobility. Dame Juliana does not offer much in the way of viable information for would-be falconers, and her list is not always easy to unscramble as some of the spellings do not allow an unambiguous definition of the bird in question, while others are clearly wrong. Vultures are not the chosen

An illustration of a falconer with his birds and accoutrements from *The Gentleman's Recreation* by Nicholas Cox, originally published in 1677. The book was in four parts, covering hunting, hawking, fowling and fishing. The falconer is attempting to organise the falcon's feet using a stick. I am told by falconer friends, that this section of the drawing is a copy from a 16th century book by George Turberville, but that Cox's version has the process illustrated incorrectly.

falconry bird of Emperors, three Peregrines of differing types are mentioned, and most dukes would not be at all content to be given a Bustard as a sporting bird (though it is now generally considered that 'bustard' is a misspelling of 'bastard' and refers to a hybrid falcon). Dame Juliana's list must surely not be considered as an indication of which noble had, or wished to have, which bird. It was, rather, a list of avian raptor hierarchy set against the hierarchy of English nobility, a list which would likely have been seen as crucial in the class-steeped society of the time. And so we read that a King must have a Gyrfalcon, for a Prince it must be a Peregrine and a Lady must have a Merlin. For a Young Man it must be a Hobby. Usually the list is extended to include a Kestrel for a Knave (see *Footnote 1*), but that association is not, in fact, from the *Book of St Albans*. Rather, it derives from the *Harleian Manuscripts* held at the British Museum. Collected by Robert and Edward Harley, the first two Earls of Oxford and Mortimer, in the late-17th/early-18th centuries, the relevant manuscript dates from the first half of the 15th century and replicates the list in the *Book of St Albans* (is it the original source or is it *vice versa*?), but adds the famous line about the Kestrel for a Knave. The latter became the title of a well-known book by Barry Hines, published in 1968, which itself became the subject of a film, *Kes*, directed by Ken Loach and released in 1969.

Section of a female Kestrel's wing showing the emargination on the anterior primaries.

General Characteristics

Kestrels share the design details of all birds – the fusion and elimination of some bones, the hollowing of others to reduce weight (but with the use of strut-reinforcing to increase strength), and a respiratory system which maximises oxygen uptake to assist flight. The 'True Falcon' wing is the classical design for rapid flight, the leading-edge outboard of the wrist being swept back and tapering to a point, a wing shape which defines falcons as fast, agile hunters, rather than soaring raptors which search for prey on the wing. However, it is not the case that the Kestrel's wing (or, indeed, the wing of any falcon) is modified solely for rapid flight. The Kestrel has ten primary feathers (plus a vestigial 11th), the longest of which are Nos. 8 and 9, with emargination on the anterior three (*i.e.* 8-10) so that a degree of wing-slotting (as seen in classic soaring birds) exists. Wing-slotting is considered to reduce drag and so enable a bird to soar, and is therefore seen in species such as vultures which spend considerable times at height while searching for food. While the slotting is less pronounced in falcons, it is adequate to allow soaring, though this flight mode is only used for limited periods.

The primary feathers are attached to the 'hand' of the bird's wing, the secondary feathers on the bones of the 'arm'. The leading edge of the 'arm' is encased in contour feathers. The arm also carries, on its leading edge, the alula, which corresponds to the human thumb. The function of the alula is still debated, though most experts believe that, at least in part, it performs the function of the leading-edge slats of aircraft, inducing a vortex which delays stalling at high angled, low speed flight, *e.g.* landing.

Female Kestrel coming in to land, clearly showing the extended alula. *Torsten Prohl.*

Further details on the falcon wing, particularly wing loading are given in Chapter 4.

The Kestrel's tail comprises 12 feathers, the inner feathers, Nos. 6 and 7, being about 1.5cm longer than the outer, Nos. 1 and 12. For further information on the Kestrel's tail, particularly in relation to flight-hunting, see Chapter 4.

Kestrels, indeed, all falcons, are diurnal raptors, adapted for hunting, and killing, live prey. The eyes are protected by a distinctly developed superciliary ridge which extends above and in front of the eye. The ridge and its feathers give the birds their famous piercing stare. The ridge is assumed to protect the eye from the wind and debris. Falcons also have a region of dark feathers in front and below the eye which create an equally distinctive malar stripe which is assumed to reduce glare from the sun when the bird is hunting.

The eyes are enlarged, to such an extent that they cannot be manoeuvred as human eyes can, movement being highly restricted, to about 2-5° only (see, for instance, Jones *et al.* (2007)). The bird must therefore move its head to change its line of sight. Falcons also have two *foveae* (from the Latin *fovea* for a pit) or positions of visual acuity, rather than the one in the human eye. One fovea is positioned to give forward binocular (stereo) vision, the other being positioned at an angle of about 40° to the axis of the bird and is probably associated with the tracking of prey.

Above   Tail of a juvenile male Kestrel. One feather has already been replaced with a feather of adult male form. The tail is rounded (but may appear wedge-shaped when the two central feathers are moulted). In the tail above, feathers 1, 2, 11 and 12 were 15.5cm; feathers 3, 4, 9 and 10 were 16cm; feathers 5 and 8 were 16.5cm; and feathers 6 and 7 were 17cm.

Below   This male Kestrel has moulted a grey, barred tail feather to an unbarred one, but is also replacing a grey barred feather with a similar one.

Head of a female Kestrel, showing the relative size of the eye, the tomial tooth and the curvature of the upper mandible.

The upper mandible (maxilla) of the beak is decurved and hooked at the tip for tearing at the flesh of prey[5]. The tomia (cutting edge) of the maxilla have a distinct notch which creates the tomial tooth used to sever the spine of the prey by biting the base of the neck. This manner of killing differs from that of hawks which tend to kill by squeezing their prey with their talons, death occurring by asphyxiation or by bodily penetration, with damage to

---

[5] It would be assumed that the shape of the Kestrel's (indeed all raptor's) beak is an evolutionary adaptation to diet. But work by Bright *et al.* (2016) has shown that about 80% of the shape variation in raptor beaks arises primarily from the integration of beak and skull, *i.e.* changes to beak shape result in predictable changes in the skull and vice versa. The major adaptive difference in terms of diet is size.

vital organs causing death by multiple body traumas, or by a combination of the two. The muscles which control the Kestrel's beak are large, giving them a powerful bite, the better to overcome occasional large, well-armed prey such as rats and Weasels (*Mustela nivalis*). Use of the bill to kill is inherited rather than learned, appearing to be automatic when faced with live prey. That said, adult falcons overcome this auto-function during the fledgling stage of their offspring's development as they occasionally deliver live prey for the young to deal with themselves. However, instances of adult Kestrels delivering live prey to chicks are rare and speculative, and while prey is often cached, there is no evidence of Kestrels storing live prey as has been recently observed in Eleonora's Falcon. In their remarkable observations in Morocco, Qninba *et al.* (2015) noted the falcons pulling flight and tail feathers from prey and then placing the prey in rock 'caves' for later consumption, and even pushing prey into rock crevices from which they could not escape.

In common with other falcons, Kestrels have hooked claws (talons) on their toes which are used to seize prey in an unremitting grip which may be fatal, even if the main purpose is to ensure and maintain capture. The toes also have a ratchet-like tendon which means that once prey is gripped no muscular effort is required in maintaining the grip, a useful attribute when prey must be carried away to a nest or secure place for consumption. Interestingly, the Kestrel has the lowest ratio of middle toe length to tarsus length (67-68%) of the four UK breeding falcons: the ratio for the other three are 100% (Peregrine), 95% (Hobby) and 80% (Merlin). The variation of these ratios has led to some interesting hypotheses, for instance that avian prey specialists have longer toes than the ground-feeding Kestrel as an aid to the grabbing of prey in the air, and that the relatively longer tarsus of the Kestrel (in relation to body length, excluding tail length) is a possible adaptation for hunting in long grass.

Distribution

Falco tinnunculus, the Common, or Eurasian, Kestrel, is, as noted in Chapter 1, the most widespread of the kestrel group. The Common Kestrel is a chestnut falcon, a little larger than the Hobby, but more significantly larger than the Merlin. Kestrels primarily hunt rodents, taking them after a short flight from a perch or by dropping on them from a stationary, flight-hunting (see Chapter 1 for a definition) position a short distance above the ground (see Chapter 4). But while primarily mammalian hunters[6], Kestrels will take a very wide range of other prey, this dietary flexibility allowing the species to have adapted to a variety of habitats. Interestingly, while the Kestrel is assumed to be a 'lowland' species, a bird of farmland and low moorland, the work of Lehikoinen *et*

---

[6] Recent work on mice has found that if the rodents are infected with *Toxoplasma gondii*, a protozoan parasite, of which they are a secondary host, they lose their fear of cat urine (Ingram *et al.*, 2013). As cats are the primary host of the parasite this suggests the protozoan is influencing the behaviour of a natural feline prey to aid its own life cycle, a remarkable, if somewhat alarming, finding. It makes one wonder whether, if Kestrels could read, they would currently be crossing their talons in the hope that some other parasite could persuade rodents to lose their fear of clear skies.

The Common Kestrel

al. (2018) suggests the population in the alpine areas of central Europe is increasing, perhaps as a consequence of global warming.

Falco tinnunculus is the most numerous Old World falcon and until the early years of the new century Kestrels were probably the most numerous of British raptors, but a decline in population – see Chapter 11 for more details – coupled with an increase in the population of Common Buzzards means the species is possibly no longer the nation's most populous diurnal raptor.

Adult male Kestrel.

Adult male Kestrel.

Plumage

Adult Male

The back and wing-coverts are chestnut, spotted with teardrop/arrowhead black spots. The primaries are dark brown/black, edged paler or buff, this being noticeable on the perched bird or the flying bird if viewed from above. The head is slate-grey, with a darker malar stripe and pale buff/cream throat: the contrast between moustache and throat becomes more pronounced with age. The tail is blue-grey with a broad, black sub-terminal and a narrow white terminal band. The underparts are buff, spotted with dark teardrops. The under wing-coverts are paler buff/cream, much paler on the undertail and vent. The underside of the primaries is pale grey/cream, with a dark brown/black tip and bars. The iris is dark brown, the bill blue-grey with a darker tip. The cere, tarsi and feet are yellow, the claws black.

Adult Female

The upperparts, including the head and nape, are chestnut with more abundant, darker spotting than on the male. The malar stripe and throat are as for the male, though the stripe is usually much less well-defined. The primaries are as the male. The feathers of the tail are more rounded than those the male. The tail is chestnut, with darker barring continuously along the length, but with broad, dark sub-terminal and pale buff terminal bands. The underparts are a darker buff than the male, with more spotting. The undertail is continuously darker barred, as is the underwing. The iris is dark brown, the bill blue-grey with a darker tip. The cere, tarsi and feet are yellow, the claws black.

Adult female Kestrels.

The Common Kestrel

It is often suggested that the unbarred grey tail of the male can be used to distinguish the sexes if the male's head is not seen, but some males show pale banding and in some it is very distinctive, while some females have grey, not chestnut tails. The easier distinguishing feature (if the male's head is shrouded) is the richer coloration of the female's back and upper wings, with heavier spotting and barring on the secondary coverts.

Above Adult Kestrel pair at a woven basket nest in southern Scotland. The male, on the left, has a distinctly barred tail and even shows brown colouring towards the tip. The female, on the right, has a barred grey tail. If only the tail was visible the female would likely be assumed to be male, while the male's tail might cause confusion. With sincere apologies for the poor quality of the photos, which are from a trail camera assumed, wrongly, to have been set up to provide crisply-focused images.

Below A leucistic female trapped on the Isle of Bute. The female was breeding and had a brood of three which included two very pale chicks and one which was normal colour. *Gordon Riddle*.

Nestling

The first down is sparse and white. The second down (seen at about 8 days) is buff/grey on the upperparts, paler on the underparts. Feathers appear at about 14 days. The down is almost gone, and all feathers apart from the primaries and the tail, are fully grown at around 28 days.

Juvenile

Juveniles resemble adult females, but are usually paler with heavier streaking and barring on both upper and underparts, and buff fringing to the primaries. The iris is dark brown, the bill blue-grey with a darker tip. The cere is pale yellow, the tarsi and feet are yellow, the claws black. Historically juveniles were considered extremely difficult to sex accurately (*e.g.* Village, 1990), but Dijkstra *et al.* (1990b) were 99% accurate in the sexing of 193 nestlings (191 correct sexes verified by trapping after first moult) using the tail coverts, which were grey in males, with narrow, pointed crossbars, but brown in females. Riddle (2011) agrees with this method of differentiating the sexes, provided the chicks are at least three weeks old and have wing lengths of more than 138mm. However, Village (1990) is less convinced, noting that in his work 16% of juvenile males had brown, barred coverts (and so would have been classified as female by Dijkstra *et al.* One more issue is tail colour, as some observers still maintain that it is possible to sex juveniles by tail colour (see photos opposite).

Why juvenile males resemble adult females, a phenomenon found in many dichromatic species during their second calendar year (*i.e.* first potential breeding year), is a question that has taxed researchers for years and has resulted in two main theories. The female-mimicry hypothesis suggests it aids juvenile males to mate by deceiving adult males into believing that they are females and, therefore, not competitors for mates. The status-signalling hypothesis suggests the similarity allows adult males to differentiate low status males, again reducing competition. This second hypothesis assumes that adult male Kestrels can distinguish sex despite the plumage similarity. In their study of the Kestrels in Finland (Hakkarainen *et al.*, 1993) the research team noted that juvenile males tended to choose breeding sites closer to adult males and that this increased both their mating success and the success of breeding, particularly if they mated with adult females. This suggested that plumage mimicry was an adaptive feature to enhance breeding potential, but still did not differentiate between the two hypotheses. To aid with distinguishing the two, the Finnish team showed captive adult male Kestrels both adult female and female-like juvenile males. If the two were shown simultaneously the adult males preferred the females, but if the two were shown separately the adult males were unable to distinguish them. Hakkarainen and colleagues conclude that adult male Kestrels are not good at distinguishing sex. However, they note that the inability to detect sex is a male trait, female Kestrels being

Above These photos are of the tails of the last two, of five, chicks to leave the nestbox which forms the basis of the study in Chapter 8. The desperate attempt of one of the birds to regain the box when it realised it could not join its siblings outside the barn resulted in the last two landing on the floor – see photos in Chapter 8. The pair were rescued and replaced in the box (from which they both fledged the next day). The colouring of tails and coverts would leave many observers to conclude that the two were male and female (as would the weights and overall sizes of the two).

Below Three near-fledgling Kestrels. The three were from a brood of five and are watching a sibling take its first flight. Except in the hand, there is almost no discernible difference in the plumage of young males and females.

Juvenile Kestrel preening. Preening begins early, with down being removed as the first feathers grow, and is then maintained throughout the bird's life.

much better at distinguishing the sex of juvenile birds. While the Finnish study provides good evidence for the female-mimicry hypothesis, recent studies on an unrelated species (Morales-Betancourt and Castaño-Villa, 2018 researching the White-bearded Manakin (*Manacus manacus*)) have found that the UV reflectance of juvenile males, which strongly resemble adult females in visible light, differed significantly, *i.e.* the sexes were dichromatic at UV wavelengths. If that were also to be found in Kestrels, then it would potentially negate female mimicry – the plumage similarity of juvenile males and adult females is therefore likely to be debated into the future.

Moult

To maintain feathers in good condition Kestrels preen often. They also occasionally bathe, Shrubb (1993a) observing one adult which deliberately sat at the top of a tree, taking a shower in pouring rain. However, bathing seems to be rare behaviour, the Dutch group at the University of Groningen (which carried out a series of studies on captive and wild Kestrels as we shall see in the following chapters) making no mention of it. There are no recorded observations of dust bathing in the literature, but in autumn 2018 I was told of one, a prolonged dust bath by a male Kestrel. The male was one of a pair which had bred in 2017 (see Chapter 8) in a barn in Hampshire (S. Newberry, pers. comm.), and bathed in dust which collected at the entrance of the barn in which the nestbox was positioned.

Above   Female Kestrel bathing. *Frank Leo.*

Right   Bedraggled female. It would be fun to think that the bird has deliberately stood in rain in order to bathe, as Shrubb (1993a) records. But more likely it was the need to deliver prey to her brood in pouring rain. The shot was taken as part of the study reported in Chapter 8.

However, despite the fastidious care birds take in ensuring their feathers stay in good condition, feathers are damaged, raptor feathers probably sustaining a higher level of damage because they may add collisions, struggles *etc.* to the normal wear-and-tear experienced in take-off and landing. In addition, feathers are subject to the attention of a myriad of parasitic feather mites which degrade both the vanes and quills (as well as infesting the skin and subcutaneous tissues, blood *etc.*). In a review of the parasitic mites of falcons and owls Philips (2000) found a total of 21 families infesting falcons (and 17 infesting owls), including at least ten species in the feathers and nasal cavities of Kestrels, or which were found subcutaneously: see Chapter 11 for more detail of a Kestrel's potential parasitic burden.

To maintain good feather condition, all the feathers of adult Kestrels are replaced once annually, the requirement to maintain flight ability in order to hunt meaning that replacement is over an extended period. Adults moult from May onwards. Females moult earlier than males, beginning a few days after laying the first egg, with first-year females starting a few days earlier than older females (Village, 1990). Village notes that earlier moulting is also seen in non-breeding first year females. Village does not suggest that the moult is triggered by calendar time, though an experiment by the Dutch University of Groningen group with captive Kestrels (Meijer, 1989) showed that day length affected both reproductive and moult sequences. While some Kestrel pairs were subject to natural day length, for two other cohorts day length was altered using artificial lighting. For each of these groups from 1 December onwards the day length was increased, to 17.5 hours in one case (*i.e.* 17.5 hours of day, 6.5 hours of night) and 13 hours in the other (13 hours of day, 11 hours of night). All Kestrels were fed 'normal' daily rations. In the first cohort, of 20 pairs, with manipulated day length, egg laying began on 29 December, though some females did not lay until early February. In the second cohort, of 21 pairs, laying began on 15 February. In all cases moulting was synchronised with laying.

Another team at Groningen University studied the energy requirements of the moult in Kestrels (Dietz *et al.*, 1992). Their work included measuring the moult rate for various feather groups for an 'average' Kestrel in natural conditions (Fig. 2).

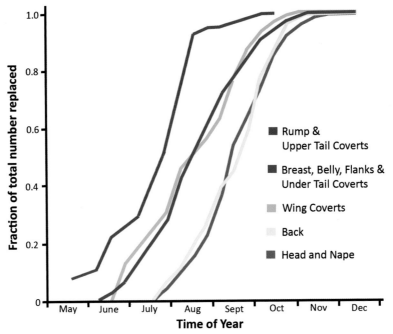

Figure 2. Average moult rate of flight feathers (expressed as a fraction of total feather length) and body feathers (expressed as a fraction of total number) replaced as a function of time. Redrawn from Dietz *et al.* (1992).

Dietz and co-workers also measured the mass of the feathers of four non-moulting Kestrels, Table 1, and studied the feather mass replaced as the moult progressed, Fig. 3.

	Female 1	Female 2	Male 1	Male 2
Primaries	3.59	3.64	3.51	2.93
Secondaries	1.67	1.58	1.58	1.33
Rectrices	2.39	2.18	2.28	1.84
Body Feathers	17.50	13.46	11.07	9.67
Total	25.15	20.86	18.44	15.86

Table 1. Dry mass (g) of flight and body feathers of four non-moulting Kestrels.

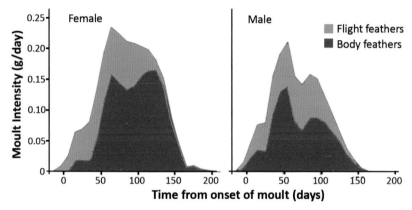

Figure 3. Average moult intensity (expressed as dry feathers in g/day) as a function of time from onset of moult. Redrawn from Dietz et al. (1992).

The Dutch team were then able to calculate the energy required to synthesis feathers, assessing this as approximately 106kJ/g (dry weight). To this must be added the energy equivalent of the feathers themselves (23kJ/g) to produce an overall cost of moulting of 129kJ/g (dry weight). Moulting increases the Basal Metabolic Rate (BMR – the minimum rate needed to sustain life assuming an organism in an ambient temperature equal to body temperature) of the female by up to 30%. Egg laying is also energy intensive and so it is clear why females would wait until after laying had begun, suggesting that more experienced females also wait until incubation is underway before starting the moult.

On average the male's moult starts about 15 days later than his mate, delayed by the necessity of hunting to provide food for both the female and hatchlings, which reduces or eliminates the energy available for feather synthesis. The ability to hunt would also be compromised by the loss of flight or tail feathers:

while this clearly occurs during the moult, the male's hunting ability is at its most stressed during the time he is feeding himself, his partner and their hungry brood. Both males and females end the moult at about the same time, so males moult faster than females. Both sexes may exhibit an arrested moult, *i.e.* they may halt the moult, the likely cause being a shortage of prey during the chick-rearing phase, as both parent birds hunt once the chicks are about 10 days old and able to regulate their own body temperatures.

Adult moulting starts with the upper tail coverts and rump, followed by the wing coverts and underside, then the back and head. Moulting of the primary feathers follows the sequence 4-5-6-3-7-8-2-9-10-1, with two or three feathers in moult simultaneously, while that for the half-tail is 1-6-2-3-4-5 (6 being the outer feather). Moulting is usually symmetrical on both wings and both sides of the tail. During the moult, flight-hunting is considerably reduced as it is much more difficult if there are gaps in either the flight or tail feathers. Shrubb (1993a) notes that some flight-hunting Kestrels exhibit an impressive degree of feather loss or damage, but are still able to function, and this is confirmed by personal experience of falconry birds. This might imply that the moult sequence of both wing and tail is an evolutionary advantage for a flight-hunting species, but in fact the sequence is standard for falcons. The duration of a full moult is about 180 days in total, though the main moult phase, when the primaries and tail feathers are moulted, takes 136 days in females and 122 days in males Village (1990). Village also considered that the growth time for an individual primary feather was about 25 days.

Juvenile Kestrels start to moult body feathers as soon as they fledge, though the timing varies with individual birds and the sequence is haphazard. Work on trapped birds by Village (1990) showed that although some birds began their moult as early as August, most did not start until October and the majority were moulting during the coldest months (Fig. 4). Since loss of feathers increases body heat loss this seems a potentially dangerous strategy, but is perhaps required if the juvenile is hoping to breed during the coming spring. In support of this argument, neither the primary nor tail feathers are moulted during the winter, so hunting is not compromised: primaries and tail feathers are moulted at the same time as adult birds. At that time the juvenile feathers match the size of adults, the first juvenile flight plumage showing shorter wings and tails. Since the weight of juveniles is as for adults, the implication of this is that the wing loading of the younger birds is higher than that of adults, presumably an evolutionary tactic to aid the young birds when they are learning to hunt their own prey, as a higher wing loading means more stable flight. During the primary and tail feather moult, juveniles finally acquire the characteristic plumage of the adults (for instance juvenile males acquire a grey tail).

The Common Kestrel

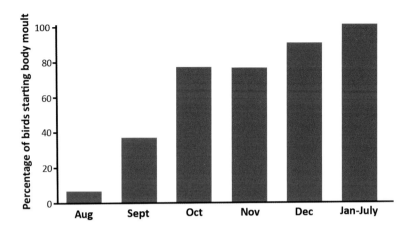

Figure 4. Start of body feather moult, by month, in first-year Kestrels expressed as a percentage of birds that had replaced at least one feather on the head, back, rump or upper tail coverts. Redrawn from Village (1990).

Dimensions and Weight

Overall length (bill tip to tail tip) 320-350mm. Wing: female 256±8mm; male 246±6mm. Tail: female 171±7mm; male 163±6mm. Bill: female 15.0±0.9mm; male 13.9±0.6mm. Tarsus: female 39.6±1.2mm; male 39.6±0.9mm. Toe: female 27.0±1.0mm; male 26.5±0.8mm. Claw: female 11.9±0.5mm; male 11.4±0.3mm.

Juvenile
Wing: female 249±14mm; male 240±12mm. Tail: female 165±13mm; male 157±12mm.

Common Kestrels (in general all kestrels) have tails which are long in proportion to both body and wing length. While this is generally assumed to be an aid to the specific hunting technique of flight-hunting, though this assumption is subject to debate, as we shall see in Chapter 4.

Males 210±20g; Females 250±30g. In both cases the weight is variable throughout the year and particularly during the breeding season. The changes in weight, and the reasons for these changes, are discussed in Chapter 5.

As part of the work carried out on Kestrels by Dutch researchers at Groningen University, which has already been mentioned and will be mentioned again, the constituent parts of two dissected Kestrels were weighed (Daan *et al.*, 1990a) – see Table 2 overleaf.

B	H	Li	K	Lu	DT	BM	LM	S	P	W	F	R	Tot
0.51	0.47	1.20	0.35	0.41	1.56	6.39	4.65	2.74	21.11	104.34	31.67	27.61	203.0

Table 2 Dry weight (g), apart from water and fat, of various parts and internal organs of a Kestrel. Data from Daan *et al.*, 1990a.

Key:
B: Brain; H: Heart; Li: Liver; K: Kidneys; Lu: Lungs; DT: Digestive Tract; BM: Breast Muscle; LM: Leg Muscle; S: Skin; P: Plumage; W: Water; F: Fat; R: Remainder; Tot: Total Weight.

From Table 2 it can be seen that, as with most homeotherms, Kestrels are mainly water and fat. Plumage is a substantial fraction of the total weight, as are the flight muscles.

Reverse Sexual Size Dimorphism

Female Kestrels are larger than males (about 20% on average). In this respect Kestrels are as other falcons (and raptors) in showing Reverse Sexual Dimorphism (RSD). While it is sometimes maintained that RSD is unique to raptors, this is not the case, the trait also being seen in Stercoraiidae (the skuas or jaegers), Scolopacidae (sandpipers) and Sulidae (boobies and gannets). It is also occasionally maintained that RSD is associated with predatory behaviour, but although the skuas are predators, neither sandpipers nor gannets are, and the Laniidae (shrikes), which predate birds, rodents and lizards, do not exhibit RSD. However, predatory behaviour is often chosen as a starting point for considering RSD because of the effect of body mass on flight, and therefore hunting, characteristics.

Andersson and Norberg (1981) suggested that RSD evolved in raptors as a consequence of three factors. Firstly, role partitioning favoured one sex being larger to guard the nest while the other hunted for the pair and their offspring. This does not, of itself explain why females are larger, but accepting this, a second factor suggested that it made more sense for the female to be the nest guard. The reasoning is that the female risked injury to her eggs during prey attacks, and that her role in egg laying predisposed her to stay close to the nest. Once her role as a guard was established, a larger female represented a more formidable opponent. The third factor suggested a smaller body mass favoured better hunting so that males of species which hunted avian prey would tend to be smaller. In favour of this Andersson and Norberg pointed out that RSD was more pronounced in species which specialised in avian prey. This is certainly true of the British falcons, the size differential of Peregrines and Merlins being greater than that of the rodent-hunting Kestrel. The position of Hobbies is more ambiguous as the

It is difficult to obtain a photo of male and female Kestrels side-by-side to illustrate the size difference. The images above were taken in 2020 with the male and female falcons perched in almost identical positions at the Chapter 8 nestbox (on different days with differing lighting conditions). The RDS of Kestrels is usually about 20% in size, a difference which seems appropriate for the 2020 pair.

size differential is relatively small, but Hobbies are chiefly insects feeders, tending to hunt avian prey only during the breeding season, though this is not an absolute as wintering birds will also take birds.

However, while these reasons for the evolution of RSD are compelling, debate over the issue has resulted in a considerable number of papers over the years, both before and after the work of Andersson and Norberg (1981). No consensus has emerged, either on the reasons for RSD or on whether it emerged in ancestral falcons because males became smaller or because females became larger (though data in support of either would be difficult to obtain). More than 20 theories have been put forward over the years, with varying levels of success in terms of both explaining the phenomenon and in achieving a measured degree of scientific acceptance. The 'most successful' theories fall into three main groups.

The first is ecologically based, the idea that the sexes operate in specific niches, in this case prey size, to avoid competition on a shared territory. Evidence in support of this theory is contradictory. Pande and Dahanukar

(2012) noted that their work on Barn Owls in India indicated that the mean mass of prey brought to the nest by males was significantly lower than that brought by females, but that males brought a greater number of prey items. Pande and Dahanukar considered that this supported an ecological basis for RSD. The work prompted an interesting debate which highlighted a lack of consensus. Olsen (2013), replying to the Indian researchers from Australia, pointed out that a study across all raptors showed that while males did indeed carry lower mass prey than females in many species, in others the mean mass of the sexes was the same and some males carried heavier prey than their larger female partners. Pande and Dahanukar (2013) agreed these comments were valid and called for more research into the nature of the relationship between RSD and foraging in male and female raptors.

In Britain, separate studies on Peregrines, which show a high degree of RSD, had also come to different conclusions. In Cornwall, Treleaven (1977) found no difference in the mean size of prey delivered to the nest by the two parent birds, while Parker (1979) in Pembrokeshire, and Martin (1980) in the Lake District found that the prey delivered by males was significantly smaller than the prey of females. In a more recent study, Zuberogoitia *et al.* (2013), studying Peregrines in northern Spain, found no difference in the size of prey delivered to the nest. Other studies have noted that in single chick broods the male does much of the provisioning of the youngster, while in broods of three or four the female brings more of the prey: as she is larger, she is able to carry larger prey to the nest which would be advantageous. This adds a further argument to the ecological theory, positing that females are larger so that they can carry larger prey to the nest. Further evidence comes from studies which note that male Peregrines consistently bring small prey items to the nest even if they are killing larger prey, for instance grouse, which they feed on. The ecological theory therefore has much to recommend it – but it does not explain why females are larger than males rather than vice versa and has failed, to date, to muster a consensus among researchers.

The second RSD theory is behavioural, suggesting one of three possibilities. The first suggests females are larger in order to dominate their male partners, maintaining the pair bond and ensuring food deliveries. The second that larger females can outcompete conspecifics in competition for males. The third that as males compete for females, smaller males can make superior (more agile) display flights and so are more attractive to potential mates.

The third RSD hypothesis is physiological – either because smaller males, being more agile, are better hunters (an extension of the second, behavioural, hypothesis as flight agility indicates probable worth as hunters and, therefore, providers for their mates and broods), or that larger females can produce larger eggs and/or clutches and so enhance breeding success. One aspect of the latter idea is the 'starvation hypothesis' which posits that larger females are better able to withstand food shortages during the breeding season. A physiological basis for RSD is certainly supported by the observation of Newton (1979) that in female falcons, body condition is critical for reproduction and that females are at their heaviest during the laying phase of the breeding cycle and remain heavy (though weight is lost) during incubation and the early nestling phase. Although it is in the reproductive interests of the male to feed his mate and so increase her weight, the increased size of the female would also allow her to dominate her mate, 'bullying' him into hunting. In a study of Tengmalm's Owl (*Aegolius funereus* – now more correctly called the Boreal Owl) Korpimäki (1986a) found that the weight dimorphism in owl pairs was positively correlated with the laying date of the first egg, *i.e.* pairs in which the weight difference was larger bred earlier than pairs where it was smaller. As early breeding tends to produce larger clutches and more fledglings this supports the starvation hypothesis. However, Korpimäki found no evidence that heavier females were better nest guards. Korpimäki also found that male wing length was correlated with laying date, suggesting that females were selecting shorter-winged males as they were likely to be more agile fliers and, therefore, better hunters.

Further evidence for a physiological aspect to RSD came from the work of a Norwegian group. Slagsvold and Sonerud (2007) studied seven species of owl, hawk and falcon (the latter being Merlin and Peregrine) and noted that the ingestion rate of prey was variable, and that this might influence dimorphism. Ingestion rates were higher for smaller prey and for mammalian rather than avian prey. Mammalian prey was also ingested faster by raptors which were primarily mammal feeders. The researchers argue that taking larger prey increases the feeding time for chicks, which both ties up the feeding parent and heightens the risk of the chicks chilling. Males, the main providers, should therefore be selected for taking smaller prey which would favour smaller size. Females, on the other hand, would benefit from being larger in order to better cover the eggs and chicks, and to survive a winter shortage of smaller prey. In a later study (Sonerud *et al.*, 2013, Sonerud *et al.*, 2014a) on the Kestrel, the Norwegian team noted that video evidence suggested that the assumption that females delivered larger prey to the nest was flawed. It seemed that females intercepted larger prey items the male was carrying, part consuming the prey and then feeding the remainder to the chicks, while the male delivered smaller, more easily handled, prey items directly to the chicks once they could handle these. The change from the male passing food to the female and directly provisioning the chicks occurred first for insect prey, then lizards, mammals and birds, and occurred earlier for small mammalian prey than for larger items. The female's ability to intercept prey deliveries depended on her size (allowing her to bully the male) while the male's ability to return quickly to hunting also depended on his size, lending support to the contention that ingestion rates may have been influential in the development of RSD. The Norwegian team followed up this study with another (Sonerud *et al.*, 2014b) covering nine raptor species which showed similar results. From their study they noted that the size difference in sex should increase as the size of prey taken increased, from insects through reptiles to mammalian and then avian prey. Because of the range of weights seen in UK falcons it is difficult to assess with certainty whether the suggestion of Sonerud *et al.* (2014b) translates to the UK's breeding falcons. It is the case that female Peregrines are much larger than males (by around 50%), but for the three smaller falcons (Merlin, Hobby and Kestrel) the size differential is 20-35% depending on individual weights: Sonerud *et al.* (2014b) would suggest that the mammal-hunting Kestrels show a smaller disparity than the avian hunting Merlins (and the Hobbies, which as noted above, are primarily insect feeders outside the breeding season, but are avian hunters when breeding).

Fig. 5 is a pictorial representation of the three RSD hypotheses. The figure is from the work of Krüger (2005) who collected data on 237 species of Accipitridae (hawks), 61 Falconidae (falcons) and 212 Strigiformes (owls), and used comparative studies of 26 variables (both physiological – body mass, wing length *etc.*; behavioural – breeding system, displays, *etc.*; and ecological – habitat, range size *etc.*) to investigate potential correlations with RSD. Krüger found different levels of correlations between the three raptor forms, but in all three the dominant correlation supported the third of the hypotheses mentioned above, *i.e.* that small males evolved to efficiently hunt small agile prey. Krüger's findings suggest that the trend for smaller males occurred before falcons began to attack larger prey, which would explain why males are relatively smaller in larger falcon species. Fig. 6 shows the variation in RSD, defined as the cube of wing length ratios of the two sexes, across the studied species.

Krüger's work was a step forward in that it offered an explanation for why males had become smaller than females (rather than for females to become larger than males). In more recent work Pérez-Camacho *et al.* (2018) collected data from 75 raptor species and scored four factors linked to RSD to create a mathematical model of their interaction. The factors were prey agility, the structural complexity of the hunting habitat, territorial behaviour and territory size. The analysis again suggested that the evolution of RSD strongly favoured a reduction in male size rather than an enlargement of females, and that as prey agility and the complexity of the hunting habitat increased (each of these factors influencing territorial behaviour and territory size) the degree of RSD would increase. In combination, the work of Krüger (2005) and Pérez-Camacho *et al.* (2018) appeared to narrow the choices in terms of RSD theories, but within two years a further paper (Schoenjahn *et al.*, 2020) suggested that another physiological factor might be involved, requiring females to be larger than males so they could better defend eggs and chicks against predators and so enhance breeding success. Schoenjahn *et al.* suggest their methodology overcomes one difficulty of all research on RSD, in that it distinguishes between whether the identified benefits of RSD derive from the proposed causal process(es) or are incidental. Whether the academic community accepts the idea as definitive or just another option for an ongoing debate is, itself, a matter of debate.

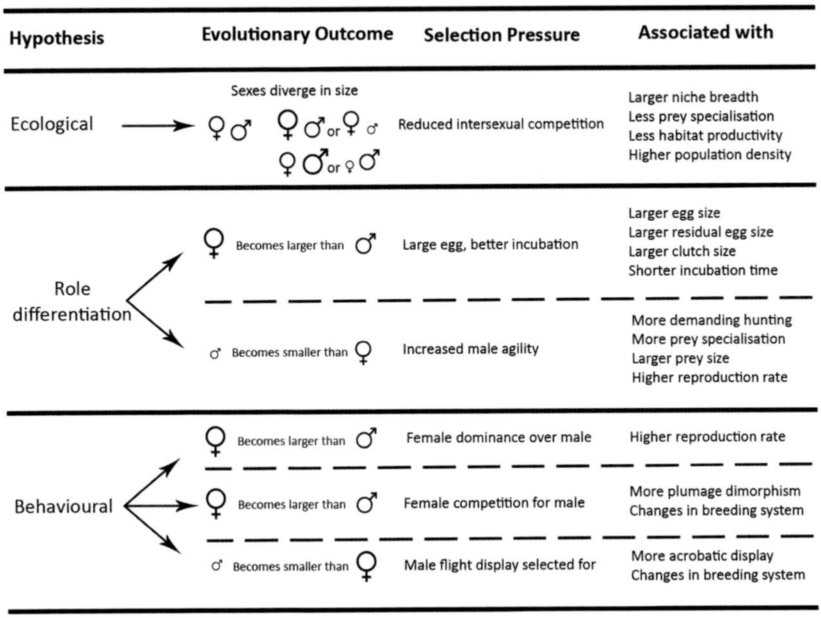

Hypothesis	Evolutionary Outcome	Selection Pressure	Associated with
Ecological	Sexes diverge in size: ♀♂, ♀♂ or ♀♂, ♀♂ or ♀♂	Reduced intersexual competition	Larger niche breadth Less prey specialisation Less habitat productivity Higher population density
Role differentiation	♀ Becomes larger than ♂	Large egg, better incubation	Larger egg size Larger residual egg size Larger clutch size Shorter incubation time
	♂ Becomes smaller than ♀	Increased male agility	More demanding hunting More prey specialisation Larger prey size Higher reproduction rate
Behavioural	♀ Becomes larger than ♂	Female dominance over male	Higher reproduction rate
	♀ Becomes larger than ♂	Female competition for male	More plumage dimorphism Changes in breeding system
	♂ Becomes smaller than ♀	Male flight display selected for	More acrobatic display Changes in breeding system

Figure 5. The three main hypotheses proposed to explain RSD, with their predicted outcome in terms of sex differential size, and the selective advantage of, and main correlations of, the resulting RSD. Redrawn from Krüger (2005).

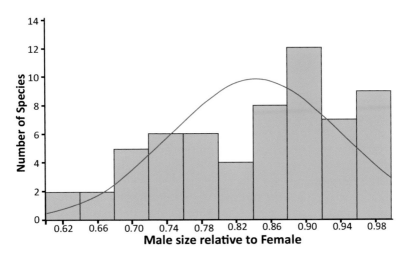

Figure 6. Distribution of RSD in the Falconidae. The histogram includes the 61 species recognised as forming the family at the time Krüger carried out his analysis. The lowest value (0.61) is for the Bat Falcon, the highest (0.99) for the Black Caracara (*Daptrius ater*). The red line is a normal distribution. The distribution is skewed from normal, indicating a shift to a relatively small male size across the falcons. Redrawn from Krüger (2005).

Habitat

Kestrels are primarily birds of open country – farmland, heath, grassland, scrub and low-lying moorland, as well as features such as river banks and woodland edges – but also do well in cities, taking advantage of any open area which might offer the possibility of rodent prey, or taking city-dwelling passerines. They also frequent other human-defined landscapes such as road verges, railway embankments, canal edges, and areas of felled forestry where they take rodents from the newly-created scrub before new tree growth closes out the area as a hunting ground. Despite the extensive range of the species they do not inhabit tundra, forest-tundra or taiga. While in parts of their range Kestrels are found at considerable heights (up to 3,000m or higher in the Caucasus) in Britain they rarely breed above about 500m: from personal experience in Scotland, Kestrels breed in areas normally associated with Merlins, nesting at similar heights to their cold-climate cousins on weatherswept moorlands. However, while Kestrels as a species show a high degree of flexibility in habitat, the study of Avilés *et al.* (2001) in south-west Spain showed that the clutch size and breeding success of Kestrels in pastures was higher than in cereal cropland. This finding was echoed in a study in England (Garratt *et al.*, 2011) which found that Kestrels preferred recently cut grassland for hunting. (For further details of the work of Garratt and co-workers across a spectrum of habitats see text and Figs. 40 and Figs. 41 in Chapter 4). A similar study (Shrubb, 2003), but encompassing a larger number of Kestrel territories (461 or 462) over a period of years (1997-2002), also identified a spectrum of habitats in Wales, covering coastal sites (91 or 92 territories), farmland (127), hill grazing (100), forest (5) and urban/industrial areas (21).

Habitat, and its influence on Kestrel hunting is considered further in Chapter 4, while land management for habitat conservation beneficial to Kestrels is considered further in Chapter 11.

Voice

Kestrels are vocal birds, particularly during the breeding season, but personal experience suggests Walpole-Bond (1938) was incorrect when suggesting the species was mute during November and December, though calls are much rarer in those months. The most common call, the alarm call, is usually described as '*kik-kik-kik*' or '*kee-kee-kee*'. The call, used by both males and females, is most often heard when the birds are disturbed at the nest or if the nest is approached, but the birds also use it during any aggressive event, *e.g.* harassing another raptor or a trespassing corvid. The number of syllables is most often 2-4, but is variable, with longer calls being heard. Longer calls are usually associated with more aggressive situations, the pitch rising so that the call becomes much shriller. Occasionally a single '*kee*' is heard, particularly from fledglings exercising their wings and voice at the same time.

Kestrels are vocal birds. Here, male and female discuss who will take the next session of egg covering.

The common call is the Kestrel's communication with the wider world, but a second call is more private, a gentler, trilling '*wheee*' heard when the Kestrel pair are together and when they greet each other at the nest. The female also uses the call when begging for food from the male, the call then being more of a nasal whine. A shorter, higher-pitched, more clipped version, usually written '*clip*', '*kit*' or '*tsick*', is also used by the female when feeding her brood, and by the male when he arrives at the nest with food.

The calls of nestlings are variations on adult calls, but they also have a plaintive *cheep* most often heard during the first days after hatching when they are cold and in need of brooding, or later when they are hungry. In a study on American Kestrels, Smallwood *et al.* (2003) found a number of different calls of nestlings, but were unable to use voice as a means of distinguishing nestling sex. The authors noted that by the age of 16 days the chicks were producing calls akin to those of adults. It is likely that vocal development in the Common Kestrel follows a similar path.

Kochanek (1984) analysed the calls of three pairs of wild Kestrels, both adults and young, by sonogram and concluded that females made 11 different calls, males nine and young Kestrels five. While these were all variants on the basic calls noted above, Kochanek's work does suggest that the bird-to-bird communication of the Kestrel is rather more complex than it first appears.

3 DIET

Kestrels may prepare their food, by plucking birds and removing the fur of mammals, especially when feeding chicks. However, while as many feathers as possible are removed from avian prey, adult falcons may also swallow rodent prey whole. All meals will therefore involve the ingestion of feathers and fur. These items are added to indigestible parts – teeth and bone, claws and bill, and the chitinous parts of insects – which cannot easily pass through the digestive tract. This detritus is formed into a mass within the bird's gizzard and regurgitated at regular intervals, regurgitation having the added advantage of purging the upper section of the tract. Known as casts by falconers, but as pellets by others, the masses are cylindrical, Kestrel pellets being 20-30mm long, 12-17mm in diameter (Village, 1990), though Riddle (2011) suggests 20-40mm x 10-25mm in his absorbing book on his own experiences with Kestrels. Pellets weigh 1.2-1.5g, are usually pale grey when dry, and are rounded at one end, pointed at the other. When cast they are slimy with mucus which often means they stick to branches close to nest sites. Pellets are normally regurgitated at dawn and as this is when the bird is at its roost,

Top   Screen grabs from a video shot by the external camera set up for the study reported in Chapter 8. The process of disgorging a pellet starts with the falcon opening its beak wide and moving its head up and down as spasms in its digestive tract bring the pellet to its mouth. The pellet is then simply dropped from the mouth. The photos above were taken at 05.23 on 6 June 2019.

Above   Screen grabs from a video shot at 05.35 on 13 April 2020, the camera being in b&w, IR mode at that time. The adult female Kestrel ejects a pellet which lands on the platform at the front of the nestbox.

or at the nest in the breeding season, this aids recovery for analysis, careful separation of the contents revealing the consumed prey.

Kestrels also occasionally ingest grit or small stones (officially termed gastroliths, but called *rangle* by falconers). Such behaviour is well-known among herbivorous birds, the grit grinding the forage, but is less well understood for falcons. Though gastroliths may assist in the grinding of food items, they may also aid the cleaning of the digestive tract of mucus or the grease of prey, or aid shedding of the gizzard lining. While not fully understood, the necessity of rangle for the health of falcons has probably been known since the beginnings of falconry: Symon Latham, whose books on falconry, published in the early 17th century, are still considered to be among the best ever produced on the topic, wrote '*washed meat and stones maketh a hawk to fly*'. Though well-known in falconry circles, the use of gastroliths by wild falcons is assumed, but rarely observed: Albuquerque (1982) did note it in a female Peregrine overwintering in Brazil, the falcon taking material from a sand store. Albuquerque noted the behaviour several times between January and March.

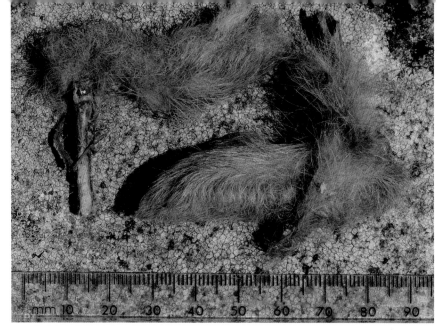

Rabbit feet found in the nestbox of the study of Chapter 8 after the chicks had fledged in 2019.

Kestrels are flexible in their diet, hunting earthworms and insects, and taking avian and reptilian prey, but are primarily predators of Microtus voles and other rodents. Pellet analysis has shown that more than 20 mammal species are taken by Kestrels across their vast range, the majority, as expected, being small mammals (mice, voles and shrews). In the absence of voles or mice, the birds will feed on anything readily available on the ground which they believe they can overpower, taking young Rabbits (*Oryctolagus cuniculus*), leverets, and the young of rats and squirrels. Village (1990) reports seeing a male Kestrel attack a Brown Hare (*Lepus europaeus*) leveret. The falcon was able to keep hold of the animal, but was unable to kill it. The female of the pair arrived and dispatched the hare with several bites at the base of the neck. However, the female was barely able to take off with the kill and Village (displaying an admirable enthusiasm for science, if unchivalrously depriving the falcon of her meal) forced her to leave it: he measured it at 120g, about half the likely weight of the female (about 250g). In the breeding study detailed in Chapter 8, on one occasion the male Kestrel (probable weight 210g) brought a young rabbit to the nestbox platform. The exhausted falcon paused for several minutes before attempting to lift the rabbit over the lip into the box. He tried several times, but failed on each, eventually giving up and flying off with the prey. Over time he returned carrying sections of the rabbit which he had obviously dismembered. Rabbits weigh 30-35g at birth and grow rapidly. To have been above ground the rabbit could have been comparable in weight to the male Kestrel: full-grown rabbits weigh 1.2-2.5kg.

Opposite

Above   A collection of pellets below a corvid stick nest occupied by Kestrels.
Far left   Dissected pellet of a male Kestrel with a number of rodent teeth.
Near right top   Rodents skulls from Kestrel pellets.
Near right bottom   Juvenile Kestrel pellet with numerous beetle wing cases.

Common Shrew (*Sorex araneus*) left behind after the chicks had fledged from a nestbox in southern Scotland. To the left of the shrew is a grisly reminder of another prey item.

Kestrels also take shrews not, apparently, being put off by the foul-tasting mammals as most other potential predators are, though the photographic evidence of nestboxes suggests that shrews are only eaten if more appealing prey is unavailable.

Moles (*Talpa europaea*) have also been found – both as remains, in pellets, and as prey items at nests – these invariably having been taken after periods of heavy rain when the animals had been forced to abandon waterlogged burrows. Uncorroborated reports exist of Kestrels taking Weasels (*Mustela nivalis vulgaris* – the central/western European sub-species), usually young animals. However, in the breeding study described in Chapter 8, photographic evidence of the male Kestrel bringing an adult Weasel to the incubating female was obtained. Weasels are fierce and well-armed, and the male falcon needed to be very precise in its first attack as there would probably have been no time for a second before the weasel had inflicted significant damage. Korpimäki (1985b) notes the taking of a northern sub-species of the Weasel (*Mustela nivalis nivalis*): further details are given later in this Chapter.

The dreadful quality of this screen grab from the 2017 video study which is discussed in Chapter 8 led to the decision to replace the external camera with one which lived up to its advertised capabilities. Nevertheless, it is proof of the male Kestrel having brought a Weasel to his mate.

Diet

Once taken, prey is dispatched with the standard falcon bite at the base of the neck or on the head. In a somewhat gruesome experiment in northern Italy Csermely *et al.* (2009) offered mice (in an enclosure which simulated the local environment: the rodents were contained within the enclosure, but not tethered) to both captive and wild Kestrels. In all cases the falcon took a short exploratory flight above the victim, presumably assessing capture possibilities, before swooping onto it. The only difference between the two cohorts was that wild Kestrels flew away with their prey, taking it to a perch where it was consumed, whereas the captive Kestrels killed and ate the mouse at the point of capture.

Insects are also taken, both larger beetles and much smaller insects (*e.g.* ants), these sometimes being taken by ground foraging after both successful and unsuccessful rodent kills. In a study of Kestrel pellets from the Pentland Hills, south of Edinburgh, the elytra (wing cases) of the Dor Beetle *Anoplotrupes stercorosus*, several ground beetles (including *Carabus problematicus*, the Bronze Carabid (*Carabus nemoralis*), Violet Ground Beetle (*Carabus violaceus*), and the scarce, but highly distinctive, bright green *Carabus nitens*), various Pterostichus and Agonum Ground Beetles, and Burying Beetles (including the common Black Carrion Beetle (*Silpha atrata*)) were found among Field Vole jaw bones and lizard tail fragments. In a study in southern Alicante, Spain, Orihuela-Torres *et al.* (2017) found insects in 89.9% (by occurrence) of 571 Kestrel pellets. Of other prey (by occurrence) 7.5% of pellets including the remains of birds, 2.5% of mammals, and 0.08% of reptiles (one reptile being the Ocellated (or Jewelled) Lizard (*Timon lepida*), an exotic-looking blue-spotted green lizard which can grow to lengths of over 20cm (Fig. 7).

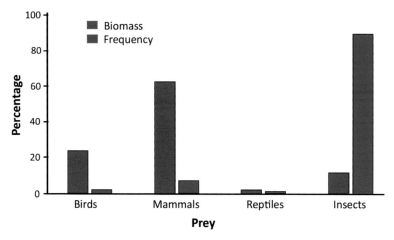

Figure 7. Diet of Kestrels in southern Spain. While mammals formed only a small fraction of the total prey by occurrence, they constituted 62.3% of the biomass, with insects forming only 12.3% despite being the majority of captured prey. Of the insects, the bulk (55%) were grasshoppers, beetles making up 13.3%. Redrawn from Orihuela-Torres *et al.* (2017).

A female Kestrel delivers a Vivaporous Lizard to her brood hatched in an old shopping basket in southern Scotland.

The Spanish researchers suggest the reason for the difference between the diet of the studied Kestrels and their northern cousins was climatic, with a relative abundance of insects and a lower density of small mammals in southern Spain. The dominant mammal in the Kestrel diet was the Mediterranean Pine Vole (*Microtus duodecimcostatus*). Interestingly, the insects consumed by the falcons included a recent invasive species, the Red Palm Weevil (*Rhynchophorus ferrugineus*). In an earlier study across five years in Spain (the Alto Palancia region of Valencia, east-central Spain) Gil-Delgado *et al.* (1995) found that grasshoppers formed the bulk (83.0%) of the diet of nestling Kestrels: overall insects formed 96.4% of the diet, mammals representing only 2.6%. (The remainder was 0.7% avian, 0.3% reptile.) Dragonflies are also recorded as having been taken, but in general Kestrels lack the aerial agility of Hobbies and do not take these or other fast-flying insects often, though Kestrels in Africa do take newly-emerged swarming termites.

A high insect fraction of the diet is not restricted to Spain. Shrubb (1993a) collated data from studies across Mediterranean Europe noting that vertebrate prey constituted only 14% of the Kestrel diet (voles being 15% of this fraction) with invertebrate prey comprising the remaining 86%.

In warmer parts of the species range lizards may form the main vertebrate prey, but even in Britain lizards are taken (both Vivaparous Lizard (*Lacerta vivipara*) and Sand Lizard (*Lacerta agilis*)), as are Slow Worms (*Anguis fragilis*)

The female Kestrel of the study reported in Chapter 8 delivering one of four Slow Worms (*left*) she captured and a Common Frog (*Rana temporaria – right*) to her brood in 2019. The video stream includes the head and front legs of the frog, but the screen grab was too blurred for reproduction. However, the long back leg is unmistakable.

and, though much more rarely, snakes. Amphibian prey includes newts and frogs (Deane, 1962), while Village (1990) records seeing a Kestrel feeding on a Common Toad (*Bufo bufo*): as the toad was 'road kill' Village believes the bird was able to bypass the poisonous skin to reach the harmless flesh beneath. Earthworms are frequently eaten – indeed, Kestrels have been seen following tractors ploughing fields prior to seeding[1] – and slugs are also consumed, while rarer menu items include fish and crabs (Village, 1990 and references therein).

Perhaps the most unusual dietary addition is that observed by Rejt (2005) while studying urban-nesting Kestrels in Warsaw. Rejt noted that unhatched eggs disappeared from the nest as soon as the last chick had hatched, and that continuous video monitoring of one nest showed an adult taking an unhatched egg and caching it, then feeding the contents to the hatched chicks a few hours later. The observation implies that infertile eggs may be routinely fed to nestlings.

Avian prey is also taken. Kestrels are not adept bird hunters, Cresswell (1993) noting that in a study of the birds taken by four raptor species during three winters at an estuarine site in East Lothian, Scotland, a pair of Sparrowhawks (*Accipiter nisus*) took 753 birds, two Peregrines took 393 and two Merlins took 223, while a pair of Kestrels took just 5, one of the pair taking no birds at all. But Kestrels will take both adult and fledgling birds, and chicks, and these may even form a significant part of the diet of fledgling Kestrels. The birds taken are species which share the Kestrel's preferred habitat – in open-country Meadow Pipit (*Anthus pratensis*), Skylark (*Alauda arvensis*) and Starling (*Sturnus vulgaris*): in urban environments House Sparrow (*Passer domesticus*) and other

---

[1] While it is the case that Kestrels will follow tractors towing ploughs to take earthworms, the falcons will follow all sorts of farm machinery to seize rodents disturbed by their passage. Especially popular seem to be combine harvesters which are often accompanied by a rodent-seeking Kestrel.

Above　　Male Kestrel plucking a captured bird. *Torsten Prohl.*

Below　　The male delivering a plucked bird to the chicks in 2019 in the study described in Chapter 8.

Plucked and partially eaten prey found in a nestbox after the Kestrel chicks had fledged. The long back claw identifies the bird as a Meadow Pipit. The box was on moorland south of Edinburgh, a habitat shared by Kestrels and Merlins, the latter specialising in catching the pipits.

town passerines – though many other species have also been noted, including waders and the young of much bigger species such as ducks and gulls. Village (1990) records a male Kestrel killing a Turtle Dove (*Streptopelia turtur*) and then flying off with it in a series of short, presumably, exhausting flights. Given that the falcon may have weighed only about 190g, while the dove was likely to be around 120g if a juvenile (the kill was in July) or more if an adult, this shows both notable strength and determination by the falcon. This was also shown in the breeding study detailed in Chapter 8, when the female Kestrel brought a Jackdaw (*Corvus monedula*) to the box. While, again, the bird was likely to have been a juvenile, it would probably have weighed around 200g. At that stage of the breeding cycle the female's weight was close to its annual minimal and so she may not have weighed more than the prey.

Further evidence of Kestrel's determination when hunting is not lacking: Gentle *et al*. (2013) record a (failed) attempt at predation of House Sparrow nestlings from a nestbox, while Ponting (2002) reported seeing a Kestrel grab a Canary (*Serinus canaria*) through the bars of a cage suspended outside an apartment block on Tenerife. The Kestrel held the bird with one foot and plucked it through the bars, eventually letting go of the dead bird and flying to perch on another cage. The owner of the second cage appeared and scared the falcon. To complete the picture of predation of typical caged birds, Riddle

(2011) records finding feathers at one site which could only have come from an escaped Budgerigar (*Melopsittacus undulatus*).

Kestrel chicks and eggs, and the feathers of an escaped Budgerigar. *Gordon Riddle.*

Shrubb (1993a) notes that Kestrel prey has included Collared Dove (*Streptopelia decaocto*), Stock Dove (*Columba oenas*) and adult Lapwing (*Vanellus vanellus*), while other records include Snipe (*Gallinago gallinago*) all of which would have been a handful (as it were) for the falcons. In The Netherlands, Cavé (1968) also noted Snipe among the avian prey of Kestrels hunting on a Dutch polder (reclaimed land). Other surprising species among a list of 28 in total from the Dutch study were Coot (*Fulica atra*), Moorhen (*Gallinula chloropus*), juveniles of Black-tailed Godwit (*Limosa limosa*), Mallard (*Anas platyrhynchos*) and Grey Partridge (*Perdix perdix*). Cavé also notes juvenile Black-headed Gulls (*Larus ridibundus*), though only in one of his five years of observation. Personal observation includes a juvenile Black-headed Gull being taken from a nesting colony by a bold adult Kestrel, who was pursued by a screaming gull flock: for several days after the attack, the, now-fledged, juvenile Kestrels who had been fed the young gull were relentlessly harried by adult gulls.

Mead and Pepler (1975) record Kestrels at a Sand Martin (*Riparia riparia*) colony, but note that the falcon was not especially skilled at taking the martins, those that were taken probably being juveniles[2]. Riddle (2011) also mentions an incident on a North Sea oil platform where a Kestrel, having consumed a good fraction of the small birds taking refuge, retrieved a Goldcrest (*Regulus*

---

[2] Rich (2016) recorded (and photographed) a Kestrel attempting to predate a House Martin nest under the eaves of a house roof in Sardinia, flying to it and peering in. The editorial comment accompanying the report notes that similar behaviour had been seen before.

A female Kestrel has caught a mouse, and after dispatching it is preparing to take it away.

It is mid-winter and the female Kestrel from the previous page has decided to eat her mouse close to where she caught it.

regulus) which had dropped onto the water, and later retrieved several bodies of recently drowned passerines in similar fashion. Messenger *et al.* (1988) also observed a Kestrel at Seaforth, Merseyside attacking a group of Leach's Storm-petrels (*Oceanodroma leucorhoa*) catching one about 15cm above the water.

In a study of urban Kestrels in a town in north-east Slovakia, Mikula *et al.* (2013) noted the birds taking Swifts (*Apus apus*) and bats. Although some prey were taken in opportunistic surprise attacks, the technique favoured by the falcons for both prey types was to wait close to, or at the mouth of, ventilation shafts in the hope of taking prey as they emerged from roosting. The authors suggest 'sit-and-wait' as a new technique to be added to the hunting repertoire of Kestrels. The bats and Swifts had arrived in the town before the Kestrels, and were roosting and breeding in the numerous ventilation shafts and crevices of older buildings. The Kestrels, arriving several years later, had needed time to develop the new hunting technique, Mikula and co-workers noting that not all observed Kestrels were adept at taking this 'new' prey. However, those that were, reared, on average, one more fledgling than those that did not (a mean of 3.5 fledglings rather than 2.5).

Riddle (2011) also records the attempted taking of Hobby chicks, the Kestrel being fought off by the adult Hobby, the two raptors eventually locking talons as they fought. Before leaving avian prey, it is worth noting a fascinating study in Finland by Huhta *et al.* (1998). Using a black marker pen (!) to dull the plumage of Great Tits (*Parus major*) the Finnish researchers showed that hungry Kestrels were no more attracted to the unmarked, bright tits than to the duller ones, which, if they were able to understand, would doubtless be of some comfort to our more colourful small passerines.

Being crepuscular as well as diurnal, Kestrels take bats. One study has already been mentioned above (Mikula *et al.* (2013)). In another, in southern Spain, Negro *et al.* (1992) note predation of Pipistrelles (*Pipistrellus pipistrellus*), the Kestrels hunting these in winter, preferentially choosing fine weather, when the bats emerged from their roosts earlier and in greater numbers. The Kestrels' technique was either to circle above the bat flock before diving on an individual, or by perch-hunting, a short fast flight taking a bat from the flock. By analysis of its pellets, one radio-tracked male Kestrel was found to have a daily diet comprising 30%-60% of bats over a nine-day period. The male hunted bats only in good weather: in poor weather the Kestrel roosted early. Duquet and Nadal (2012) also studied bat hunting, on this occasion in France. The researchers list a total of 11 bats taken by avian predators. Although Hobbies took the highest percentage (39%), Kestrels were responsible for 20% of identified bat kills.

In general Kestrels do not take carrion, but records of them doing so exist (*e.g.* Dickson and Dickson, 1993) and it may be that for some individuals or in some circumstances it is a more common behaviour than most observers consider. Kestrels also exhibit kleptoparasitism, the pirating of prey, from other species, a behaviour which is common among predatory species (Brockman and Barnard, 1979). Kestrels will take prey from other raptors such as Sparrowhawks and other Kestrels, and from other species, *e.g.* Magpies (*Pica pica*): both Village (1990) and Korpimäki (1984a) note that Short-eared Owls (*Asio flammeus*) are frequent victims of piratical attacks, probably because both species are diurnal vole hunters and so more likely to share hunting areas and times. Interestingly, in a study in western France, Fritz (1998) noted that kleptoparasitism of Short-eared Owls was more prevalent when the wind was lighter and so unfavourable for flight-hunting. Fritz estimated that one successful piratical attack compensated for 50% of the reduction in hunting yield on days when the wind was too light for flight-hunting. Overall, Fritz found pirate attacks were successful against the owls 50% of the time. Interestingly, Fritz also noted that pirating was most successful when a pair of Kestrels operated in conjunction. The female would initiate the attack, both falcons then attacking in turn until the owl dropped its prey. The male would then dive after the prey, taking it before it hit the ground, while the female would stay below the owl, preventing it from any attempt at retrieval or revenge attack. Such attacks show a high degree of co-operation and emphasise the flight agility of the smaller male.

Piratical attacks on Barn Owls are also known, while Shrubb (1993a) notes that Kestrel attempts to steal prey from Weasels have been recorded. However, Kestrels are also the victims of piracy, Combridge and Combridge (1992) noting a male Red-footed Falcon forcing breeding Kestrels to give up prey destined for their nestlings. Kettle (1990) records a juvenile female Red-footed Falcon attacking a juvenile Kestrel sat on a fence post to pirate a vole. Given the Red-footed Falcon was smaller this was a bold strategy, the more so as it dropped onto the back of the Kestrel, sinking its talons into it and dragging it to the ground. In the fight that followed the Red-footed seems to have been lucky to survive, but took the vole when the Kestrel flew off. In Scotland, Wildman *et al.* (1998) observed a Red Kite (*Milvus milvus*) robbing a Kestrel of a Brown Rat (*Rattus norvegicus*) on the ground, while Prŷs-Jones (2018) records a pair of Red Kites pursuing a Kestrel with prey, one kite following the falcon's every twist and turn, while the second flew behind and 5m below. Eventually the Kestrel dropped its prey which the lower kite retrieved in mid-air. But the kites were not satisfied, the one without the prey most definitely not, as it continued to pursue the falcon. This incident is interesting for two reasons, firstly as it illustrates further cooperative behaviour in piracy, and secondly because it implies predatory behaviour: that possibility is considered again when Kestrel foes are explored in Chapter 10.

Both *et al.* (2013) also recorded a curious incident in The Netherlands where a Peregrine attacked a Kestrel in the air, forcing it to drop its prey which the larger falcon then caught in mid-air and landed. However, the Peregrine then discarded the prey (a Common Vole (*Microtus arvalis*)). As Both *et al.* note, a vole is only a snack for a Peregrine, but it still odd that it did not eat it. In this case, it seems, the Peregrine reacted instinctively to the Kestrel rather than being driven by hunger. Both *et al.* also recorded other successful Peregrine pirate attacks on Kestrels.

Piratical attacks by Kestrels on Hobbies have also been noted (*e.g.* Huitzing (2002) who saw a Kestrel pirating a juvenile Hobby), though the reverse seems more common, presumably because Hobbies have superior aerial agility. Cresswell (1993) observed the mobbing of raptors, noting that if this was by prey species then it occurred to emphasise that the hunter has been seen and so should leave the area. This confirms the results of Pettifor (1990) who found that the mobbing of Kestrels caused them to move greater distances between foraging positions, with flight-hunting falcons being mobbed more frequently than those perch-hunting. But Cresswell also noted piratical attacks, in both directions, by raptors (including Kestrels) and corvids, and concluded that much of what was assumed to be mobbing between raptors with species of comparable size, *i.e.* other raptors or corvids, is in fact piratical behaviour.

One of the more unusual prey items for a Kestrel. This female has caught a European Eel (*Anguilla anguilla*) on a damp meadow in the Merdja Zerga National Park, Morocco. *Torsten Prohl.*

Juvenile Kestrel with a Common Shrew. Given a choice nestlings will leave shrews, but once fledged they are less fussy..

Composition of the Diet

Analysis of Kestrel pellets has allowed a reasonably accurate examination of the diet of Kestrels across a range of habitats, though the poor survival of feathers and the difficulty of identifying the skeletal remains of birds may mean that unless claws or bills are found the avian content is underestimated.

Village (1990) notes that studies of Kestrel diet in temperate regions indicate that voles are the most important prey. In his own studies, undertaken in an area of grassland near Eskdalemuir, Scotland, and in mixed and arable farmland areas in east-central England, Field Voles (*Microtus agrestis* – also known as the Short-tailed Vole, a name favoured by Village) were the most common prey, occurring in over 90% of the pellets collected in Scotland and around 75% of those from the English sites. Shrubb (1993a), collating data from studies in both Britain and Scandinavia, noted that overall for those countries vertebrate prey accounted for 56% of the total diet (75% being mammal, 22% bird, and 3% reptile or amphibian): voles comprised 45% of the mammalian prey. It therefore seems probable that voles form the greater part of the Kestrel diet where they are abundant, and this was certainly borne out in a study in The Netherlands where voles (in this case the Common Vole) contributed over 90% of the Kestrel diet in all months except June, when the Common Shrew formed about 10% (shrews being more active on the surface at that time), and December when songbirds formed about 10% (Masman *et al.*, 1988a).

With such an attrition rate it might be supposed that voles would alter their behaviour, particularly their breeding behaviour, but as there are few, if any, areas in which vole populations are not predated this is not an option. To confirm this Klemola *et al.* (1998) set up an experiment in Finland, in which caged voles were placed beneath active Kestrel nestboxes and, as a control, under empty nestboxes. Although the voles were not subject to predation (no attacks on the cages were recorded), they were subjected to the begging calls of hungry Kestrel nestlings, the excreted scats of those nestlings and the comings-and-goings of adult birds. There was no alteration in the breeding behaviour of the voles in comparison to those not subjected to such obvious signs of the presence of predators, female voles becoming pregnant with equal probability in both the active box and control cages. The body weight of the voles showed no decrease during the experiment suggesting that the presence of the Kestrels was not unduly stressing the rodents. In a rather more gruesome experiment, carried out under licence from the Finnish Ministry of the Environment, Hakkarainen *et al.* (1992) recreated natural conditions in the laboratory to see if Kestrels had a preference for prey species. They found that the falcons had no preference, but that the probability of a Sibling Vole (*Microtus levis*)[3] being captured was much higher. Hakkarainen and co-workers did not find a satisfactory explanation for this result, though they did note that Sibling Voles tended to spend more time in cultivated fields than their Field Vole cousins who preferred the longer grass of hay meadows. The researchers also noted that the home ranges of Sibling Voles overlapped much more than those of Field Voles and that they were consequently more social: this meant that the Sibling Voles aggregated more readily and so offered a greater target for a hunting Kestrel. However, later work by Koivula *et al.* (1999b) has offered a different suggestion as we shall see in Chapter 4.

---

[3] The Sibling Vole has a complicated taxonomic history having been originally classified as *Microtus rossiaemeridionalis*, then *M. epiroticus*, and currently *M. levis*. Occasionally called the East European Vole, the species differs from the Common Vole only at the chromosomal level, though there are minor physical differences in the genitalia. The Sibling Vole is found in eastern Europe and central/southern Finland. Interestingly, there is also a small colony on Spitsbergen, an island of the Svalbard archipelago. It is believed the vole was accidentally introduced there some time between 1920 and 1960 when hay was shipped to the Russian coal mining colony of Grumantbyen as horse fodder: the horses were used in the mines. The voles were first noticed in 1960 and are the only permanent terrestrial mammal on the island (though some Arctic Foxes (*Alopex lagopus*) are probably resident and Polar Bears (*Ursus maritimus*) are frequent visitors). That such an animal, which remains active throughout the winter, has survived is remarkable.

Studies elsewhere in Finland (Korpimäki 1985a, 1985b) have confirmed that northern Kestrels are primarily a predator of mammals, but have also revealed some interesting aspects of the species diet. Korpimäki (1985a) confirmed that the majority of the prey was mammalian, but that Finnish Kestrels hunted in different habitats when seeking avian rather than rodent prey. Rodents were hunted in open fields, while 71% of avian prey were forest species, implying that Kestrels find forested areas an easier environment for hunting birds. One interesting outcome of the work was that males, in both rodents and birds, dominated the catch: Korpimäki conjectures that this was likely due to the greater activity of males in the spring. By studying pellets and prey remains, Korpimäki (1985b) noted a total of 11 mammal species comprising 72.1% of the diet by biomass. A noteworthy 29 bird species comprised 27.0% of the diet by biomass. The remaining 0.9% comprised amphibians (a single Common Frog), reptiles (a single Viviparous Lizard) and insects. Of the mammals the majority were voles, though the percentage varied with the cyclic nature of vole numbers. Mammals included both the Brown Rat and the northern sub-species of the Weasel (*Mustela nivalis nivalis*). Occasionally called the Pygmy Weasel, the sub-species is significantly smaller than its British cousins (males about 166mm long, 54g, against British males 202mm, 73g: females are smaller in each case), but would still be a formidable foe if the attacking Kestrel were to fail to kill at the first attempt.

Of the birds identified by Korpimäki (1985b), the most commonly taken were Redwing (*Turdus iliacus*), Fieldfare (*T. pilaris*), Chaffinch (*Fringilla coelebs*) and Phylloscopus warblers. Korpimäki noted that if a chick was to die in the nest it would usually be eaten by its siblings. Korpimäki also noted that the dietary width (*i.e.* the number of prey species appearing in the diet) of breeding Kestrels increased as summer progressed. Korpimäki conjectured that this increase was influenced by several factors. One was a change in the local environment, an increase in vegetation cover meaning that voles, the primary prey, became less accessible. Another was the increased food requirements of the Kestrel nestlings, while a third was the availability of fledglings of other species, young rodents and insects. Extending this study of dietary width, Korpimäki looked at the diversity of prey with latitude, *i.e.* comparing the diet of Kestrels from across their European range. What was clear was that the further north Kestrels bred the greater their dietary width. As already noted above when considering the diet of Kestrels in southern Europe, at that latitude the bulk of the prey is insect, with a limited range of mammalian and avian prey species (even though these provide the majority of the biomass). However, in Scandinavia, the Kestrels find a much broader range of species to hunt.

In a separate report covering the insect species taken by Kestrels in western Finland, Itämies and Korpimäki (1987) noted that beetles represented 75% of all the insects taken, though the range of species was very wide, including bees and ants. The Finnish researchers also noted that the percentage of insects in the diet declined with both clutch size and the number of nestlings, reflecting

Willow Warbler, Varangerfjord, northern Norway. It was surprising to find these tiny warblers not only this far north so early in spring, but also so abundant. As Kestrels were preparing to breed it was interesting to note that, consistent with Korpimäki (1985b), the warblers were part of the falcons' diet.

the fact that while insects are abundant, their energy content is low and the cost of catching and transporting them is high, making them unattractive prey when there are many hungry bills to feed.

Studies of the Kestrel diet in The Netherlands confirm a high vole fraction (Cavé, 1967 and particularly Masman *et al.,* 1988a), as does that of Shrubb (1993a) in Britain. In the data of Village (1990) the variation in the percentage of pellets containing vole remains in Scotland throughout the year was minimal, as it was in English mixed farmland, though the percentage dropped in English arable farmland area from mid-summer, reaching about 55% in mid-winter before rising again as spring approached. Interestingly, in Village's Scottish study area about 50% of pellets (on average) during the five-year period contained only vole remains, that percentage dropping to about 10% in the farmland study areas, reflecting the larger number of habitats in the latter and, hence, the greater number of rodent species. Also of interest was the fact that Wood Mice (*Apodemus sylvaticus*) were entirely absent from the Scottish pellets, but were found in English farmland in summer. As Wood Mice are nocturnal their absence from Scottish pellets is understandable: their presence in English pellets suggests that the mice moved into arable fields as crops become available and were then more easily discovered in summer, particularly during harvesting when machinery disturbed them, making them visible to waiting Kestrels. This result was confirmed in the studies of Kestrels nesting in a barn in southern England (see Chapter 8) where mice were not seen in the prey fed to the chicks until late in chick rearing. In Ireland, where there are no Field Voles, Kestrels take Wood Mice, but mostly prey on birds (Fairley and McLean, 1965, and Fairley, 1973). Fairley and McLean (1965) also note the taking of House Mice (*Mus musculus domesticus*). In his study of the Kestrel diet in English farmland, Village (1990) noted that there was some suggestion that Kestrels took voles when both voles and mice were available, but did not consider that the evidence was sufficient to imply genuine selectivity. In general, the results of his study indicated that prey selection was based on abundance.

Bank Vole. Though the species is preyed upon by Kestrels, the rodents are chiefly nocturnal in summer and so are less often seen in the prey list of breeding Kestrels (see, for instance, the study reported in Chapter 8). They are more prevalent in the Kestrels' winter diet.

By contrast to the essentially constant fraction of voles in the pellets analysed by Village (1990), the occurrence of bird remains was seasonal, peaking in June/July, the time when broods fledge, reflecting the relative ease with which Kestrels can catch young, inexperienced birds. Village noted beetle remains in 50-60% of farmland pellets but, perhaps not surprisingly, in a smaller fraction of Scottish grassland pellets: in each case the percentage of pellets containing insect remains increased during the summer, before falling off as winter approached. The earthworm content of the pellets was surprisingly high, the annual average for the farmland study area being around 50%. The percentage was much lower (a few percent only) in the Scottish grassland. In each case the percentage was higher in the late winter/early spring period, reflecting the likelihood of unfrozen, but waterlogged ground forcing worms to the surface. In the farmland area, the percentage then fell, before rising again in autumn, as rainfall increased (Fig. 8).

The data of Fig. 8 are consistent with that of Shrubb (2003) for Kestrels in Wales, and with Shrubb (1993a) for areas of England, but the latter adds interesting data on the variation of mammalian prey during the year, noting that Bank Voles (*Myodes glareolus*), Harvest Mice (*Micromys minutus*) and Wood Mice were taken more often in winter when they became less predominantly nocturnal, while rats were more frequently taken in spring and autumn as they moved to, and from, the fields. The data of Fig. 8 are also consistent with that from a study in a boreal area of Norway (Steen *et al.*, 2011) in which video monitoring of 55 nests sites allowed the composition of prey delivered to nestlings to be analysed. Steen and co-workers noted that 60% (by number) of prey items were voles, with birds (14%), shrews (12%), lizards (9%), insects (3%), frogs (0.4%), unidentified prey forming the remainder.

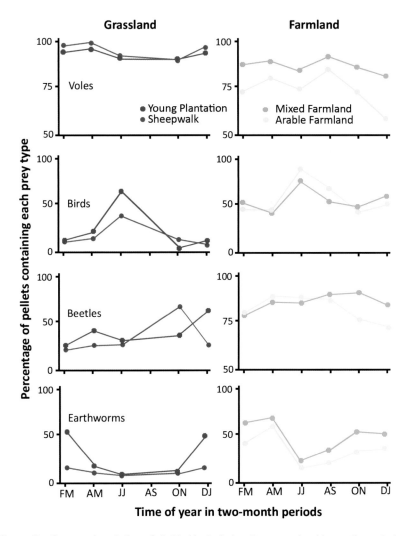

Figure 8. Seasonal variation of diet in Kestrels hunting grassland in south-central Scotland and farmland in east-central England. Each dot is the percentage of pellets gathered in a two-month period that contained the particular prey type. Redrawn from Village (1990).

A female brings in a Common Shrew (*above*) and the male brings in a Field Vole (below) for their chicks. in each case the adults have arrived from behind the nestbox to the obvious surprise of one chick. Normally prey was brought in over the author's hide allowing the chicks to see what was coming and to start calling loudly, an unwitting alarm that it was time for photography.

Female Kestrels bring in a Wood Mouse (*above*) and a Field Vole (*below*) for their chicks.

Shrubb (1993a) also noted the difference in diet between male and female Kestrels: males took about twice the fraction of insect prey that females secured, but this was reversed for earthworms. Males took more birds than females (though this, perhaps, may have been accounted for by the number of chicks taken by a male feeding a female and nestlings), but it was invariably the female Kestrels that took reptiles. These data are also consistent with that of Village (1990) and of a personal study in Hampshire (see Chapter 8), though the conclusion must inevitably be tempered by the fact that prey delivered by the female to chicks need not necessarily have been caught by her.

Village (1990) also compared the diet of adult and juvenile Kestrels (Fig. 9) noting that the fraction of avian prey in the pellets of younger falcons was reduced, while that of insects increased, clearly indicating the relative ease of capture of insects. Interestingly, the fraction of voles in the pellets was similar for adults and juveniles, supporting the intuitive suggestion that the skills required to catch voles are less demanding than those required to catch birds.

Yalden (1980) studied the diet of urban Kestrels in Manchester. The diet comprised birds (76% by weight) and mammals (22%) with traces of insects and earthworms. The avian prey chiefly comprised passerines (58% by number, 64% by weight) of which the majority (about 90%) were House Sparrows. The next largest fraction was pigeon (Feral (*Columba livia var*)), which comprised 4% by number (10% by weight) of the diet. Goldfinch (*Carduelis carduelis*), Dunnock (*Prunella modularis*), Starling, Blue Tit (*Parus caeruleus*) and Pied Wagtail (*Motacilla alba yarrellii*) were also found, though in all cases Yalden is cautious to claim certainty of identification because of the difficulty of identifying bird remains in pellets. More recent studies of an urban Kestrel diet have been made by Romanowski (1996) in Warsaw. Romanowski studied two urban sites, one close to the centre of the city, an open area with parkland, the other a heavily built-up area where the falcons nested on the flat roofs of houses. In the central area birds comprised only 12% of the breeding season diet, House Sparrows being the major prey species, with mammals forming the bulk (79%), and insects the remaining 9%. However, in the built-up area birds comprised most of the diet (57%) with mammals accounting for 38% and insects 5%, consistent with the Yalden data. The differences seen in Warsaw between the 'true' urban and 'open' urban areas were also well-illustrated in a study in Berlin where Kübler *et al.* (2005) observed the dietary differences in various city areas. The team defined three areas – the City with its high concentration of housing, a Mixed Zone, with apartment and industrial buildings, interspersed with parkland and gardens, and the Outskirts, with detached housing and some industrial buildings, but a higher fraction of parkland and gardens. The team then collected 2144 pellets (split 619, 695 and 830 between the three zones) and identified four prey types – birds, insects, mice and shrews – in seven combinations. The differences in diet are shown in Fig. 10.

A similar result to that of Kübler *et al.* was found by Boratyński and Kasprzyk (2005) in north-central Poland. In this case the authors assumed that they

Diet

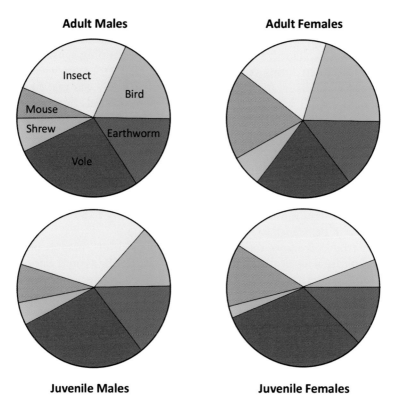

Figure 9. Variation of diet between juvenile and adult males and females during one winter on farmland in east-central England. Redrawn from Village (1990).

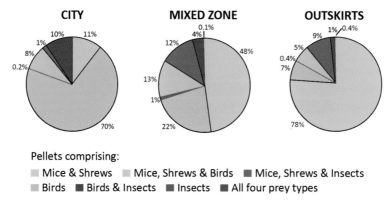

Figure 10. Diet of Kestrels from an analysis of over 2,000 pellets collected in three urban areas (as defined in the text) during the breeding season. Redrawn from Kübler et al. (2005).

would find a preponderance of avian prey in the urban Kestrel diet and were surprised to find that mammals comprised the overwhelming bulk. It was only when they observed the hunting range of the falcons that this apparent anomaly was explained: the Kestrels' range was extending across a large area outside the town limits comprising suburbs and farmland where rodents were readily available. Another interesting urban study was made by Riegert and Fuchs (2011) who studied the diet of male Kestrels during the winter in České Budějovice, a town in southern Bohemia of the Czech Republic. Winters are harsh in the town, yet insects remains were found in almost 16% of analysed pellets. The bulk of the falcon diet was rodents (67.5%), chiefly voles, with only 6.5% of pellets including bird remains. However, the town had many days of snow cover during the winter, Riegert and Fuchs finding that as the fraction of days with snow cover rose the proportion of rodents and birds altered significantly, with bird remains climbing to around 40% as vole fragments in pellets fell to about 20%.

While Fig. 10 shows the greater dietary breadth of 'rural' as opposed to 'urban' birds, it is also a clear example of the flexibility and opportunistic nature of the Kestrel's diet. An even more obvious example of this was found by Carrillo *et al.* (2017) in their study of the Kestrel diet on the island of Tenerife. The island is home to the sub-species *F. t. canariensis* which inhabits a range of habitats. Across the range insects formed the bulk of the prey by occurrence, though in the arid southern island the endemic Gallot's Lizard (*Gallotia galloti*) was the primary prey both in terms of occurrence (for both adult falcons and chicks) and biomass. Carrillo *et al.* also compared the diet of the island's falcons with that of other southern breeding Kestrels (Fig. 11).

As can be seen from Figs 10 and 11, Kestrels have the flexibility to switch prey if their territory does not provide adequate numbers of their preference

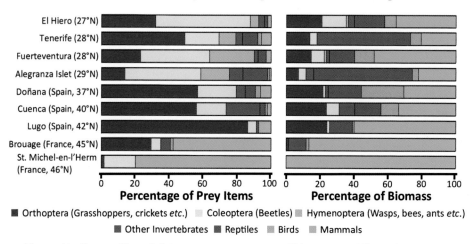

Figure 11. Composition of diet as prey occurrence and biomass, at different latitudes. The first four entries are islands of the Canaries. Data sources for the individual entries are given in Carrillo *et al.* (2017). Redrawn from Carrillo *et al.* (2017).

(*i.e.* rodents), but will always take a higher fraction of rodents if a full spectrum of prey is available.

Hunting Strategy and Food Caching
Kestrel hunting techniques are dealt with in detail in Chapter 4, only some general comments being made here.

The main hunting techniques are flight-hunting and perch-hunting. In the latter the falcon chooses a perch at a height which allows a wide scanning area and watches for movement on the ground or in the air. Once prey is detected the Kestrel will then fly directly to it or may flight-hunt above ground prey briefly. Generally, while perch-hunting the falcon will move to a new perch every few minutes. Kestrels will also soar, but this is not a frequently used technique as it requires specific weather conditions, *e.g.* rising thermals, but thermals which do not carry the bird too high or too fast as either would make spotting rodents more difficult. Updrafts can, of course, also aid soaring, but are usually more useful for flight-hunting. Kestrels will occasionally use a combination of soaring and flight-hunting to move across the landscape, the flight-hunting pauses allowing a better scan of likely-looking areas. The success rate of hunting Kestrels is considered in Chapter 4.

It is no surprise to discover that birds know the feeding times of their prey. In studies in The Netherlands (Rijnsdorp *et al.*, 1981, RUG/RIJP, 1982) the researchers noted that voles emerged from their burrows to feed in two-hour cycles (Fig. 12 overleaf), and that the arrival of Kestrels to hunt corresponded with these probable emergence times. The number of Kestrels hunting decreased over the two-hour feeding time as a successful bird would catch prey and fly off to consume it, perhaps not returning until the next vole feeding period. The Dutch noted that Hen Harriers (*Circus cyaneus*) saved about 90 minutes of flight-hunting daily by adopting this temporally-linked behaviour (corresponding to about 12% of their daily energy intake). It must be assumed that Kestrels benefit from a similar reduction.

While Fig. 12 indicates a two-hour cycle of vole activity, a similar cycle was not seen in the study reported in Chapter 8 where peak capture rates were seen in the early morning (9-11am) and the evening (around 8pm), with no peaks at other times. This suggests that in southern England both voles and Kestrels are not as strictly time-limited in terms of activity as the Dutch observations suggest, with a cyclical nature of activity overlaid by a more random pattern of foraging. Kestrels usually consume invertebrate prey on the ground, though larger insects may be lifted into the air, and even partially consumed in the manner of Hobbies while flying to a nearby perch. Mammals and birds are taken to a perch: birds are plucked before the meat is taken in a succession of torn-off morsels. Mammals are eaten in similar fashion, sometimes after an attempt to remove the fur. Once they are large enough juveniles at the nest often consume rodents brought in by their parents by swallowing them whole. This behaviour is probably to avoid competition by siblings, though hungry

The Common Kestrel

Male Kestrel starting to eat its prey on the wing. Invariably the falcons, and other birds of prey, eat the brain first as it is rich in lipids and other essentials. It is noticeable when watching falcons feeding their young that the head is almost always consumed before the young are fed during the early period of chick-rearing, though when the chicks are capable of handling the prey the head may be left on.
Torsten Prohl.

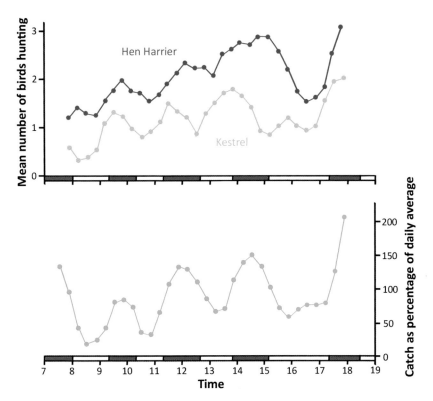

Figure 12. The upper curves show the daily average number of Hen Harriers and Kestrels hunting over a field of Lucerne (Alfalfa) in The Netherlands during a period of the breeding season (running mean of three 20-minute periods). The lower curve shows the number of voles trapped by the research team as a percentage of the daily average catch (running mean of three 20-minute periods). In each figure the red blocks on the time axis indicate the periods of maximum vole activity as indicated by catch rate. Redrawn from RUG/RIJP (1982).

adults will occasionally do the same thing or will consume the whole prey in one meal. In general, though, adults discard the internal organs, the grass-filled stomachs of voles seeming to be especially distasteful (and pointless in view of the available energy content).

All falcons cache food, and the behaviour is not restricted to wild birds as falconry birds will also cache despite their meals being delivered regularly. Since cached food will, if discovered, be eaten by the discoverer, falcons are furtive when caching, moving the food if (presumably) the bird believes it has been watched or if it believes the hiding place offers inadequate protection from theft. Once food is hidden, the falcon will observe the cache site from a short distance, presumably to fix it in its memory. Caching is used when prey is abundant, particularly if the falcon catches several prey items, for instance if taking chicks from a nest, if catching several juveniles in a flock, or if catching prey which is more abundant at certain times of day (*e.g.* bats as they emerge at dusk, or certain seabirds which fly far out to sea to feed, returning to their roosting/breeding sites at nightfall). It is also used in winter, allowing the bird to benefit prior to or after enduring the long hours of darkness: it may also be useful if winter weather prevents hunting. Caching may also be used occasionally as a form of food pass, a male caching prey in full view of a female, then departing.

Kestrels tend to cache larger prey items, particularly if smaller prey items have already been consumed. Masman *et al.* (1986a) provide interesting data on the weights of voles (in this case Common Voles as the study was carried out in The Netherlands) eaten, fed to chicks or cached which indicates (Fig. 13) that adult birds tend to consume smaller voles, feeding larger specimens to the chicks and caching the largest. In their study of the energy intake of Kestrels (see Chapter 5) the Dutch group noted that it was clear the birds were heavily dependent on cached food during days when the weather prevented hunting, a fact which probably explains the reason for caching larger prey. The observations of the Kestrels allowed the Dutch team to retrieve the cached items, identifying the species and weighing the prey before returning it to the cache. An interesting outcome of the weighing was that subsequent observation of adult Kestrels eating the cached prey allowed the Dutch to plot time of consumption against prey weight (Chapter 5 *Fig. 43*).

Village (1990) reports two occasions on which Kestrels discovered young rats, as a result of nest disturbance by harvesting on one occasion, juveniles making early explorations from the nest on the other: the Kestrels were able to

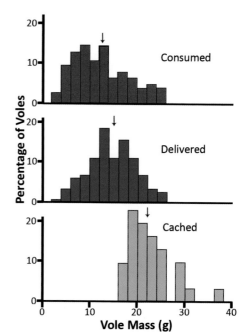

Figure 13. Frequency distribution of body mass of voles consumed, delivered and cached by a male Kestrel provisioning himself, his mate and their brood during April and May. The arrows indicate the mean mass of each population. Redrawn from Masman *et al.* (1986).

take five or six in the space of a few minutes, caching them all. Such caching would be particularly useful if there are chicks to feed. Although it appears to invite theft by terrestrial scavengers, most caches are on the ground (Table 3). This is perhaps because invariably Kestrels are hunting on open ground with few possible above-ground cache sites, but might also be due to cached food being most frequently retrieved on the same day or early the following day, reducing the time frame for, and therefore the likelihood of, theft. Caches are usually on a prominent tussock or at the base of a post, but despite the bird retrieving the prey within a relatively short period, Kestrels seem to lack the legendary memories of Corvids (particularly Jays (*Garrulus glandarius*)) and may fail to find their caches.

Cache Position	Percentages of 45 identified caches
Percentage on the ground	
at base of post, pole or tree stump	22
tuft of grass or straw	33
clod of earth	16
stone	2
Total percentage on ground	**73**
Percentage above ground	
on buildings	5
on fence posts	5
on straw stacks	11
in trees or shrubs	6
Total percentage above ground	**27**

Table 3. Kestrel food cache sites. Tabulated from data in Village, 1990.

While caching obviously has benefits in times of prey abundance, it is also seen in winter when it would allow the bird to profit from an evening meal before enduring the long, cold hours of darkness with no opportunity to hunt. Coupled with the possibility of poor weather the next day which might make hunting difficult, this would make winter caching a survival strategy, something seemingly borne out by the work of Rijnsdorp *et al.* (1981) on Kestrels in The Netherlands (Fig. 14 overleaf). However, Rijnsdorp *et al.* also noted that caching an evening meal allowed the bird to hunt most efficiently by reducing body weight during the day which could be useful during the breeding season: the Dutch team suggest that putting off eating until the evening would result in an energy saving during flying of 7% of the daily subsistence energy. However, it is also possible that for male falcons, the need to feed chicks during the

The Common Kestrel

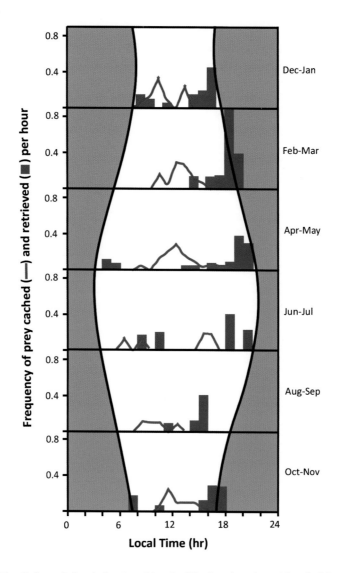

Figure 14. Daily variation in food caching (red line) and cache retrievals (blue histograms) for Eurasian Kestrels in The Netherlands. The figure clearly shows the prevalence of late evening retrievals in both the breeding season and winter. The shaded areas are night-time. Redrawn from Rijnsdorp *et al*. (1981).

day might explain why meals are taken in the evening. Rijnsdorp *et al.* note that female Kestrels do not concentrate their feeding during the evening when breeding, reinforcing the idea that male evening feeding is a weight reduction strategy: during the early phases of breeding female Kestrels do not hunt and do not, therefore, need to minimise their daytime weight.

Diet

In winter a more logical explanation for food caching and evening consumption would seem to be a survival strategy, consistent with the data of Masman *et al.* (1986a) during observations of wild Kestrels and calculations of energy intake by the birds. This work allowed the energy input of the falcons to be established, both in terms of the variation during the day, and month-by-month throughout the year (Fig. 15).

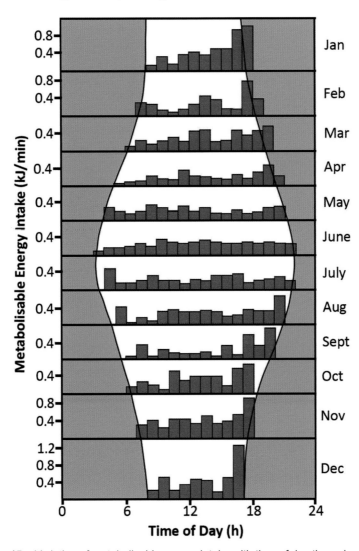

Figure 15. Variation of metabolisable energy intake with time of day throughout the year. As can clearly be seen, while during the summer months the energy intake is more or less constant during the day, from September to March the intake increases towards the end of the day to aid the bird in coping with the rigours of night-time. The shaded areas are night-time. Redrawn from Masman *et al.* (1986a).

Fig. 15 clearly shows that in winter months the Kestrels' energy intake rose in the period immediately prior to nightfall, whereas it remained essentially static during the days of summer. It is also consistent with the data from another study by the Groningen University group (Dijkstra *et al.*, 1988). Using electronic weighing scales mounted at the entrance to nestboxes used as roosts, and later for breeding by five pairs of Kestrels, the researchers were able to weigh male and female falcons at various times during the day and through the breeding cycle. The results are shown in Fig. 16.

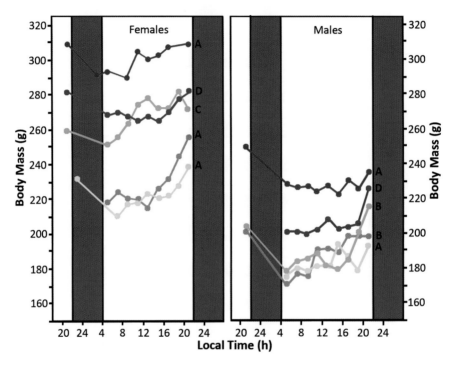

Figure 16. Variation of the weight of 5 female and 5 male Kestrels during the day at various stages of the breeding season. 'A' is during courtship, 'B' is during incubation, 'C' is while chicks are less than 10 days old and 'D' is when chicks are more than 10 days old. Shaded areas are night-time. Redrawn from Dijkstra *et al.* (1988).

The fluctuations in weight of both sexes are consistent with other studies by the Dutch team (see Chapter 5), but what is striking is the weight loss of the birds during night hours. Given that the study was carried out during the breeding cycle when night-time temperatures are benign in comparison to those of winter, it is not surprising that the Kestrels would wish to eat prior to nightfall in winter: if spring/summer nights can result in weight losses of *c.*10%, winter losses might well be catastrophic without an evening meal.

Some indication of that is given in data of Masman *et al.* (1986a) taken from 125 observations of the energy intake of three Kestrels during four winters in The Netherlands (Fig. 17).

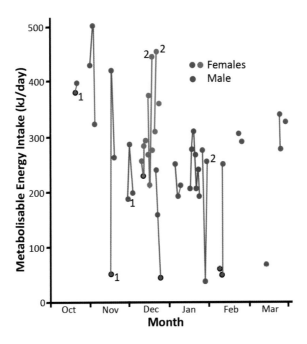

Figure 17. Day-to-day energy intake of two female and one male Kestrel during the winter months. Lines connect consecutive days of observation. Days of bad weather (rain or high wind) likely to have inhibited hunting are marked with black circles around the data points. '1' indicates days when prey was caught, cached and then retrieved by the falcon. '2' indicates when prey was caught and cached but not retrieved. Redrawn from Masman *et al.* (1986a).

As can clearly be seen, the effect of bad weather is to dramatically reduce the energy intake of the falcon. Equally clearly, Fig. 17 shows the value of consuming a cached meal on those occasions by allowing the bird to replace its energy losses.

4   Hunting

Avian flight is complex, the mathematics used to explore it no less so. The mathematical problems are also compounded by the fact that flight is more than mere wing shape and beat frequency, flying being energetic and so involving biology as well as physics. For interested readers, the books of Pennycuick (2008) and Videler (2005) represent good starting points. Here I briefly discuss only the basic features of falcon flight.

In a classic paper on the flight of avian predators, Andersson and Norberg (1981) defined six flight characteristics important in the capture of prey – linear acceleration in flapping flight; maximum speed in horizontal flapping flight; terminal dive speed; maximum rate of climb; angular roll acceleration; and turning ability. The last two define the manoeuvrability of the bird[1]. What Andersson and Norberg found was that of the six, only terminal dive speed was improved by body mass, all the others varying with the inverse of mass, *i.e.* smaller birds fare better. Higher body mass allows a stooping falcon to accelerate faster under gravity, the bird closing its wings to reduce its profile and so reduce drag. However, despite what is occasionally seen written, stooping falcons cannot fold their wings completely unless they dive vertically: in any high angle stoop the wings must be partially open to provide the lift necessary to maintain the stoop angle. But while a high body mass increases stoop speed, it reduces the possible rate of climb: acceleration and horizontal speed also decline with increasing body weight. It is therefore not surprising

---

[1] In another classic paper, on the influence of natural selection on the shape of birds' tails, Thomas and Balmford (1995) define two aerodynamic parameters which influence flight patterns. These are 'agility', defined as maximum rate of turn (usually associated with high speed flight) and 'manoeuvrability', defined as minimum turning radius (which occurs at low speeds). The two parameters are essentially the same as those that define 'manoeuvrability' in Andersson and Norberg (1981).

A female Kestrel departs after delivering prey to her brood.

that stooping is most usually associated with the larger falcons, though the smaller falcons will also utilise it as a hunting technique, with Kestrels being known to stoop occasionally[2].

These considerations of flight ability against body mass might imply that a low body mass is an advantage to a falcon whose hunting technique does not include regular stooping, but mass is not important only in terms of acceleration under gravity. Falcons must be able to stop their prey and then carry it away (while eating at the point of capture is acceptable to an adult bird, breeding requires prey to be returned to the nest), both of which require a body mass which is at least comparable to that of the prey. Secondly, as already noted, there is more than physics to falcon flight, and while body mass is undoubtedly important (and wing loading (see Box 1) perhaps more so) the power input to flapping flight and into a quick take-off and fast acceleration are important in surprise hunting. Physiology must therefore be considered, initial power output being high, favouring short, fast attacks followed by longer resting periods, as may be seen in Kestrel perch-hunting and in their hunting of avian prey.

---

[2] That Kestrels are not designed for regular stooping as a hunting technique is apparent from the bone structure of their arms (*i.e.* their wings). In an investigation of the difference in bone strength of Peregrines and Kestrels (and Sparrowhawks and pigeons for comparison) Schmitz *et al.* (2018) found that the bone mass of the arm and shoulder girdle of Peregrines (which regularly stoop on prey) were significantly higher than those of Kestrels (and the other investigated species), as was the mineral density and strength. The wings of Kestrels are not adapted for regular high-speed stooping.

Box 1 Wing Loading
Andersson and Norberg (1981) noted that wing loading was important in the two characteristics which affect the 'manoeuvrability' (see *Footnote 1* p90) of an avian predator. Wing loading is defined as the ratio of body weight and wing area and is generally given in units of g/cm^2. The variation across the avian world is remarkable, varying from around 0.2 for hummingbirds to 2.3 for members of the auk family which are heavy, but have short, stubby wings used for both flight and underwater 'flight', water being a medium 1000 times denser than air. The wing loading of birds of prey is generally low, at about 0.3, reflecting the need to carry prey, but does vary with prey type. Wing loading also varies with hunting technique. Harriers and kites, which spend long periods in the air searching for prey which is, in general, small and slow-moving, but (relatively) abundant, have lower wing loadings than those, such as hawks and falcons, which spend less time in the air and attack faster and heavier prey which is less abundant. Falcons feeding on avian prey (*e.g.* Peregrines with a wing loading of about 0.5) have high wing loading which underpins fast flight. However, high wing loading decreases manoeuvrability and falcons which feed on insects (*e.g.* Hobbies with a wing loading of 0.30-0.35) have lower wing loading, (*i.e.* are more agile in the definition of Thomas and Balmford (1995) – see *Footnote 1* p90), and can make tighter turns, the radius of turn being proportional to the wing loading.

In his classic book on falcons, Cade (1982), Tom Cade collected wing loadings from various sources. Most of these sources used an equation of Greenewalt (1962) to give wing area and combined this with the weight of an individual bird. From this, and the average weight of Kestrels, Cade calculated wing loadings of 0.19 g/cm^2 (male) and 0.22 g/cm^2 (female), noting that as Kestrels are heavier in winter the loadings would then rise to 0.26 (male) and 0.29 (female). While the actual numbers Cade quotes are subject to discussion, the fact that wing loading varies with the season is clearly correct as the weight of an individual bird varies significantly – as we shall see in Chapter 5 when considering the weights of both male and female Kestrels during the breeding season. Wing area also varies during each wing beat which means that assessing it from photographs of flying birds is difficult. Assessing it from trapped birds is also difficult as live birds do not react well to having their wings stretched. Even falconry birds, which, when feeding, are willing to have their wings somewhat extended by their owners (with whom they are, of course, familiar) are steadfastly intolerant of having the wing fully extended. The best solution is to work with carcasses, when both weight and wing area can be accurately measured. In his work on assessing the various physical parameters which influence the flight characteristics of birds Pennycuick (2008), having recognised the importance of wing loading in assessing fundamentals such as wing beat frequency, noted a method for accurately assessing wing area on a carcass. (Fig. 18).

Hunting

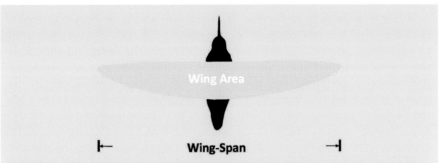

Figure 18. Assessment of wing area. The weight of a gliding bird is balanced by the pressure difference between its upper and lower surfaces multiplied by the wing area. While the wing provides most of the area, the pressure differential operates across the body as well and so the relevant portion of body area must be included. In practice, working with a carcass, the area of one wing, extended to include the body to the spine, is measured and then multiplied by two. The same procedure is used for a live bird, though, as noted in the text, producing full extension may be problematic and requires patience. Redrawn from Pennycuick (2008).

Despite the difficulty of calculating wing area on a living bird, it has been attempted. Videler *et al.* (1988a) quotes wing loading values of 0.27g/cm^2 (for a single 160-163g male, range 0.270-0.274) and 0.31g/cm^2 (for a single 185-192g female, range 0.303-0.316) for wild Kestrels captured for experimentation. For this book I measured the wing loading of *Kai*, a male bird trained by Lydia Newberry. Extending the wing as far as the bird would allow (for which, in comparison to some other falconry birds, *Kai* was very amenable), photographing it, then superimposing a grid on the photograph I calculated a wing area of 630cm^2 with an estimated error of 5%, the area more likely larger than smaller. *Kai*'s weight on the day was 165g (including a 3g radio-finder which was always attached during flights). The wing loading was therefore 0.26g/cm^2, range 0.25-0.27 which compares favourably with the Dutch values. As a 'domestic' bird, *Kai*'s wing loading would not vary with the season.

Measuring the (half) wingspan of the male Kestrel *Kai*.

While it is possible to list the characteristics which define raptor flight, analysing flight mathematically is, as noted, difficult. The mathematics of fixed-wing aircraft, *i.e.* the interaction between the wing and the air, is now well-understood, particularly with respect to the vortices generated and their effect on preferred wing shape and performance. In gliding flight, the closest avian form to a fixed-wing aircraft, the vortices have been studied for the Kestrel (Spedding, 1987a) and allow a reasonable understanding of the behaviour of the wing. However, the avian wing need not maintain the 'simple' geometry of a fixed-wing aircraft: birds can morph their wings by moving individual feathers in three dimensions allowing a vast number of potential geometries. Combined with highly sensitive wind/air pressure sensing this wing morphing allows amazing active flight control even if the wing is not flapping. Flapping flight brings even more complexity as the bird's shoulder rotates during a complete up/down stroke and, as has recently been shown, the left and right wings may move differently during the complete wing beat to allow for alterations in wind speed across the body due to gusting[3].

Spedding (1987b) produced beautiful images of the vortices in flapping flight (Fig. 19), but the aerodynamics of flapping remain far from clear. And if the physics is poorly understood, the underlying mathematics cannot be established.

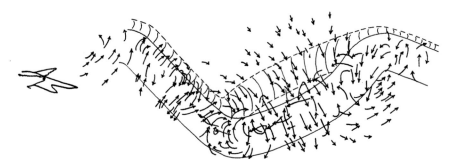

Figure 19. The airflow behind a Kestrel in flapping flight reconstructed from stereo-photographs. The arrows represent 3D velocity vectors caused by one downstroke and the subsequent upstroke projected onto the plane of the page. The continuous lines are the vortex behind the left wing: the dashed line represents the right wing. From the position of the bird it appears that either the wing beat amplitude was not symmetrical, or the wake was convected after shedding, or a combination of the two. From Spedding (1987b) reproduced by kind permission of *The Company of Biologists*.

---

[3] To investigate the effect of gusting Ravi *et al.* (2020) implanted electrodes into the flight muscles of Ruby-throated Hummingbirds (*Archilochus colubris*) and exposed them to gusting wind in a wind tunnel. While hummingbirds are very small and so will likely be more influenced by gusting than larger birds, there is no reason to doubt that all birds can morph their wings individually to counteract its effects.

The difficulty of mathematically analysing the flying bird led Colin Pennycuick (whose death was announced during the writing of this book), a leading expert on the modelling of avian flight, to develop a theory based on defining the clearly important parameters – wingspan, wing area, mass *etc.* – and combining these using dimensional analysis rather than pure theory to produce a model which could evaluate the basic characteristics – wing beat frequency, minimum power speed *etc.* – for any bird. Below, that model is used to define these characteristics for the Common Kestrel, together with details of an inertial measuring unit (IMU) designed and constructed for attachment to falconry birds to check the validity of the modelling.

Flapping Flight
While Kestrels are best known for flight-hunting they must, of course, use other forms of flight – flapping, soaring and gliding – to reach positions where flight-hunting is possible. Kestrels also hunt insects and avian prey, occasionally by cruising and waiting for the opportunity to make a surprise attack. With avian prey this will involve a stoop or a powered dive, gathering the speed necessary to both overtake and stop the prey before killing it. Kestrels will also perch-hunt: spotting an unwary bird they may attack in fast, level flight, again gathering the speed necessary to catch and stop the prey.

Wing Beat Frequency
A fundamental parameter of bird flight is wing beat frequency. While to an extent the frequency at which a bird beats its wings is controlled by the bird itself, all bird observers will have noted that smaller birds flap at a higher rate than larger ones, and that for birds of the same size, those with smaller wings flap faster than those with larger wings. From his work on the flight dynamics of birds Pennycuick (2008) derived an equation which relates the obviously important variables of a bird to derive what may be termed its 'natural wing beat frequency':

$$F = M^{3/8} S^{-23/24} A^{-1/3} g^{1/2} \rho^{-3/8} \qquad (1)$$

where:
F is the wing beat frequency;
M (kg), S (m) and A (m^2) are the mass, wingspan and wing area (see Fig. 18) of the bird;
g is the acceleration due to gravity;
and ρ is the density of air (kg/m^3).

In their experiments with Kestrels in laboratory conditions, a Dutch team at the University of Groningen (Videler *et al.* (1988a, 1988b)), were able to measure wing beat frequency in still air. For their experiment the Dutch team captured two wild Kestrels, a male (which they named *Jowie*) and a

Male Kestrel. *Torsten Prohl.*

female (*Kes*), and trained them to fly between two handlers (in exchange for small portions of mouse) stationed at each end of a straight corridor at the University. The corridor was 142m long, 3.4m wide and 2.4m high, and the birds were flown along it either for distances of 50m or 125m many times during each daily flying session. The mouse portions offered to the bird were minimised to maximise the number of flights per session, with the female averaging 94 flights, the male 78 flights. The total weight of mouse fed to the birds during each session also ensured that their average weight remained essentially constant during the experiment. The experiment lasted about 6 months and overall *Kes* flew more than 1200 times, *Jowie* more than 1000 times. On some flights small weights were attached to the Kestrels to simulate the carrying of prey to allow investigation of the effect on wing beat frequency and cruising speed: weights of 31g or 61g were used, these based on observations of prey cached by wild Kestrels. At the end of the experiment the two birds were released back into the wild.

Wing beat frequency was measured by analysis of film taken of the flying birds at 200frames/sec (fps). The speed of the birds was measured using a system of infra-red lights and cells mounted in the corridor's ceiling and floor respectively, the cells triggering a quartz clock each time the bird broke the light beam. The Dutch also investigated the flight mode of the Kestrels. They found that the flight of each of the birds followed an almost precise pattern. The falcon descended a few centimetres from the handler's fist, then flew level for about ⅔ of the distance to the second handler before changing to gliding flight prior to a final swoop up to the handler's fist. The experiments noted that the Kestrels achieved cruising speed after about 20m.

The mean wing beat frequency for *Kes* was 5.52Hz (range 5.24-5.78Hz) (with a body mass in the range 185g-192g during the experiment) and for *Jowie* was 5.87Hz (range 5.59-6.14Hz) with a body mass of 160-163g. The imposition of a 31g mass did not increase the mean wing beat frequency of *Kes*, though the measured range did increase (5.10-6.10Hz). However, adding weights increased the mean frequency for *Jowie* (5.98Hz, range 5.88-6.14Hz). Adding a 61g mass increased the mean wing beat frequency of both falcons – *Kes* (mean 6.19Hz, range 6.10-6.25Hz), *Jowie* (mean 6.25Hz, range 5.95-6.54Hz). Both birds increased their wingspan and wing area while carrying the additional loads (though *Kes'* wingspan increase was minimal). The inclination of the two falcons and of their tails relative to horizontal increased with loading. Further details of the flights are noted in Box 2 (p98).

Over recent years the availability of electronic components for attachment to birds to track and analyse flights has improved to the point where Inertial Measuring Units (IMUs) can be constructed of sufficiently small size and weight to fly on relatively small species. With the assistance of interested colleagues such IMUs were constructed to investigate the flight characteristics of falconry birds. The availability of captive, but free-flying, Kestrels has allowed these IMUs to be flown to investigate some aspects of the flight of the species such as wing beat frequency, cruising flight and flight-hunting.

The first IMU comprised a gps unit which ran at 1Hz, a tri-axial accelerometer, gyroscope and magnetometer unit which ran at 200Hz, and a barometer, running at 20Hz, which backed up altitude data from the gps satellites: while satellite altitude data are excellent if many satellites are in view, the data become less reliable if fewer are seen. Having learned from the behaviour of this unit, a second was constructed with tracking at 18Hz, barometer at 50Hz and tri-axial units running at speeds from 200Hz to 1.6kHz depending on the species being flown. Both IMUs are powered by small LiPo batteries which give flight times from about 25 minutes to an hour depending on battery size (*i.e.* mAh). In both IMUs data are stored on an onboard flash drive. The IMU is retrieved after each flight, data being downloaded for analysis. Each IMU measured about 33x18x8mm: the first weighed about 8g including battery, the second 3g plus the battery (0.7-1.2g, depending on power). Each requires a 1g metal spring which allows clipping to a standard harness worn by almost all falconry and demonstration birds primarily to allow the attachment of a location device in case the bird goes missing (Box 2 overleaf).

Box 2   Using IMUs on Birds
Size and mass were, of course, a critical element of the design. The mass of a unit has a direct effect on the bird as it represents an immediate increase in weight without the advantage of an increase in fitness which might otherwise compensate in the longer term. In the previously mentioned studies by the Dutch team at the University of Groningen, male and female falcons were loaded with 31g or 61g of lead (approximately the weight, and double the weight, of the voles which wild Kestrels hunt). Unloaded the falcons flew at speeds close to the V_{mr} (maximum range speed – see below) predicted by theoretical models, but as the payload increased speeds declined, approaching the V_{mp} (minimum power speed – see below) predicted by the same models – these models are explored further in the text. As the payload mass increased, the decline in speed was accompanied by an increase in wing beat frequency and an increase in tail inclination. The payloads carried by the falcons were large. The male weighed 160g, the female 190g, so 31g represented 19% and 16% of body weight respectively for male and female, while 61g represented 38% and 32%.

In the wild, smaller alterations in body mass are likely on a daily basis. Hambly *et al.* (2004), looking at the variation in flight energy costs and flight speeds in the Cockatiel (*Nymphicus hollandicus*) – an endemic Australian species also known as the Quarrion – noted that daily weight changes of 5-10% were normal. Hambly and co-workers noted declines in flight speed of the birds with payloads of 5%, 10% and 15%, but then a reversion to control flight speed (*i.e.* 0% payload) at 20%. The decreases in speed for lower payloads were initially small (of order 1% at 5% payload, 3% at 10% payload), but rose to 7% at 15% payload, a statistically significant difference). The reversion to normal flight speed at the higher payload was accompanied by a significant increase in wing beat frequency: the Cockatiels were changing their behaviour rather than simply reducing their speed. The energy cost of flying increased with payload, but were not significantly higher than an unloaded flight for any payload. While this is surprising, Hambly *et al.* note that Cockatiels regularly increase their body mass by 20% prior to migration, suggesting that the observed behavioural change at the higher payload was an inherent strategy.

While falcons differ from Cockatiels in many ways, the results of Hambly *et al.* suggest that provided the mass of a unit added to a falcon was minimised the effect on flight characteristics would be limited. This was confirmed by a study by Pennycuick *et al.* (2012) using Rose-coloured Starlings (*Sternus roseus*), but a separate issue was noted in that work, namely interference with the air flowing over the back of the bird. In the work of Pennycuick *et al.* the payload (a dummy transmitter) projected 6mm above the back of the bird. This increased the bird's drag coefficient (as measured in a wind tunnel) by 50%, but the addition of an angled aerial increased the drag coefficient by almost 200%. Pennycuick and co-workers then simulated the effect of the mass and

Hunting

drag coefficient increases by considering the actual flight of a Barnacle Goose (*Branta leucopsis*) which had been fitted with a satellite transmitter to study its migration. The simulation suggested that the effect of increased mass due to the transmitter would have been small, but the effect of the increase in drag coefficient would have reduced range and decreased energy reserves on arrival.

A related concern was raised by Sodhi *et al.* (1991) who wondered if the radiotagging of urban-breeding Merlins in Saskatoon was affecting breeding success or survival, particularly as other studies (on game birds) had noted reductions in both, as well as abnormal behaviour in tagged birds. The tags used by Sodhi and co-workers weighed 4g, representing 2.4% and 1.6% of the body weight of male and female Merlins respectively. In a study in which the breeding success of tagged and untagged males was measured, there was no difference between the two groups. In addition, survival rates of both males and females, as measured by the return rates of the birds the following year, were indistinguishable. By observation there was also no discernible behavioural difference between tagged and untagged birds, though those fitted with tags (either to the underside of tail feathers or legs) tended to peck at the tags for a period and to preen frequently immediately after release. In conclusion Sodhi *et al.* considered that the small radiotags were having no behavioural effect on the Merlins.

While considerations of breeding success were irrelevant in the studies carried out on falconry birds for this book, it is heartening to note that all the experiments reported above suggest that in the short term the use of tags on birds do not significantly affect flight performance.

The UK restricts attachments intended for wild birds to units weighing preferably less than 3% and definitively below 5% of body weight. For captive birds this limit is somewhat artificial as the IMU is removed after each flight: in the case of the Kestrels used to collect the data described below the unit represented no more than 2-3% of normal flying weight.

Using the IMU, the wing beat frequency of the captive Kestrel, *Kai*, was measured in natural conditions, *i.e.* with the bird flying outdoors, wind speed being measured using a standard anemometer. To allow comparison with the data of the Dutch team, a 26g weight was attached to *Kai* on some flights. This weight was comparable to the largest vole weight and average mouse weight measured during the comprehensive breeding season data obtained at a Kestrel barn site (see Chapter 8).

Allowing for the IMU mass, and assuming the wing area calculated above and an air density of $1.225 kg/m^3$, Pennycuick's equation (Equation 1, p95) gives a wing beat frequency of 5.4Hz for *Kai*. In a specific test flight on the day that weights were added to the bird, *Kai*'s unweighted wing beat frequency was 5.6Hz (Fig. 20 overleaf). This measured frequency agrees well with both the

Kai, the male falconry Kestrel used in the experimental flights.

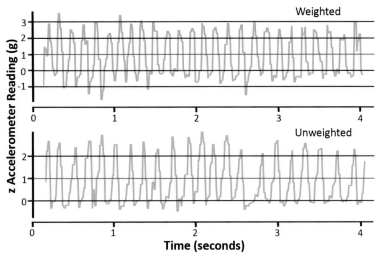

Figure 20. z accelerometer readings during a typical 4 second period of level flight by the Kestrel *Kai* with and without weights added to simulate prey.

Dutch experiments and the Pennycuick equation. The two flights were made on a day with little wind (a gentle breeze of about 0.5m/s) within about 6 minutes of each other across 120m of gently sloping paddock. The zero level on the graph is at 1g as the bird is in level flight. The downstroke of the wing generates lift and speed, while the upstroke also generates lift, but decreases the speed. The wing is held rigid during the downstroke, but is part-folded during the upstroke, the primary feathers rotating around their longitudinal axis to create slits through which air can pass. The difference in the wing beat, as well as the frequency, is clear in the two sequences. When weighted, Kai beat his wings both harder and faster, But also with a different pattern between upstroke and downstroke.

Pennycuick's equation for a mass increase of 26g gives a wing beat frequency of 5.7Hz. *Kai*'s weighted wing beat frequency was 6.1Hz (Fig. 20). Again, the frequencies are comparable with the Dutch experiment, though the difference from theory is increased. It is possible that this was in part due to the nature of the weighting. Whereas hunting Kestrels manoeuvre rodent prey in the same way that an Osprey holds captured fish – as a torpedo – to lessen drag, two tear-drop 13g fishing weights were attached to *Kai*'s jessies and these hung in space below him.

The wing beat frequencies for an unweighted bird observed in *Kai* and the two Dutch Kestrels *Kes* and *Jowie* (*i.e.* approximately 5.4-6.0Hz are in close agreement with those measured on migrating Kestrels above Switzerland, southern Germany, Mallorca, Malaga and southern Israel (Bruderer and Boldt, 2001, Bruderer *et al.*, 2010) who measured a range of 4.9-5.5Hz, corrected from altitude to mean sea level, using high-speed photography. From the work I have carried out, what is noticeable is that an individual bird,

The Common Kestrel

having reached adulthood and therefore having a (more or less) fixed wing area and span will have a fixed (perhaps preferred would be a better word) wing beat frequency for a given body mass, and will use that frequency at all times, varying the power input per beat or utilising the wind to vary flying height and speed.

Take-off

Two take-offs of *Kai* during the day which provided the data of Fig. 20 are shown Fig. 21.

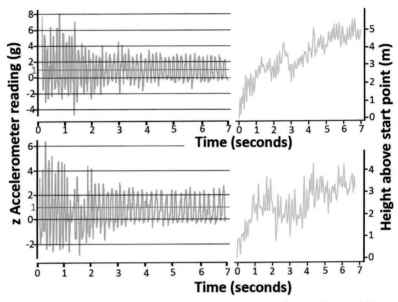

Figure 21. z accelerometer data and barometer height data for the first 7s of flights by the Kestrel *Kai* with (upper figure) and without weights added to simulate prey.

Again, the effort of lifting additional weight is clear in the accelerometer readings, in wing beat frequency and shape and in the height rises. Carrying the extra load the Kestrel rose 5m in the first 7s. In the unweighted flight the bird initially rose at the same rate, probably because it was not aware that the weights had been removed (the time between the two flights was only about 6 minutes). The Kestrel then levelled as it realised it was not carrying the weights, before rising again, but less steeply. Presumably the steeper rise while carrying weights reflects the need to gain height early to allow a margin against turbulence. Note that the barometer readings also reflect the wing beats – not surprisingly as the device is incorporated into the same unit as the tri-axial. In each case the accelerometer data indicates the effort of take-off.

Interestingly, despite *Kai* taking off from an outstretched arm (*i.e.* at about 1.5m), in each of these flights, and in almost all flights, he rose in the air rather than allowing gravity to aid gathering speed during an initial short downward 'dive'. Kai's behaviour is at odds with that noted by the Dutch experimenters with *Kes* and *Jowie*, and so it is possible that *Kai*'s curious take-off technique was idiosyncratic. However, photography of wild Kestrels suggest that both take-off methods are seen in the same bird on occasions and so may reflect local conditions.

The same female Kestrel leaving the nestbox after delivering prey on different days, clearly showing that jumping up or falling down on take-off are interchangeable. The photographs have been reversed to allow reading left to right.

For the weighted flight *Kai* rose 3m in 1.3s, achieving a forward speed of 10.7m/s. Assuming no energy input from the wind, *i.e.* assuming both kinetic and potential energy gains were entirely due to the falcon, *Kai*'s energy output was 17.5J equating to a power output of 83.1W/kg for the 162g bird. In the next Chapter the metabolic rate of a Kestrel is considered by reference to more of the superb series of experiments carried out at Groningen University in The Netherlands. That work found that the energy cost of flying for a Kestrel was 62W/kg. In a classic work on bird flight, the Russian ornithologist Viktor Dolnik (Dolnik, 1995) noted that the energy of flight was 12xBMR, *i.e.* 12-times the Basal Metabolic Rate of a bird. He also noted that in take-off the multiple increased to 16x. In the case of *Kai*, weighted with the equivalent of a 26g prey, the difference (83W/kg *cf.* 62W/kg) is 34% suggesting an increase from 12x to 16.1x, in extremely good agreement with Dolnik's value. Given the hunting requirements of a male Kestrel during the breeding season, provisioning both his partner and their brood, it is no surprise that the extra energy required results in weight loss, as we shall see in the next Chapter.

Cruising Speed

A bird in flight generates lift and thrust by pushing air downwards and backwards. From basic aerodynamics the power required to generate lift varies with speed, while the power to overcome drag, and therefore generate thrust, varies with the cube of speed. These theoretical considerations predict that the power required for a bird flying horizontally at a range of speeds will follow a U-shaped curve, *i.e.* the power output required to fly slowly, and to fly quickly, will be higher than for intermediate speeds. Proof of this theory is complex, but early experiments by Tucker (1968, 1972) suggested that the theory holds. Flying birds in a small wind tunnel, the bird fitted with a mask to measure respiratory rate, Tucker (1968) measured the breathing rate of two Budgerigars (weight 30-40g) at various speeds and produced a classic U-shaped curve (Fig. 22), as did Bundle *et al.* (2007) for two different bird species.

Over the three decades subsequent to Tucker's experiments, other researchers measured the power .v. speed curve for differing species. Dial *et al.* (1997) used *in vivo* bone-strain measurements of the pectoralis muscle in three Black-billed Magpies (*Pica hudsonia,* weight about 175g) flying in a wind tunnel with a range of flight speeds from 0m/s (*i.e.* hovering) to 14m/s (50.4kph). The resulting power curve resembled the theoretical U-shape, the birds requiring almost three times the power to hover in comparison to minimum power flight speed, but was flattened at higher speeds. *i.e.* the range over which the power was minimal was greater than might have been expected. Tobalske *et al.* (2003), integrating *in vivo* measurements of the pectoralis muscle with quasi-steady-state aerodynamic modelling, measured curves for Cockatiels (weight about 90g) and a dove species (weight 150-160g, see Fig. 23), while Askew and Ellerby (2007) used *in vivo* pectoralis muscle measurements to produce

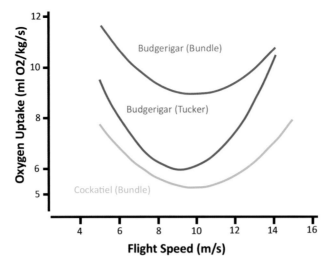

Figure 22. Variation of oxygen uptake with flight speed for Budgerigars (Tucker, 1968) and Budgerigars and Cockatiels (Bundle et al., 2007). The Tucker data were extrapolated by Bundle et al. to fit their range of flight speeds. Bundle et al. considered the difference between the Budgerigar data in the two data sets was due to the differing experimental set-up. Redrawn from Bundle et al. (2007).

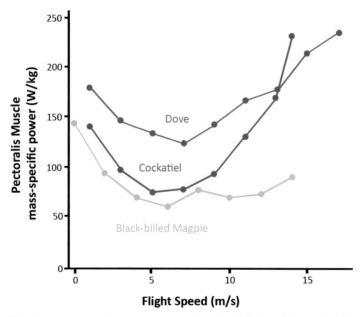

Figure 23. Power .v. speed curves measured for Cockatiels and doves by Tobalske et al. (2003), and for Black-billed Magpies by Dial et al. (1997). Note that the dove used in the experiment of Tobalske et al. (2003) was identified as a Ringed Turtle-Dove (*Streptopelia risoria*), but in Tobalske (2007), which replicated the comparison figure, was identified as a Mourning Dove (*Zenaida macroura*). While the curve for Cockatiels in the work of Bundle et al. (2007 – Fig. 22) is very similar, there is a clear off-set in terms of the speed at the minimum of the curve – 6m/s in Tobalske et al. (2003) against 10m/s in Bundle et al. (2007). In the later paper by Tobalske et al. (2007) the researchers suggest the two distinctly different experimental methods (measuring mechanical power and respiration) results in differing flight efficiencies across speeds and that in future it is vital to understand the difference when making comparisons.

curves for Budgerigars and Zebra Finches (*Taenopygia guttata*, weight 12g) which had been trained to fly in a variable-speed wind tunnel.

Fig. 23 shows data from species differing in weight by a factor of 5, suggesting to some researchers that true U-shaped flight curves were applicable only to lighter bird species. With the Zebra Finch also showing a U-shaped curve, the weight factor rises to 10, but again the finches are very light. Tucker (1972) produced data on two 300g Laughing Gulls (*Larus atricilla*): the curve in that case was similar in having a minimum point (at about 6m/s) and an increasing power output at higher speeds, but the experiment did not include lower speeds. Hudson and Bernstein (1983) flew five Chihuahuan Ravens (*Corvus cryptoleucus*), mean weight 480g, fitted with masks in a wind tunnel, but only obtained data over a limited speed range (8-11m/s). The power requirement increased linearly throughout that speed range, a result which is consistent with a curve which has a minimum of <8m/s, but hardly conclusive proof that U-shapes are 'real' curves across a forty-fold range of avian body weights. The debate over the shape of the power .v. speed curve continues.

But assuming a true U-shaped curve applies to the mechanical power required for flight, the minimum point of the curve represents a bird's most economical speed, *i.e.* where the energy used per unit flight time is a minimum (the minimum power speed, V_{mp}). The curve also allows a second speed to be calculated, a speed at which the ratio of lift to drag is a maximum. This is the speed at which the energy required to travel a unit distance is a minimum (the maximum range speed, V_{mr}) – Fig. 24. V_{mr} is of critical importance to migrating birds.

Pennycuick (2008) provides an equation for the calculation of V_{mp}, this again depending on the mass and wing dimensions of the bird, as well as the frontal area the bird presents to the air and the body's drag coefficient,

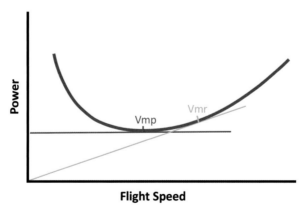

Figure 24. Theoretical mechanical power curve (power .v. flight speed) for a Eurasian Kestrel. The minimum of the curve reveals the minimum power speed, while the asymptotic line from zero to the curve reveals the maximum range speed.

together with air density, gravity and the induced power factor. To keep itself airborne the bird must power its wings. In an ideal world, all the downthrust power of wing movement would perform this task. In reality, the wing is performing two jobs simultaneously, both maintaining the bird airborne and creating forward thrust. The induced power factor is a multiplier on the ideal induced power to allow for this real world effect.

Pennycuick's equation is:

$$V_{mp} = \frac{0.807 k^{0.25} M^{0.5} g^{0.5}}{\rho^{0.5} S^{0.5} A_b^{0.25} C_d^{0.25}} \qquad (2)$$

where M, S, g and ρ are as in Equation (1), and:
k is the induced power factor;
A_b is the bird's frontal area (m²);
C_d is the bird's body drag coefficient.

The induced power factor usually differs from 1.0 by a surprisingly small amount, lying between 1.1 and 1.2. Spedding (1987a) suggests k = 1.04 for a gliding Kestrel, but for flapping flight the value would be higher: Pennycuick suggests 1.2.

In an earlier paper, Pennycuick (1988) provided information for the calculation of A_b and C_d. Using data obtained from both museum specimens and living birds he produced an empirical equation for A_b:

$$A_b \text{ (m}^2\text{)} = 8.13 \times 10^{-3} M^{0.666}$$

from which A_b = 2.6x10⁻³m² for *Kai*.
Pennycuick (2008) suggests a value of 0.1 for C_d.

With these data Pennycuick's equation predicts a cruising speed of 9.5m/s.

Using the IMU detailed above, the cruising speed (V_{mp}) of *Kai* was measured in natural conditions. On a day with a steady wind of 0.2m/s gusting to 0.6m/s, flying with the wind *Kai* averaged 10.2m/s (9.4-10.0m/s wind corrected). For comparison, the Pennycuick equation was applied to the dimension data of *Jowie*, the Dutch male Kestrel which is of comparable size to *Kai*. In the still-air corridor of Groningen University *Jowie* flew at 9.3m/s.

To allow further comparison with the data of the Dutch team, similar weights were attached to *Kai* on some flights. On the same relatively calm day, weighted by 26g *Kai* flew at 10.7m/s. This result seems odd at first glance as it would, perhaps, be anticipated that the falcon would fly slower when carrying additional weight, but the Pennycuick equation predicts that V_{mp} varies with the square root of total mass and so a weighted bird would indeed fly faster[4]. The equation predicts a speed of 10.2m/s and a speed of 10.1m/s for *Jowie*. The argument would be that a bird carrying prey would need to flap its wings faster (which both *Kai* and *Jowie* did) to stay aloft and this would mean a higher speed. But in the Dutch experiments (Videler *et al.* (1988a, 1988b)) both male and female Kestrels slowed while carrying extra (31g) weights, and slowed further when the weight was increased to 61g. The speed differential between the two sexes (about 0.4m/s, with *Kes* being faster) remained constant when the two birds carried 31g, but disappeared when both flew with 61g weights. The disparity between the two experimental birds, and between the birds and the Pennycuick theory probably relates either to the bird's ability to alter its wing area in flight, or to variations in the drag coefficient due to differences in the way the weights were attached in the two tests. As speed varies with the inverse of the drag coefficient an increase in drag would reduce speed so that the theory predicts opposing factors and it will depend in each individual case which factor wins out.

Videler *et al.* (1988b) also studied the Kestrels' attitude during cruising, using high-speed photography (200fps) to note the position of the tips of the bill, tail and wings (Fig. 25) in the x and z planes (*i.e.* the direction of flight and perpendicular to the direction of flight).

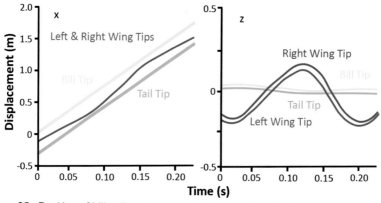

Figure 25. Position of bill, tail and wing tips in the x and z directions of a male Kestrel flying at minimum power (cruising) speed. Redrawn from Videler *et al.* (1988b).

Fig. 25(x) shows that the bill and tail tips move linearly in the x direction. Cruising not only means constant speed, but that the falcon neither loses nor gains height, the wing tips showing little deviation from a straight line. Wing tip movements are not entirely linear as the tip is pulled towards, and away

---

[4] Pennycuick's equation has wing beat frequency varying not only with the square root of mass, but the inverse square root of wingspan and, of course, a larger (heavier) bird would be expected to have a longer span. Therefore, when comparing different birds (male to female or species to species) rather than an individual bird carrying different loads it is not simply the case that wing beat frequency varies with the square root of mass.

from, the body at the start of, and end of, the downstroke: minimum power cruising really does mean extremely efficient travelling. Fig. 25(z) confirms the minimal movement of bill and tail tips, and the symmetric movement of the wings. The maximum wing tip velocity measured during the experiment was 10m/s (36km/h, c.30mph), this occurring at the start of the downstroke. The minimum wing tip velocity at any point in the complete wing beat was about 5m/s (8km/h, c.15mph). Videler et al. (1988b) also calculated that the maximum acceleration of the wing tips occurs at the start (acceleration) and end (deceleration) of the downstroke and reached an extraordinary ±250m/s (25.5g), the downstroke taking only 0.08s (the upstroke taking 0.1s, i.e. a total time of 0.18s, equating to a wing beat frequency of 5.6Hz (Fig. 26).

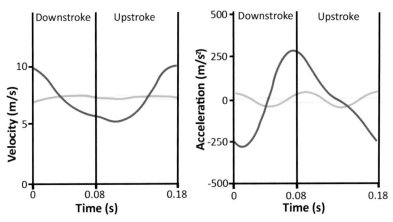

Figure 26. Velocity in the x direction and acceleration in the z direction of the bill (yellow), tail tip (green) and right wing tip (blue) during one complete wing beat of a male Kestrel in level flight at minimum power (cruising) speed. Redrawn from Videler et al. (1988b).

Soaring and Gliding

Despite the adaptations for flying it is energy intensive and birds utilise both spatial and temporal changes in wind velocity to gain height and, therefore potential energy, by soaring. The energy can then be exchanged for kinetic energy as the bird loses height. Potential energy can be exchanged quickly or more gradually. For a stooping falcon, the exchange is quick, a near-vertical dive resulting in fast acceleration as prey is attacked. A more gradual loss results in gliding, the bird exchanging potential energy for position as it traverses the landscape in its search for food. There is a limit to how slow a bird may glide as ultimately the wing is no longer able to generate the lift required to stay aloft, and the bird stalls. Again Pennycuick (2008) defines equations which allow V_{bg}, the best glide speed, i.e. the speed at which the bird covers the greatest distance for a given height loss, to be calculated. For a 206g Kestrel this speed is 8.4m/s. At that speed the glide angle is 3.9°.

Soaring utilising spatial changes related to the vertical gradient of wind velocity due to frictional losses close to the Earth's surface and the potential existence of updrafts created by local topography, while temporal changes arise from gusting as consistently homogenous winds are rare. In general, the use of the differential speeds of air layers is termed dynamic soaring as it is usually associated with seabirds (particularly albatrosses), while the use of spatial and temporal changes induced by land topography is termed slope soaring. In each case the bird first travels across the wind, gathering energy in the way that sailing boats do while tacking. The bird then turns downwind, gathering further energy as it is pushed along by the wind. The bird then turns across the wind again, before turning into the wind and converting kinetic energy to potential energy as it rises against the air flow. This is a necessarily simplistic explanation as in practice the conversion of kinetic to potential energy is complicated by the fact that kinetic energy varies as the square of velocity and velocity is varying, and drag must be included for a full understanding (and drag is also variable with time as the bird's attitude alters).

As well as using gliding to increase the distance of travel, Kestrels also utilise 'bounding flight' in which periods of flapping are interspersed with periods of gliding, a flight pattern which is frequently seen in passerines. In their study of captive Kestrels already mentioned above, the Dutch researchers found that their 'corridor flying' Kestrels flapped for about ⅔ of the distance between handlers, then glided, gently losing height before a short swoop upwards to the handler's glove and mouse portion reward.

Although Kestrels undoubtedly soar and glide when they are able, they are not renowned for these flight patterns, and no direct data was obtained with the IMU on either.

Migration Flights
Pennycuick (2008) also sets down an equation for the calculation of V_{mr}, the maximum range speed, which a bird would be expected to use (in still air) on migration:

$$V_{mr} = \frac{k^{0.25} M^{0.5} g^{0.5}}{\rho^{0.5} P^{0.25} S_d^{0.25}} \qquad (3)$$

where k, M, g and ρ are as for Equation (2) and:
P is the equivalent flat-plate area of the bird's body;
S_d is disc area of the bird.

The equivalent flat-plate area of the body is defined as $P = A_b C_d$ with the two terms as in Equation (2), and S_d, disc area produced by the wings is derived from $S_d = \pi S^2/4$, with S the wingspan (as in Equation (1)).

For *Kai*, V_{mr} = 12.1m/s which would be the expected migration speed for an adult male as *Kai* is an average size. Bruderer and Boldt, 2001 observed migrating Kestrels above several countries and measured air speeds of 12.3-13.2m/s adjusted for altitude and wind speed, which are very close to the calculated speed for *Kai* using Pennycuick's equation. Unfortunately the nature of *Kai* – a 'domesticated', falconry bird, meant that no equivalent speed could be measured using the IMU, although speeds comparable to, and above (up to 13.7m/s), the calculated V_{mr}, were measured during flights which included shallow dives.

Female Kestrel. *Torsten Prohl.*

First flight, though rather more jump than flight, by a fledgling Kestrel.

Early flight by a fledgling Kestrel. On their first flights the falcons look somewhat uncoordinated, the wing beat 'lumpy' and the undercarriage down in case of an emergency landing.

Flight-Hunting

In order to catch prey it is obvious that Kestrels must be able to see their victim and then have sufficient flight abilities to reach it before it can move to a position of safety. In this section I investigate the form of hunting most often associated with Kestrels. But first, we consider the falcon's vision.

Vision

Falconiformes differ from humans in having two foveae rather than one. Foveae are areas of closely-packed cone cells, cones being one of the three types of photoreceptor cells in the eye. In humans the eye's single fovea is set centrally. In Falconiformes there is a similar fovea, giving the birds stereo vision and an appreciation of depth as humans have. But there is also a second fovea which is more visually acute. This deep fovea is set an angle of about 40° to the main axis of the bird.

Occasionally it is claimed in the popular media that the eye-sight of falcons and eagles is far superior to that of humans, with numbers ranging up to 10x being quoted. While such figures capture the imagination and seem to be borne out by the activities of the birds, science does not support such claims. Hirsch (1982), in a study of the geometry and make-up of the eye of an American Kestrel (a male named *Argus*), considered that the visual acuity of the falcon and a human were comparable, despite the eye of the former being only half the size of a human eye, the difference explained by superior optics and closer-packed photoreceptors. Reymond (1985), working with the Wedge-tailed Eagle (*Aquila audax*), suggested the eagle's visual acuity was probably twice that of a human, but noted that its vision was attuned to the highest light levels that occur naturally, performance deteriorating more rapidly than does human sight as luminance declined: this finding confirmed an earlier study of Fox *et al.* (1976) which had shown a similar decline in falcon vision with declining luminance. Jones *et al.* (2007) supported Reymond's finding. If a factor of two improvement in vision does not sound enough it has to be remembered that human vision is itself remarkably acute so that falcon vision is extraordinarily good.

Jones *et al.* (2007) also note that as well as having more acute vision than humans, falcons have a higher flicker-fusion frequency (FFF). FFF is a measure of the ability to resolve rapid movements. Films are shown at a frame rate of about 30Hz as below this rate humans are able to resolve the change from one frame to the next and so see flicker as frames change. Some humans can resolve higher frequencies, though few are able to detect the 50Hz flicker of a fluorescent light. By contrast birds have an FFF which is at least 100Hz, with some species having even higher rates. Such rates are required by both hunters and hunted: flying through woodland at speed humans would likely collide with branches which birds easily avoid as their eyes, and, of course, brain, process visual information more quickly. While this ability is at a premium for woodland prey species and such predators as the Goshawk, falcons also need to be able to react quickly to changes in flight direction by fleeing prey or to avoid lunges by captured prey. Recently, Potier *et al.* (2020) have accurately measured FFF for Peregrines (129Hz), Saker Falcons (102Hz) and Harris Hawks (*Parabuteo unicinctus* – 81Hz). Potier *et al.* suggest a link between fast vision and hunting strategy, so that raptors chasing fast moving prey would have a higher rate than those which hunt slower prey. On that basis, the Kestrel, primarily a hunter of (relatively) slow-moving ground prey would be expected to have an FFF closer to that of the Harris Hawk.

Just as it was assumed for many years that the vision of falcons was many times more acute than that of humans, it was also assumed they viewed the world in much the same way as we do, *i.e.* that their vision encompassed the same range of wavelengths as ours does (the visible spectrum) even though it was not known if their colour vision was the same. Visible light is a very small range (in terms of wavelength) of the electromagnetic spectrum. This is not the correct forum for a discussion on whether electromagnetic radiation is particle-like, wave-like or both at the same time: it is sufficient to note that the spectrum is most conveniently considered by wavelength, and covers a range from radio, where the wavelength can be measured in metres to X-rays, where it is measured in nanometres (1nm = 10^{-9}m, *i.e.* one thousand-millionth of a metre) and beyond to γ-rays (gamma rays) with wavelengths measured in picometres (10^{-12}m, *i.e.* one million-millionth of a metre). Visible light is usually defined as covering the range 400-700nm with red light at the upper end and blue light at the lower, though it is worth noting that many people can see up to 1000nm and both children and young adults may see down to about 310nm. Radiation with wavelengths below 400nm is known as ultraviolet (the last colour of the 'rainbow' spectrum of visible light being violet) while at wavelengths above 700nm it is known as infra-red. Infra-red wavelengths are used in night-vision cameras *etc.* as the heat produced by warm objects (including warm-blooded animals) is in the infra-red part of the spectrum. Ultra-violet (UV) is beneficial to humans as it allows the synthesis of Vitamin D, but over-exposure is also harmful, causing sunburn and, potentially, skin cancers.

Not until the 1970s was it discovered that some birds could see UV light. At first UV vision was identified in very few species – Huth and Burkhardt (1972) with a hummingbird (the White-vented Violetear (*Colibri serrirostris*)), Wright (1972) with pigeons – but the list was gradually extended. It is now considered that UV vision is present in most, perhaps all, diurnal birds, but may be absent in most nocturnal species[5]. Since UV may cause retina damage, most animals, including humans, have UV opaque structures in front of the retina to minimise UV transmission. The obvious question therefore is why birds allow UV to reach the retina – what advantages accrue which compensate for the potential damage incurred?

Bennett and Cuthill (1994) considered this in a seminal paper, suggesting three potential advantages: that UV vision allowed the position of the sun to be determined even if it was not visible, an aid to homing pigeons and migrating species; that UV vision allowed sexual signalling, enhancing the visible light signalling of plumage; and that UV vision aided foraging. As evidence in support of the latter Bennett and Cuthill noted that many fruits reflect UV whereas leaves do not. Consequently, fruits are more conspicuous to an eye sensitive to UV.

Soon after the report by Bennett and Cuthill was published, work in Finland by Viitala *et al.* (1995) suggested that the ability to see UV aided Kestrels in hunting as they were able to see the urine and faeces trails left by Field Voles, these trails being visible in UV light (wavelength 320-400nm) with a peak at 340nm.

It was known that voles, and other small rodents, use urine and faeces to mark their trails and that both fluoresce in UV light (Desjardin *et al.* 1973 with House Mice). Fluorescence occurs when UV is absorbed by a substance and then emitted at a longer wavelength which may be visible. UV sensitivity having been identified, Viitala *et al.* (1995) therefore investigated whether this UV signal was being used by Kestrels as a clue to areas of high Field Vole numbers. The researchers captured 19 wild Kestrels which were then shown four laboratory areas, two of which were clean, the other two having been 'treated' with vole urine and faeces. One pair of clean/unclean areas was illuminated with UV light, the other pair with visible light, and the time a bird spent above each area and the number of times it scanned were recorded. Scanning, the particular form of behaviour where the falcon bobs its head,

---

[5] Recent work on the Emu (*Dromaius novaehollandiae*) by Hart *et al.* (2016) has suggested that UV sensitivity is likely to be ancestral in avian evolution and therefore present in all birds unless it has been lost.

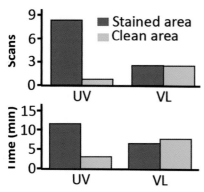

Figure 27. Mean time spent by 19 Kestrels above four areas. Two areas were stained with vole faeces/urine, two were clean. Each pair of stained and clean areas were then illuminated in ultraviolet light (UV) or visible light (VL). Time was recorded per individual minute (min). Scans were recorded per individual. The differences between time and scans above stained and clean areas under UV light were statistically significant. There was also a significant difference between time and scans under UV and VL. Redrawn from Viitala et al. (1995).

apparently measuring the distance to the area, is seen as a prelude to hunting and so indicates interest. The Kestrels were found to favour treated areas over clean areas when each was illuminated in UV light, but showed no preference when the areas were illuminated in visible light (Fig. 27). The latter was important in dismissing the idea that the Kestrels were using olfactory cues (*i.e.* they were not smelling the difference)[6].

The Finnish team then repeated the experiment in the wild, choosing areas close to 27 nestboxes erected for Kestrel occupation. These areas were then treated to create three classes: in one class artificial vole trails were formed and marked by vole urine and faeces; in a second, artificial trails were formed, but were unmarked; and in the third class the areas were clean of both trails and marking. The Kestrels preferentially hunted in the Class 1 areas (Fig. 28). (As a digression, Rough-legged Buzzards (*Buteo lagopus*) were also shown to prefer Class 1 areas, a finding confirmed by a later report, Koivula and Viitala,1999.) As it was known that mouse urine fluoresces (in faint blue light) – *e.g.* Desjardin *et al.* (1973) – Viitala *et al.* (1995) considered this was also a potential clue for the Kestrels, but believed the fluorescence was so dim that it was readily masked in daylight and so was unlikely to be offering the falcons usable information.

Later studies by the Finnish team showed that the urine of Bank Voles also reflected UV light, again with a peak at around 340nm (Koivula *et al.*,1999a) and suggested that reflectance was strongest (*i.e.* the urine tracks were brightest) in mature male voles[7]. Although Koivula and co-workers did not investigate urine frequency and pattern, this finding of a difference between

---

[6] It was long assumed that raptors had little, if any, sense of smell, but recent work by Potier (2020) has shown that olfaction might be important to both hawks and falcons. Though Kestrels were not among the tested falcons, there is no reason to believe they will not share the characteristics of the species that were.

[7] In a study on prey choice by Kestrels in western Finland, Korpimäki (1985c) noted that in both spring and autumn the raptors took more male than female voles, a difference that was statistically significant in both seasons. The team considered this to be due to the activity levels of males being higher at those times. The finding of Koivula *et al.* (1999a) that reflectance was strongest in mature male voles, suggested another reason why mature males might be preferentially predated, particularly if their urine trails were more frequent and across wider areas (as suggested by other studies) – if male voles were easier to find they would be easier to catch all other things being equal.

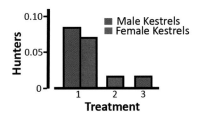

Figure 28. Number of hunting male and female Kestrels per hour (Hunters) above areas with three differing classes of treatment. The difference between the number of birds hunting above Class 1 areas, and Class 2 & 3 areas was statistically significant. Redrawn from Viitala *et al.* (1995).

the urine trails of dominant males is consistent with the finding of Desjardin *et al.* (1973), who noted that both the degree and frequency of mouse urine marking depended on the rank of the animal: dominant male mice urinated often and everywhere, subordinate males urinated infrequently and only in cage corners. Rozenfeld *et al.* (1987) found similar pattern/frequency differences in Bank Voles. In later studies the Finnish team (Honkavaara *et al.,* 2002) noted that the ability of Kestrels to detect the position of high vole activity would also be of assistance in areas where vole populations are subject to significant fluctuations – as they are in northern Fennoscandia – and so could help to explain how the falcons arrived so quickly in areas where the vole population had peaked. In a further report, the Finnish group (Koivula *et al.,* 1999b) noted that Kestrels preferred the scent markings of Sibling Voles to those of either Bank or Field voles (with no difference in preference between the latter) and preferred the trails of mature males to those of mature females or juveniles. Given that the trails of mature males were shown to glow brighter in UV, this might explain that preference, rather than that mature males were either larger or tastier (either of which could explain the preference for Sibling Voles).

As a short digression, it is worth reconsidering the potential advantages the ability to see and be seen in UV offers to those creatures that have UV vision. Studies have suggested that the selection of avian prey by a primarily avian predator is complex. In Wales, Baker and Bibby (1987) noted that Merlins took prey in relation to its abundance rather than plumage colour, but in Canada, where Sodhi and Oliphant (1993) also studied Merlin, the researchers noted that it was the time the prey species spent away from cover, *i.e.* how easy it was to spot and catch, that was the driver. So what was the value of UV vision? One clue was a study of Swedish songbirds (Håstad *et al.,* 2005) that found their plumage was UV reflecting, but at shorter wavelengths than the UV sensitivity band of the raptors: the songbirds were using a private communication band that could not be intercepted by their predators. That offered a reason for UV vision, but raised the question of why if avian prey were using frequencies not visible by their predators, rodents were not doing the same. An immediate answer was suggested by Koivula and Korpimäki (2001) who noted that for an individual vole the risk of predation from a raptor was much lower than that from mustelids which use olfactory clues rather than sight, so that UV signalling is not as dangerous as would first appear.

Further studies have indicated that the usage of UV clues is not confined to Kestrels and Rough-legged Buzzards, Probst *et al.* (2002) showing a similar preference in the Great Grey Shrike (*Lanius excubitor*) for the scent trails of Field Voles, while Härmä *et al.* (2011), found the same preference in the Pygmy Owl (*Glaucidium passerinum*), a diurnal owl, for the scent trails of Bank Voles. More interestingly, Zampiga *et al.* (2006) carried out a study to see if the use of UV was learned or innate in Kestrels. The researchers set up four experimental areas. In two, cardboard marked with vole urine could be viewed with or without a UV-blocking filter: in the other two, water-sprinkled cardboard could be similarly viewed. 44 adult and 49 juvenile Kestrels were exposed to the four areas. The results indicated that while both adult and juvenile falcons were attracted to the urine stained areas, the response of the adults was greater suggesting there were both inherent and learned components to the use of UV cues in Kestrels, juveniles having to learn by experience that what they were seeing was prey-related. Interestingly, adult Kestrels also showed an interest in the urine stained areas when there was UV-blocking filter in place: Zampiga *et al.* did not suggest what other cues the adults were using, but it seemed most likely that they were either detecting faint UV clues despite the blocking filter, or picking up fluorescence. The latter was interesting because other studies were suggesting that it was UV fluorescence rather than UV reflectance that was important to raptors.

The fact that Kestrels were seeing vole trails in UV captured the public's imagination, and at falcon displays the handlers could occasionally be heard explaining the idea, though rather than suggesting that the UV clue was guiding the Kestrel to an area frequented by voles, it was often suggested that the flight-hunting Kestrel was watching a vole that was following a trail, urinating as it went. The falcon, it was suggested, needed only to pounce at the head of the emerging UV trail to acquire a meal. Given that many of the audience would be parents with children, the idea that Kestrels spent their time watching pee trails in the grass was greeted with delight even if the explanation was very probably an over-statement.

But while the idea that UV vision was being used by a flight-hunting Kestrel was gaining acceptance among the public, it was coming under increased scrutiny within the scientific community. Working with three endemic Chilean rodents, Chávez *et al.* (2003) noted that their urine UV reflectance was as that of the north European rodents, but Kellie *et al.* (2004) did not find similar results in 13 species of endemic or introduced Australian rodents. They found no reflectance peaking in the UV, but did find UV fluorescence peaking at a wavelength of about 380nm, the wavelength of maximum UV sensitivity in raptors. When observing urine samples under a UV light the researchers were able to see faint blue fluorescence with the naked eye. More importantly, they also noted that visibility of UV was highly dependent on the UV absorption characteristics of the substrate (*i.e.* the ground on to which the urine was expelled).

The suggestion that the substrate would influence visibility was initially overlooked as researchers sought to understand why there appeared to be differences in the urine of Finnish rodents to those elsewhere. Then a new study in Finland (Huitu *et al.*, 2008) confirmed the results of Kellie *et al.* (2004), and therefore disagreed with earlier Finnish studies. But the work also found that the diet of the voles influenced the peak wavelength of urine fluorescence. If fed rye grass infected by fungal endophytes the voles lost weight, but also had the wavelength of peak fluorescence shifted so that it better aligned with the UV sensitivity of predatory raptors. Loss of weight presumably persuaded the voles to eat other vegetation – the endophytes appeared to be benefitting their host by increasing resistance to vole foraging. But the shift in UV fluorescence wavelength suggested a more profound change – that the endophytes were increasing the likelihood of foraging voles being predated, a rather less subtle way of preventing the host grass from being eaten.

Then, in 2013 further significant work was carried out by Lind *et al.* (2013) in Sweden. The team studied a fundamental property – was UV light actually reaching the retina of the Kestrels (and of Common Buzzards and Red Kites). Taking raptors which had been euthanised less than an hour before, the Swedish researchers removed both eyeballs, cut an opening in the back of the eye and used a spectrometer to measure the transmission of UV from a source placed in front of the eye. Lind *et al.* found that for each raptor the transmittance of UV fell off steeply as the wavelength of the light decreased below 400nm. For Kestrels 50% transmittance was at 379nm, and zero transmittance was reached at about 340nm. Next, Lind and co-workers collected urine and faeces samples from both Bank Voles and Field Voles and measured UV reflectance against a variety of substrates, finding that they could not replicate the data of early work on the reflectance of trail markers. In the view of Swedish team, UV reflectance from vole urine/faeces was unlikely to provide a reliable clue to a hunting Kestrel. However, the Swedish group recognised the contradiction this raised with earlier studies and noted that resolution of the discrepancy required further work.

At present, the question of whether Kestrels are using non-visible light signals to detect voles remains open. UV reflectance would make rodent urine lighter than the background, while absorption would make it darker than the background. Fluorescence would be masked by sunlight and, perhaps, lost against the background of UV reflectance from plants and other substrates. In all cases the effect of the substrate cannot be ignored. But, of course, our ideas may be biased by the assumption that Kestrel vision is largely consistent with human vision. Perhaps Kestrels (and other raptors) really are picking up signals at frequencies and with sensitivities that are, as yet, not fully understood by ourselves, as well as picking up other visual signals which we also do not, as yet, fully understand, such as the twitching of grass stems as the voles move. The only clear outcome of the work to date is that the debate will continue.

The Common Kestrel

The photos on this page and the opposite page are of a male Kestrel delivering prey to a nestbox brood. The four photographs were taken in a third of a second (*i.e.* one every 0.08s) giving a good idea of both the speed of the delivery and the wonderful flight control of the falcon.

Flight

Having considered the Kestrel's vision, we must consider the other critical aspect of Kestrel behaviour – flight-hunting. As noted previously, Kestrels have longer tails in relation to other falcons. In general, tail length decreases with increasing flight speed. However, tails increase control and stability, and since the tail also generates lift, longer tails can help a bird to fly more slowly (by decreasing the stall speed) and, in addition, improves the rate of turn. Overall it would therefore seem that the Kestrel's longer tail would be an adaptation for flight-hunting, 'hovering' being, essentially, a zero-speed manoeuvre. In their investigation of how natural selection may have influenced tail morphology for flight (as opposed to mate selection, the majestic tail of the Indian Peacock (*Pavo cristatus*) having evolved to enhance mating opportunities and is more hindrance than aid in flight), Thomas and Balmford (1995) postulated that flight-hunting birds would have longer outer tail feathers (relative to body length). However, when examining species that exhibit such behaviour (to a greater or lesser extent) the data did not support this postulate. Nine birds were considered (the Common Kestrel and eight others – Antarctic Petrel (*Thalassoica Antarctica*), Short-eared Owl, Black-winged Kite (*Elanus caeruleus*), Rough-legged Buzzard, Little Gull, (*Larus minutus*), Black Tern (*Chlidonias nigra*), Pied Kingfisher (*Ceryle rudis*) and Skylark), these being compared with similar species which do not, or rarely, flight-hunt (including the Lesser Kestrel). Data from these species showed neither a tendency towards longer tails nor any distinct difference between the pairs. Thomas and Balmford were puzzled by this as the aerodynamic model they had set up was successful in predicting other suggestions relative to tail design. They consider the likely reason for the discrepancy related to the fact that tails generate much less lift than wings so that deviations from an apparent optimum are less significant. But despite the negative outcome of the modelling, it is the case that the Kestrel has the longest outer tail feathers (relative to body length) of the nine studied birds. While the tail may not be truly optimised for flight-hunting, it is, therefore, an advantage to the bird in that mode of flight. However, it must also be recalled that Kestrels cannot hover in the way that hummingbirds do. While some observers (myself included – this is mentioned again below) consider that on occasion Kestrels may flight-hunt (and therefore hover) on perfectly calm days for extremely short periods (1 or 2 seconds), in general, Kestrel flight-hunting requires a head wind[8].

At first glance it appears that a flight-hunting Kestrel is gliding into the wind with a glide speed which is both above the bird's stall speed and equal to the wind speed so that it remains on the same spot. However, gliding is not enough, because in a true glide the bird would lose height, and that is not the case. Wing

---

[8] A recent paper on tails (Usherwood *et al.* (2020) considers the potential for tails to enhance lift and so reduce drag in gliding raptors. While this adds to the debate on the form of tails, it does not significantly aid an understanding of why the Kestrel tail is longer than that of species which also utilise flight-hunting on occasions.

This male Kestrel has lost two tail feathers, but could still hunt well enough to help raise four chicks to fledge. He has also moulted two other feathers to 'standard' adult plumage implying he is a first-year bird, which, if correct, means his achievement is even more notable.

movement is required to generate enough lift for the Kestrel to maintain height. In their study of flight-hunting Kestrels in The Netherlands, the Groningen University group, Videler *et al.* (1983), noted that the time span of bouts of flight-hunting by a female Kestrel were almost constant at wind speeds between 4m/s and 12m/s, but were significantly reduced below 4m/s and above 12m/s (Fig. 29). In later work by another Groningen team (Masman and Klaassen, 1987) the researchers compared theoretical models of the energy requirements of flight-hunting. All the models predicted that the costs would rise with wind speeds below 5m/s and above 10m/s, minimal energy costs being at 7-8m/s. The prediction was consistent both with the data from the female falcon (Videler *et al.*, 1983) and with their own observations of a male Kestrel during the winter, when energy minimisation is critical (Fig. 29).

At low wind speeds a Kestrel must generate lift and so needs to flap its wings vigorously, the amplitude of the flap being high (*i.e.* the wing-tip travels a considerable distance). The falcon's body is also kept at a high angle, sometimes almost vertical, and both the wings and tail are spread. At higher wind speeds much of the lift is generated by the wind: the wings still flap, but the amplitude of the flap is very much reduced. The body's position is also much more horizontal and the wings and tail much less spread.

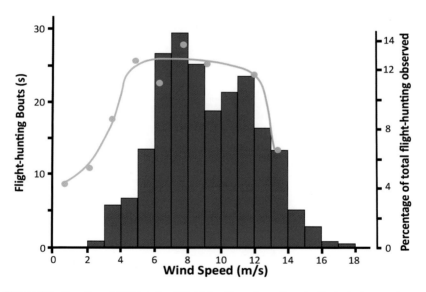

Figure 29. Duration of 794 bouts of flight-hunting of one female Kestrel observed throughout one year plotted against wind speed. Each orange dot represents the mean duration of bouts at a given wind speed. The fitted curve is approximately symmetric around a wind speed mean of 8.3m/s. At the preferred range of wind speeds the Kestrel 'hovered' for an average of 25s. Redrawn from Videler *et al.* (1983). The superimposed blue histogram is data on the percentage of flight-hunting observed in a male Kestrel during one winter plotted against wind speed. Redrawn from Masman and Klaassen (1987).

Two falconry Kestrels. The male (left) is 'hovering' in a low speed (c.3m/s) wind. The second male (right) is hovering in a higher speed (c.8m/s) wind. In each case the birds were flying on flat land, though, of course, local topography was likely disturbing true planar wind. However, with that proviso, the difference in body angle is clear. The wing spread and beat amplitude of the bird to the right is much reduced,, though the tail spread is similar in each case.

In this shot the bird is inclined below horizontal, apparently in contradiction of theory. But the bird was at cliff-top height at a coastal site with a strong updraft and an on-shore wind, and so needed to incline its body and beat its wings to maintain position. *Gordon Kirk*.

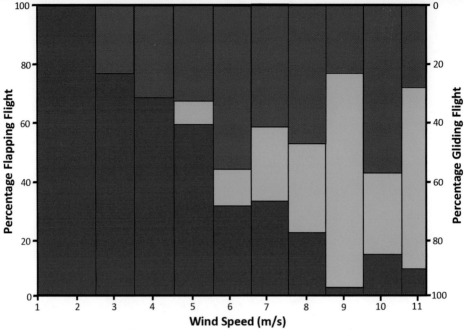

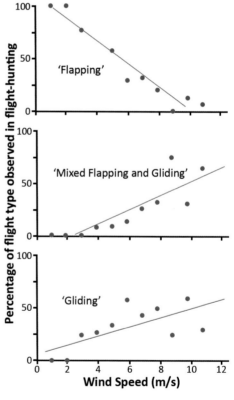

Figure 30. Percentages of flapping, gliding and mixed flight types observed in total flight-hunting as a function of wind speed. Fig. 30a (*to the right*) is a graphical representation of the variation, while Fig. 30b (*above*) converts the data into histograms, with red rectangles representing flapping, blue representing gliding, and green being mixed mode. Redrawn from Village (1983a).

In a study area in southern Scotland Village (1983a) looked at the variation of 'flapping' and 'gliding' flight-hunting at various wind speeds, the observations clearly showing the trend towards 'gliding' at higher wind speeds (Fig. 4.30a and b).

Hunting

To minimise the bird's energy input the wind must have a vertical as well as a horizontal component, so Kestrels frequently flight-hunt above slopes which face the wind, and will also utilise the updrafts created by other obstructions to the wind's progress, such as man-made structures. Village (1983a) also noted the position of flight-hunting Kestrels with respect to wind direction (Fig. 31). Kestrels prefer to flight-hunt on sloping land, using windward rather than leeward slopes, and preferring the wind to be normal, or near normal, to the slope direction.

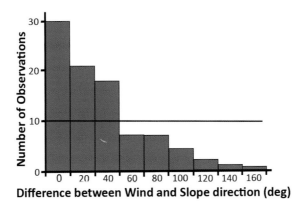

Figure 31. Slope selection by flight-hunting Kestrels. The histogram shows the frequency of observing Kestrels hunting on different slopes measured relative to the direction of the wind. The horizontal line is what would be expected if all differences were equally likely. Redrawn from Village (1983a).

As well as noting the effect of wind speed on flight-hunting duration, in their study of flight-hunting Kestrels in The Netherlands, Videler *et al.* (1983) also investigated the position of the falcon's head. During flight-hunting the falcon is watching the ground below. For maximum hunting efficiency the eyes (and, therefore, the head as the falcon's ability to move the eye within the head is minimal) must remain motionless. During the gliding phase between wing beats this can be achieved by allowing the centre of gravity of the bird to move. The centre of gravity starts close to the head, but can be allowed to move by extension of the neck. Videler *et al.* measured the potential extension on newly-dead Kestrels: about 4cm of movement could be made. So the Kestrel attains a position above the point of observation and allows its neck to extend as the wind pushes the body backwards so as to maintain the head motionless. At an extension of 4cm the bird must then flap its wings to move the centre of gravity forward again (and reduce neck extension) allowing the process to be repeated. For efficient hunting, the head must also remain motionless, or nearly so, during the wing flapping. To investigate this Videler and co-workers filmed flight-hunting Kestrels at up to 200fps. This showed that both vertical

The Common Kestrel

and lateral bill tip movements occurred during the wing beat, but that in neither case were the displacements greater than about 5mm and were often significantly less (Fig. 32).

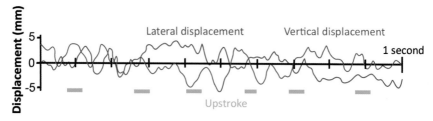

Figure 32. Vertical and lateral displacement of the bill tip of a flight-hunting Kestrel relative to nominal position as a function of time. The position of the wing upstrokes is indicated by the green bars. The horizontal axis equates to 1 second of flight time. Redrawn from Videler *et al.* (1983).

In general the topography of any Kestrel's hunting range will allow generation of updraft meaning that true planar (*i.e.* 2D) wind is unlikely even in essentially flat country because, as noted above, winds are very rarely homogenous (*i.e.* there is gusting) and friction effects at the ground surface impose changes with height. Kestrels will always, therefore, seek updrafts within their hunting ranges. But in some cases these updrafts may be so pronounced that they significantly alter the energy requirements of flight-hunting. In an interesting series of observations of Kestrels hanging more or less motionless, with minimal wing movement, in updrafts created by sea dikes in The Netherlands, Videler and Groenewold (1991) studied how this was achieved. Experiments had already been carried out (*e.g.* Tucker and Parrott, 1970, and Tucker and Heine, 1990) flying birds in a tilted wind tunnel to see how the variation in wind angle altered how the birds glided. These experiments showed that the birds (in the experiment one of three species chosen was a Lagger Falcon) altered their wingspan as the wind speed changed. Videler and Groenewold (1991) confirmed this with the free-flying Kestrels, but noted that the falcons also altered the angle of their body to the horizontal and both spread and closed their tails. The effect of these three changes was to minimise wing flapping (virtually eliminate it in some cases) and so reduce the their energy output to about 66% of that required for wing beat flight-hunting. The downside was that it took 60% longer to catch the same number of voles. This work is mentioned again when discussing the energy costs of flying in the next Chapter.

Opposite Kestrel (possibly a female, but more likely a juvenile as it was Autumn) displaying wing and tail morphing and position while hanging in the updraft caused by the wind hitting an obstruction. In this case the wind was blowing across the Somerset Levels and hitting the Mendip outcrop of Wavering Down and Crook Peak, an area famous for viewing Kestrels at that time of year, there occasionally being a string of birds across the scarp slope. With apologies for the soft focus of the images – the bird was at distance and the wind was blowing fiercely. See comments on photographs overleaf regarding head/eye position.

The Common Kestrel

During the hi-speed filming described overleaf the wind at point increased to 8m/s allowing the male Kestrel *Kevin* to perform a 'hover' which, initially, looked rather as the hanging flight illustrated on the previous page. However, in the absence of an updraft this could not be true hanging, the bird stretching its neck. in the 0.3s after the upper photograph was taken the falcon made a short wing beat, the wing tips moving up and down fractionally and the neck was reduced. Because the side-on camera was not perfectly aligned normal to the bird the stretch is difficult to assess accurately, but is certainly comparable to that suggested by the work of Videler *et al*. (1983) detailed on p127. However, as can be seen, the bill tip position moved much more than was seen in Fig. 32. This 'modified' form of hanging flight obviously produces a much less stationary head/eye position, as is also the case for the wild Kestrel in hanging flight on the previous page which might also influence the lower success rate of the technique.

Flight-hunting female Kestrel.

In order to gain further information on the Kestrel's flight-hunting technique, the IMU described earlier in the Chapter was employed together with high-speed digital video cameras to provide synchronised data between head and eye position, wing position and body position. Although the cameras could operate at very high speeds they were set at only 800fps (frames/second) as the wing beat frequency of flight-hunting Kestrels is significantly lower (2-8Hz depending on wind speed). A separate time-code generator, collecting time signals from the same satellites that provide timing for the IMU, was used to record time with millisecond accuracy for each frame of both cameras, the time being added to the metadata of each frame.

The IMU also ran at 800Hz to synchronise the camera images with output from the unit's tri-axial. The cameras were set head-on and side-on to observe eye position. Other flights were made with the cameras set head-on and tail-on. Fig. 33 is a schematic of the set-up. The bird carried the IMU described

Figure 33. Schematic diagram of the equipment set-up to investigate Kestrel flight-hunting.

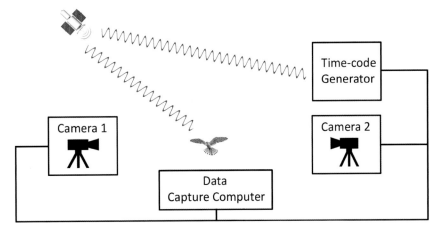

The Common Kestrel

in the text. The cameras were arranged to film combinations of head-on, side-on and tail-on simultaneously. Note that while the figure suggests a single data capture computer, this is for clarity of the schematic. In practice each camera had its own capture computer and was individually fed by the time-code generator.

The Kestrel used for the filming was *Kevin*, a 15-year-old male trained by Dave Hughes. On the day the wind was, at first, steady at about 4m/s with gusting to 6m/s, then increased in speed to a steady 7m/s with gusting to 9m/s.

Analysis of the data is not straightforward. Gyroscopes both small and lightweight enough to fly on a Kestrel, and, more importantly, cheap enough for private budgets, are susceptible to drift which accumulates errors in orientation over time. They cannot, therefore, be used to provide an accurate measure of orientation. Accelerometers and magnetometers can help by providing an absolute measure of the Earth's gravitational and magnetic fields, but are themselves subject to random errors. Using gyroscopes in combination with accelerometers and magnetometers allows computation of orientation but requires the use of complex mathematics. No attempt will be made here to investigate this mathematics or, as a consequence, to look at body orientation against wing position. A paper on the analysis of data from the IMU will be published in the near future (Madgwick and Sale, *in prep.*).

The point of the analysis is to investigate the stability of the Kestrel's eyes, in relation to the orientation and movement of the body. The former was investigated by the hi-speed cameras, while the latter was observed by the IMU which was firmly attached to the Kestrel's body between the wings. *Kevin* wore a standard harness that is attached to most falconry birds. Normally a radio-finder is attached to the harness to allow the owner to locate the bird

Measuring the separation of *Kevin*'s eyes.

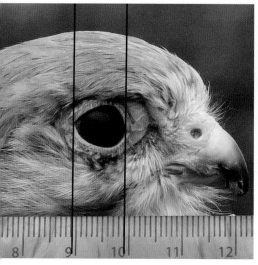
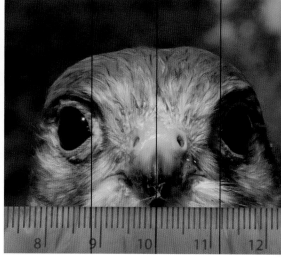

should it go missing. In our case the same type of spring used to attached the finder to the harness was used to attach the IMU. For efficient hunting the head has to remain steady. For this to happen, the body has to absorb the forces generated by beating wings and a moving tail, and the buffeting of the wind, this potentially incorporating not only a variation of head-on force due to gusting, but a potential cross-body force if the gust speed varies across the body and wings.

To calculate movement of *Kevin*'s eyes, eye and eye separation were measured prior to the flights – photographs opposite. The falcon's eye is a skewed ellipse, with a 1.0cm major axis and 0.8cm minor axis. The distance between the forward points of the eyes across the front of the head above the bill is 2.35cm.

The images overleaf are stills taken from one of the 800fps flight videos, specifically a flight with a wind speed of about 5m/s. *Kevin* was placed on one post of a marked-out flight area. Close to the centre of this a further post was positioned to be in-frame to act as a fixed marker for the cameras. The cameras were positioned so that head-on and side-on images were obtained. Dave, the falconer, was positioned equal distance from the cameras and called *Kevin* to him using a food morsel as an incentive. Over the years *Kevin* has been used many times in demonstrations and knew the drill. He flew to a position immediately above Dave and 'hovered' for as long as Dave kept pointing his fingers upward. Flights lasted only a few seconds, but during the first flights at low wind speed *Kevin* was beating his wings at 7 beats/s so that even short flights gave several dozen wing beats (and provided several hundred Gbytes of footage). The IMU's battery will power the unit for about 35 minutes at the very high data collection rate chosen. Clipping the unit into *Kevin*'s harness requires handling of the bird to which he does not take kindly. Even with the handling minimised, after the clipping the falcon needs to preen and settle before he is willing to fly. Given the time to recharge the lipo battery which powers the unit and the time taken to fly the bird, then settle it again to be ready for the another flight, the number of flights that could be made with the IMU attached was limited. A small number of flights without the unit attached were also filmed, both head-on and side-on, and head-on/tail-on.

The first flight showed that the camera positions were slightly off-angle and they were repositioned for subsequent flights. However, there were minor shifts in the wind angle on every flight and as can be seen from the images overleaf, while the side-on camera was almost perfectly aligned normal to the falcon, the head-on camera was a few degrees off centre. The still images shown above were from the second flight during which the 'hover' was the most sustained. In this flight the first two complete wing beats of the 'hover' showed greater eye and head movement as the bird adjusted his position. Eye and head positions were then held essentially constant through subsequent wing beats, the still images being at 20millisec. intervals through a 0.14s wing beat.

The Common Kestrel

Shots taken from the head-on and side-on videos during one complete wing beat lasting 140millisec. (*i.e.*7 beats/s. The IMU is visible on the Kestrel's back. The green diode indicates that the unit has a satellite lock and is collecting data.

The image to the right is a superimposition of the head-on shots above. The red rectangle represents the distance between the eyes and along the minor axis of the eye's ellipse. The green rectangle shows the maximum movement of the head during the complete wing beat. The maximum movement was 2mm.

Overleaf a superimposed wing beat is shown side-on. The red and green ellipses represent the eye and the maximum positions of the eye during the full wing beat. Again the maximum movement was 2mm.

Given the speed and acceleration of the wing tips during the upstroke and downstroke measured by the Groningen University team (see Fig. 26 on p109), even allowing for the fact that in the flight considered here the wing movement was somewhat curtailed, the ability of the falcon to control body movement such that head (and therefore eye) movement is so small is remarkable.

Hunting

Below are the head-on shots for a single wing beat, as above, superimposed through the beat. The red rectangle represents the distance between the eyes (2.35cm) and the minor axis of the eye (0.77cm). See the text for an explanation of the green rectangle.

Side-on shots for a single wing beat superimposed through the beat. The red ellipse represents the falcon's eye. See the text for an explanation of the green ellipse.

Kevin, the falconry Kestrel used in the hi-speed filming of flight-hunting. A section of the IMU, clipped into the harness the bird wears, can be seen.

Other hunting techniques
While the Kestrel is famous for flight-hunting, the technique is not the only one employed by the falcon. Kestrels also hunt from perches, choosing a perch at a height which allows a wide scanning area and watching for movement on the ground. Once prey is detected the Kestrel will then fly directly to it or may flight-hunt above it briefly. In general, in perch-hunting the falcon will move position every few minutes, taking up a new perch each time. Kestrels will also soar, but this not a frequently used technique as it requires specific weather conditions, rising thermals or updrafts created by local topography, the latter more usually being used for flight-hunting. Kestrels will occasionally use a combination of soaring and flight-hunting to move across the landscape, the flight-hunting pauses allowing a better scan of likely-looking areas.

In Britain, and across northern Europe, the principal prey of Kestrels is mammalian. But as we have seen in Chapter 3, insects may form a greater percentage of the prey even during the breeding season. In a study in South Korea (Won *et al.*, 2016) the researchers explored the difference between flight-hunting and perch-hunting. They found that flight-hunting yielded 82% mammalian prey (voles) while perch-hunting yielded 59% insect prey. While the success rate of perch-hunting was higher (*i.e.* insects were easier to catch) mammalian prey represented greater food value, which was most important when chicks were being raised. Not surprisingly therefore, breeding Kestrels spent an increasing amount of time hunting over riparian habitats (where the vole yield was higher) as the breeding season progressed (and chick food demand increased) than over other habitats (*e.g.* paddy fields) where voles were much less abundant, though insects thrived. In March, before the breeding season was underway, the Kestrels spent more time perch-hunting and hunted over a wider range of habitats, insects being relatively abundant everywhere.

In a study of Kestrels in an area of reclaimed land in The Netherlands, Rijnsdorp *et al.* (1981) also noted 'sitting' as a hunting technique. By 'sitting' the Dutch researchers referred to the falcon sitting on the ground or in sheltered positions such as in a nestbox or under the eaves of a house. Kestrels will take worms, beetles *etc.* from the ground which can also be termed 'sitting' hunting and, as with all raptors, they will also hunt opportunistically and so while sheltering from inclement weather may indeed chase after observed prey. Other interesting forms of opportunistic hunting have also been noted. Trollope (2012) reported seeing Hobbies following a steam train which was flushing dragonflies from track-side ditches, and Walker (2012), picking up on this observation, noted that his father regularly saw Kestrels following trains in the 1960s as at that time trains were all hauled by steam locomotives which regular shot jets of steam into track-side vegetation that flushed prey.

This page and opposite page, the first true flight by a fledgling hatched in a nestbox in southern England.

The success and energy costs of hunting

Flight-hunting is energy intensive in comparison to perch-hunting when the falcon watches from a perch and makes forays only when prey is spotted. Consequently, in their studies of the energetics of the Kestrel the Dutch team at Groningen University (which will be more thoroughly explored in the next Chapter) found that Kestrels perch-hunted more frequently in winter (Masman *et al.* (1988b). In summer Masman and co-workers found that flight-hunting yielded 4.7 small mammals/hour, this falling to 2.2/hr in winter. The figures for perch-hunting were 0.1/hr in summer and 0.3/hr in winter. In a separate study in The Netherlands, Rijnsdorp *et al.* (1981) found summer success rates of 2.82/hr when flight-hunting, 0.31/hr when soaring, 0.21/hr when perching and 0.07/hr while sitting. There are differences between the success data in each case, but the overall premise remains the same – flight-hunting is an efficient way of catching small mammals. Village (1983a), studying Kestrels in southern Scotland, found an alteration in the primary hunting technique with season, consistent with the Dutch studies (Fig. 34). Clearly the Kestrel is opting for a low-profit, but low-cost, hunting technique in winter to minimise energy expenditure.

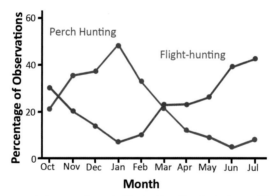

Figure 34. Seasonal changes in main Kestrel hunting techniques. Redrawn from Village (1983a).

Village (1983a) also noted that the percentage of flight-hunts increased with wind speed during spring and autumn, 'glide' flight-hunting clearly being energy efficient in comparison to 'flap' flight-hunting: in each case at speeds above about 5m/s flight-hunting was the dominant hunting technique, and at about 10m/s constituted 90-100% of all hunting. During the summer flight-hunting represented about 100% of hunting at all wind speeds, the return in terms of prey obtained outweighing energy costs when there were nestlings to feed. From Fig. 34 flight-hunting

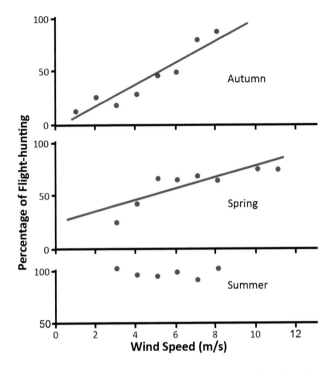

Figure 35. Variation of the percentage of all hunting that was flight-hunting in three seasons. Redrawn from Village (1983a).

accounts for only 10-15% of hunting during the winter: though winters may be windy, convective heat losses will outweigh the energy gain when the capture rate decreases. Village (1983a) investigated this variation in usage of flight-hunting in spring (March-April), summer (June-July) and autumn (October-November), noting that most flight-hunting occurred in wind speeds of 5-10m/s in all three seasons, but that the likelihood of the hunting technique was correlated with wind speed in spring and autumn, *i.e.* the change observed was statistically significant (Fig. 35).

Village does not suggest a rationale for this difference, but again it seems probable that in spring it is related to the need to provision nestlings which always requires adult Kestrels to choose the most efficient hunting technique, regardless of the energy cost, while in autumn it aids adult Kestrels to recover weight and condition after the rigours of breeding. This is consistent with the finding of Videler *et al.* (1983) in The Netherlands that Kestrels spent more time foraging while raising nestlings than at any other time of the year (Fig. 36).

One obvious question arising from the data of Fig. 36 is why Kestrels have such a low daily foraging duration in comparison to songbirds which forage almost continuously during daylight hours. We will consider this question in the next Chapter.

But while the data previously presented illustrates that flight-hunting is an efficient hunting method in terms of the number of voles caught, it does not tell us the success rate of an individual hunt. To investigate that, the University of Groningen group spent 760 days over a seven-year period observing Kestrels (a total of almost 7300 hours of observations) and compiled the behavioural data of Table 4 (Masman *et al.*, 1988a).

As can be seen, from the Table, flight-hunting is, overall, 39% efficient as a technique against mammals, and is also, surprisingly, 28% efficient in attacks on birds. Other techniques are comparably efficient, though the overall yield is much lower. Masman *et al.* (1988a) also looked at the variation of yield

Figure 36. Comparison of the foraging times for the Kestrel and the Great Tit as a function of the time of year. Redrawn from Videler *et al.* (1983).

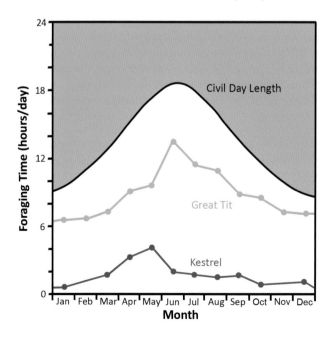

Behaviour	Flight-hunting	Flying	Soaring	Perching	Sitting[1]	Total
Observation time (hrs)	884.9	362.8	122.7	2283.0	3629.6	7283.1
Strikes: Mammals						
Total	8884	4	14	904	106	9912
Positive	3212	4	5	273	55	3549
Negative	5007	0	4	588	44	5643
Frequency[2]	10.4	0.01	0.11	0.40	0.03	1.36
Success[3]	0.39	1.00	0.56	0.32	0.56	0.39
Yield[4]	3.92	0.01	0.06	0.13	0.02	0.53
Attacks: Birds						
Total	182	57	39	193	121	593
Positive	50	1	2	47	21	121
Negative	126	16	20	145	99	406
Frequency	0.20	0.16	0.32	0.08	0.03	0.08
Success	0.28	0.06	0.09	0.24	0.17	0.23
Yield	0.06	0.01	0.02	0.02	0.01	0.02

Table 4. Kestrel Time Budget. Data from Masman *et al.* (1988a).
The total observation time 9122.9hrs, the falcons being in sight for 7283.1hrs (79.8%).

Key:
1. Sitting positions were – on the ground (1195.4hrs), in shelter (1985.5hrs) and in nestboxes (1348.7hrs).
2. Strikes/hour.
3. Number of prey captured/strike or captured/attack.
4. Number of prey captured/hour.

of male and female Kestrels throughout the year from the two predominant hunting techniques (Fig. 37 overleaf).

Masman *et al.*, 1988a also combined male and female data to compare the hunting yields across years. They found that yields were similar in different years, but that, again, the yields in different seasons varied. Hunting yield is dependent on strike (*i.e.* attack) rate and strike success. Strike success was essentially constant from year to year and from summer and winter across years, but differed markedly between the seasons (Fig. 38 overleaf). In summer there was an increase in flight-hunting yield due primarily to an increase in strike success, but the variation in perch-hunting yield was due chiefly to increased strike frequency. Most notable was the improvement in strike success from perch-hunting in early summer, the decline which followed being due to a drop in the frequency of strikes as flight-hunting took precedence.

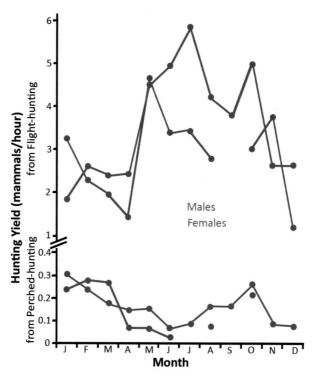

Figure 37. Annual variation in yield from flight-hunting and perch-hunting for male and female Kestrels. Redrawn from Masman et al. (1988a).

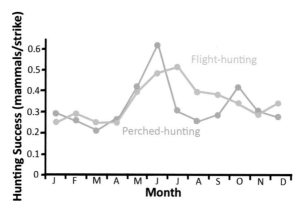

Figure 38. Annual variation in strike success for Kestrels (combined male and female data) for flight- and perch-hunting. Redrawn from Masman et al. (1988a).

What is again clear from Figs. 37 and 38 is that Kestrels transfer from a high yield, high cost hunting strategy in summer to a low cost, low yield strategy in winter.

Female (*above*) and male (*below*) Kestrels leave their nestbox after delivering prey.

The Common Kestrel

Environmental Factors

As well as observing hunting methods, the study of Rijnsdorp *et al.* (1981) in The Netherlands also noted the effect of weather conditions on hunting. As might be expected, in addition to low wind speed, flight-hunting was also inhibited by fog and rain (Fig. 39).

Kreiderits *et al.* (2016) studied the effect the weather had on urban Kestrels (in Vienna) with similar results. Low temperatures and rainfall affected the composition of the diet during the nestling phase, with the proportion of avian prey taken increasing. The weather appeared to have a greater influence on breeding performance than did the composition of the Kestrels' diet, though as the weather influences the probability of taking mammalian prey and the falcons are less efficient at taking avian prey, weather and diet are inextricably linked.

For further information on the effect of weather conditions, see also Fig. 52 in the following Chapter and Chapter 8 *Figs. 103-105*.

As well as weather conditions, flight-hunting will also be influenced by ground cover as it is the availability of prey to the hunting falcon rather than population density as such that is key to hunting success. Garratt *et al.* (2011) studied the effect of different vegetation on Kestrel hunting preference by close observation of seven pairs breeding in north-east England during the period 2006-2008. Five habitat types were identified in the home range of the falcons and the use made of these, in comparison to the habitat-type availability, were assessed – Fig. 40. The researchers also investigated the prey captures in the differing habitats, comparing the observed number with that expected if habitats were equally favourable to the hunting falcon (Fig. 41 overleaf).

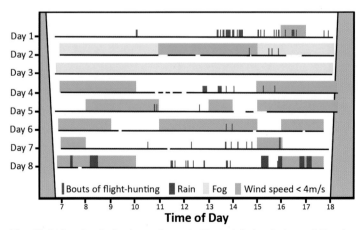

Figure 39. Flight-hunting behaviour of a male Kestrel during 8 days of October. Fog, rain and wind speeds below 4m/s clearly reduced the incidence of flight-hunting. Redrawn from Rijnsdorp *et al.* (1981).

Hunting

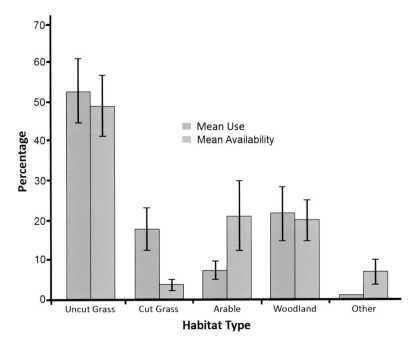

Figure 40. Comparison of mean habitat use (percentage of foraging in each habitat type by each Kestrel pair) and mean availability (percentage of habitat type in each Kestrel pairs' home range). Black dumbbells represent one standard deviation. Redrawn from Garratt *et al.* (2011).

The habitat types were defined as:
Uncut grass – comprising short grass (<20cm), long grass (>20cm), meadow (>50cm), field margin and rough buffer categories.
Cut grass – comprising recently cut (within one week) hay crops, and mowed meadow strip categories. All about 5cm or less in height.
Arable use – comprising arable crops and ploughed fields.
Woodland – comprising conifer, broadleaf and mixed woodland, and young plantations with trees up to a maximum of 4.5m high.
Other – comprising water, reed swamp, anthropogenic features and unknown categories.

Figs. 40 and 41 indicate that Kestrels prefer grass cut within the last two weeks, meaning grass blades less than 5cm in length. Garratt and co-workers found that this preferred grassland resulted in a higher take of mammalian prey (4.36 mammals to every invertebrate), with uncut grassland reducing the ratio to 1.73:1. Birds were infrequently taken during the study (9% of all prey in comparison to 48% mammals and 17% invertebrates: 26% of the taken prey was classified as 'unclear'). The only habitat in which birds played a significant role was woodland where they comprised 24% of the taken prey. While grass cutting allows the flight-hunting Kestrel a clearer view

The Common Kestrel

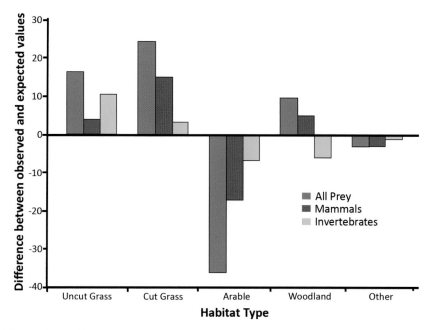

Figure 41. Difference between observed and expected prey captures by six Kestrel pairs across five habitat types in north-east England. Bird captures were scarce and included in 'all prey'. The figure indicates that grassland is the most important Kestrel hunting habitat, arable land the least favourable. Redrawn from Garratt *et al.* (2011). The habitat types are as for Fig. 40.

of its potential prey, it is not obvious that mammals living in the uncut grass will still be available for hunting after cutting. To explore the behaviour of the prey, Garratt *et al.* trapped mammals (mice, shrews and voles) prior to, and after, cutting. Not surprisingly, they discovered that most mammals left the area, but some (20-27%) remained if the cut grass stayed where it lay: the fraction halved if the cut grass was removed. In a report which took data from the same areas of study as Garratt *et al.* (2012), Peggie *et al.* (2011) looked at the time it took different species of bird which sought prey on cut grass to leave the area after cutting. The researchers found that for almost all species, within 10 days or so the number of birds using the regrowing grass had halved, but for the Kestrel, it took only four days for hunts to decline by 50%.

Vegetation management and its influence on Kestrel numbers is considered again in Chapter 11 when the population of the falcon in Britain is examined.

Female Kestrel demonstrating her flying ability. *Torsten Prohl*.

5   Food Consumption and Energy Balance

Village (1980) derived the daily time budget of Kestrels in southern Scotland by observation, and then computed the daily energy expenditure using formulae which had been developed for expressing the energy input for various avian activities – Fig. 42.

Village's work was extended in an outstanding series of experiments by the Dutch Groningen University group. This work, which has already been mentioned in previous chapters, included an accurate study of the energy intake and expenditure of free-living Kestrels, researched in the field, backed up by some excellent laboratory experiments on Kestrels trapped in the wild.

Masman *et al.* (1986a) observed wild Kestrels on 375 days over a seven-year period, during which time the falcons consumed 1944 prey items. By weighing, then replacing, 43 cached items which were later retrieved and eaten by the birds, the time taken to consume a prey item of known species and weight was calculated – Fig. 43.

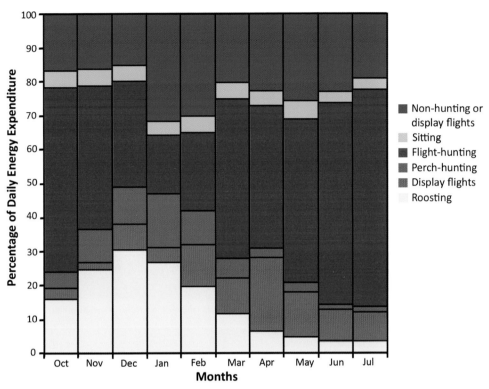

Figure 42. Percentage of daily energy expenditure for the period October to July for male and non-breeding female Kestrels. Redrawn from Village (1980).

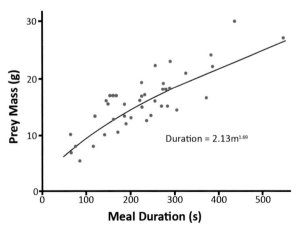

Figure 43. Meal duration for different vole masses. Each red dot represents an individual vole retrieved from a cache and weighed, then replaced and subsequently eaten. The equation was derived by least-squares fitting. Redrawn from Masman *et al.* (1986a).

Breeding puts a huge strain on the energy balance of both male (*above*) and female (*below*) Kestrels.

The Dutch team fitted a curve to the data (as indicated on Fig. 43) and used this relationship to estimate the weight of prey of the same species when observing Kestrels consuming prey on other occasions. For other species, average weights were taken from trapped specimens (mammals) or estimated (birds). Of the 1944 prey items, 1822 (93.7%) were mammals: most were unidentified voles and shrews (66%), Common Voles representing 31% of identified prey, with Common Shrews representing the remainder. The non-mammalian 6.3% of the total prey was avian, chiefly unidentified songbirds or juvenile waders, Starlings making up 14% of identified species. These data were then used to assess the energy intake throughout the year for both male and female birds (Fig. 44).

Energy intake is highest during the breeding season (April-July) and lowest during the moult (August and September) for both sexes. These peaks and troughs are as would be expected, the peak intake rate for the female corresponding to the period of egg laying and the first few days after hatching when she spends most of her time brooding and feeding the chicks. While the female may eat little during the latter period (mostly skin and the intestines of prey brought by the male, the intestines probably to provide water) her energy output is also minimal. For the male the same period corresponds to his being the sole provider of food for the family. Energy intake for both sexes declines during the moult when feather loss makes flight-hunting more difficult. In terms of vole consumption, the daily average throughout the year of both male and female is about four rodents. The male also needs to catch about eight rodents/day to satisfy the energy requirements of brood feeding, the female needs six rodents/day during egg laying. The mean energy intake during winter is similar for both males and females, reflecting the limited difference in size of the sexes.

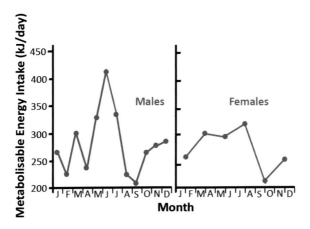

Figure 44. Monthly mean daily metabolisable energy intake for male and female Kestrels. Redrawn from Masman *et al*. (1986a).

The Common Kestrel

A female Kestrel in southern Scotland attempts to get some sleep while brooding her boisterous chicks hatched in a old shopping basket. But the attempt is short-lived.

Having established energy intake, the Dutch researchers now needed to estimate energy expenditure. Historically this has been calculated using a 'time budget', *i.e.* observing the activity of a bird during a day – the time spent flying, preening, loafing *etc.* – coupled with an understanding of the energy required for the bird to fuel its vital processes and to feed, and estimating the total energy expenditure of each. This method requires knowledge of the energy requirements of each activity, particularly of flying which is clearly the most energy-intensive. Masman *et al.* (1988b) therefore inferred an equation for total energy expenditure:

$$E = B + T + A + H + S \text{ (kJ/day)} \qquad (4)$$

where:
B is a basal component (the Basal Metabolic Rate or BMR), the energy required to keep a thermoneutral (*i.e.* requiring no energy expenditure to combat the effect of an ambient temperature lower than body temperature), fasting bird alive;
T is the energy required for thermoregulation (*i.e.* to overcome heat loss due to the difference between ambient and body temperatures). The Dutch calculated not only heat loss for a fully-fledged bird, but also the additional heat lost by a bird in moult (Tr). Moulted feathers increase heat loss by a surprising amount, particularly if the ambient temperature falls. Masman *et al.* note that while the heat losses for moulting and non-moulting birds are similar at an ambient temperature of 20°C, at 10°C the heat loss increases by almost 50% for the moulting bird;
A is the energy required for activity, which was sub-divided as Ab (activities other than flying), Af (flight) and Ah (flight-hunting);
H is the energy required for feeding (*i.e.* digestion);
and S is the energy required for tissue synthesis (*e.g.* feather synthesis during moulting). As noted in Chapter 2, the Dutch team (Dietz *et al.*, 1992) were able to accurately measure this energy requirement.

With S already known, all that was needed was an understanding of the other components in the equation. Some can be estimated using standard laboratory methods such as calorimetry, but others required more ingenious methods. To calculate T the Dutch used, amongst other things, a heated taxidermy Kestrel mounted on the roof of their building and calculated the heat loss in various weather conditions. The highest energy requirements are those associated with flight. As already noted in Chapter 4, the Groningen team trained Kestrels to fly along corridors in exchange for food titbits. By balancing the titbits to ensure the birds' weight was kept the same, the energy input for a given flight mileage could be calculated (Masman and Klaassen, 1987). The results from these laboratory tests were then checked against tests on captured live birds which were injected with heavy water. This 'double-labelled water' method of energy expenditure calculation uses a specific form of heavy water, combining Deuterium (heavy hydrogen, a hydrogen atom with a nucleus of a proton and a neutron, rather than the standard atom, the nucleus of which comprises a proton only) and O18, an isotope of oxygen with 10 neutrons rather than the 8 found in the most abundant (99.8%) form. Each of the atoms (D and O18) is easily detected by mass spectrometry: the isotopes are non-radioactive, and non-toxic in the quantities used. A given amount of heavy water is injected into the subject's blood (in this case a captured live Kestrel) and at the same time a blood sample is taken. The method works because of the body's partial conversion of inhaled oxygen to exhaled carbon dioxide. As a molecule of the latter has two oxygen atoms the body needs extra oxygen atoms to complete the chemical equation and obtains these from water molecules. O18 can therefore leave the body through exhalation as well as through water loss (urine, sweat), whereas D atoms can only leave through water loss. Blood sampling after a given time (in the Dutch case by recapturing the Kestrel) allows a comparison of the concentration of D and O18 with that from the initial blood sample. The difference in concentrations allows the metabolic rate of the Kestrel to be accurately measured. The advantage of the test is accuracy, the disadvantage being that total energy expenditure is measured rather than the expenditure of specific activities.

As with T, the component A in Equation (4) was sub-divided into the requirements of flying and flight-hunting, and for incubation and brooding. The daily energy expenditure of male and female Kestrels could then be calculated – Fig. 45.

The data of Fig. 45 for male Kestrels is comparable with the data of Fig. 42, though the individual components of the histogram differ. However, each clearly indicates that flying is energy intensive, Masman *et al.* (1988b) calculating that it requires about 62W/kg, though interestingly they found that the cost of directional flying and flight-hunting were similar. In terms of energy expenditure when flying, Kestrels compare favourably with the sprinter Usain Bolt, who, when winning his 2012 Olympic gold medal developed a maximum power output of 2619.5W or 29.1W/kg as he weighed 90kg. Since flying is usually more sustained than sprinting, a more reasonable comparison might be with the power output of a competitive cyclist. In winning the 2018 Tour de France, Welshman Geraint Thomas produced power outputs which peaked at 774W during 30s sprints and were sustained at 300W during longer periods (up to 4 hours). Thomas weighed 67.6kg during the race and so was developing power outputs of 11W/kg (sprint) and 4.4W/kg (sustained). Of course, Kestrels are a better aerodynamic shape for flying than Usain Bolt is for running or Geraint Thomas is for cycling – the best estimate being that Bolt used 92% of his energy output to overcome drag, a smaller percentage being required by the cyclist. But despite Kestrels (and other birds) being both built for it, and good at it, flying is hard work.

Food Consumption and Energy Balance

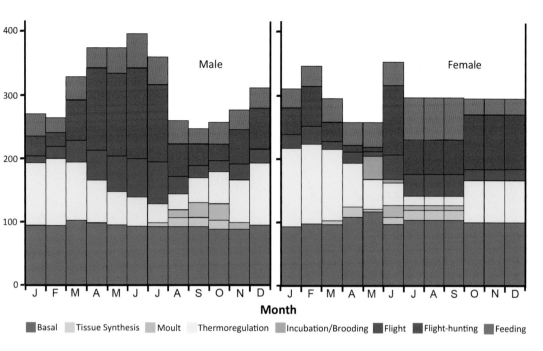

Figure 45. Monthly average energy expenditure, in kJ/day, throughout the year for male and female Kestrels. Redrawn from Masman *et al.* (1988b).

Masman and Klaassen (1987) found that the energy requirement of flight is about 16xBMR, *i.e.* sixteen times the energy required to keep the bird alive. This value differs from that suggested by Dolnik (1995) who predicted 12xBMR (as mentioned in Chapter 4): the Dutch researchers were able to identify a significant portion of this difference as being due to uncertainties in flight speed and aerial behaviour of the Kestrels, and in experimental methodology. A major part of the energy output is used in keeping the bird airborne. It is therefore worth noting that the disparity between the power outputs of Kestrels and athletes is very significantly reduced if the cost of staying airborne is eliminated. Masman and Klaassen found that if the falcon could utilise updrafts to 'hang' in the air, *i.e.* using the wind to eliminate the need to overcome gravity, the flight energy cost was reduced to 11W/kg which compares very favourably with the power outputs of both Bolt and Thomas. While at first glance pure hanging flight seems an unlikely scenario, in a study of flight-hunting Kestrels in The Netherlands Videler and Groenewold (1991) calculated that a bird's use of updrafts generated by winds rising over a sea-defence dyke allowed it to utilise hanging flight 90% of the time with a reduction in energy input to about 33%. However, there was a downside – to sustain hanging flight, the Kestrel had to maintain a specific position and the time taken to catch the same number of voles increased by 60%.

The Common Kestrel

Having now calculated both energy input and energy expenditure, the Dutch could compare the two. The comparison is most instructively considered in terms of the variation in body weight, but not as a function of calendar time, rather by considering the phase of the falcons' life during the breeding and non-breeding periods of the year – Fig. 46.

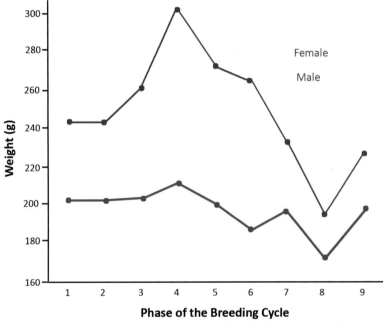

Figure 46. Variation of body weight in male and female Kestrels in relation to phase of the breeding cycle. Drawn from data in Masman *et al.* (1986a).

Phases of the breeding cycle are:
1. Wintering unpaired.
2. Wintering paired.
3. Courtship feeding.
4. Egg laying.
5. Incubation.
6. Nestlings below the age of 10 days.
7. Nestlings above the age of 10 days.
8. Dependent juveniles.
9. Post-reproductive moult.

Fig. 46 again indicates that for females, energy expenditure is at a minimum during courtship when much of her food is provided by the male[1]. In this

---

[1] Nelson (1970) feels the term 'courtship feeding' is not appropriate to describe the exchange of food between male and female birds. Although in its very earliest form it is likely that the giving and accepting of food cements the pair bond, Nelson considers the continued instances of food passing are as much related to the two birds ensuring that the female acquires the reserves necessary for egg laying and, consequently, reproductive success, and that it is therefore in the interests of both that the male continues to pass food to his mate after the initial 'courtship' phase.

phase she can store body reserves in readiness for egg laying. Village (1980) measured body weight in both males and females during the courtship/egg laying/incubation period, his data confirming the data from the Dutch study.

Once egg laying has begun, the female's body weight falls. Most striking is the sharp fall during the incubation and chick raising phases. At this time the female is unable to hunt to feed herself and most of the food provided by the male is given to the chicks to aid rapid growth. The female's weight loss during this time is about 33% from the preceding peak, and about 20% of the annual average. Even after the female can resume hunting, her body weight still declines because of the demands of the nestlings: only when the chicks become independent is the female able to replenish her body reserves. By contrast, the male's energy expenditure remains high throughout the breeding phase, the need to feed both the female and, later, the chicks requiring him to hunt for prolonged periods. Masman *et al.* (1986a) studied the energy requirements of the male Kestrel during this period of intensive feeding during the first ten days of chick life – Fig. 47.

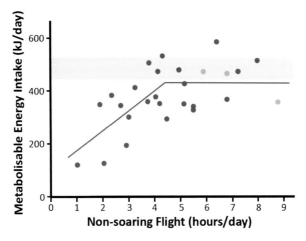

Figure 47. Daily metabolisable energy intake as a function of non-soaring flight (*i.e.* flapping flight + flight-hunting) for male Kestrels in Phases 6-8 of the breeding cycle (as defined in Fig. 46). Each red circle corresponds to one day's observation. Green circles correspond to days in which the food delivered to the chicks was reduced by the researchers, forcing the male to hunt for longer. See text for an explanation of the shaded area and red lines. Redrawn from Masman *et al.* (1986a).

Masman *et al.* (1986a) arbitrarily chose to break the data of Fig. 47 into two groups at a daily flight time of 4.6 hours, creating the two red lines on the Figure. The break is justified by the data, but the horizontal red line also sits close to the value of 445kJ/day (the lower edge of the shaded area) for a Kestrel of mass 192g, derived from *Kleiber's Law*. The law is named after the Swiss agricultural biologist Max Kleiber who, in the 1930s, first set down a relationship between the upper limit of daily energy intake set by the digestive tract and body weight. The upper edge of the shaded area is at a value of 522kJ which derives from a relationship set down by Kirkwood (1983) which defines the limit on metabolisable energy intake: the Dutch reasonably assumed that natural intake rates must lie between these values. Fig. 47 suggests that provided a male Kestrel spends less than about 4.6 hours of flight-hunting each day, his food consumption is adequate in supplying the energy requirements of hunting. But beyond 4.6 hours the male is unable to process food into energy fast enough to compensate for the energy outlay of hunting: despite consuming perhaps half his body weight in voles, the male would lose weight. While the value of 4.6 hours is obviously specific to a given male (in this case a falcon with a starting weight of 192g), the observed spread of male Kestrel body weights implies that for most males, flight-hunting for more than 5 hours will involve a loss of body weight.

Fig. 47 therefore appears to go some way to answering the question posed in the previous Chapter of why Kestrels do not follow songbirds in foraging throughout daylight hours – they cannot process food quickly enough. But that is not a complete picture as further work by the Groningen team (Masman *et al.*, 1989) showed. In this experiment the work of adult male Kestrels was increased in two ways. In broods younger than 10 days once the adults had satisfied the hunger of their brood, one or two hungry chicks from another nest were exchanged for fed chicks. In other nests with chicks old enough to feed themselves, food was removed immediately on delivery so that the chicks

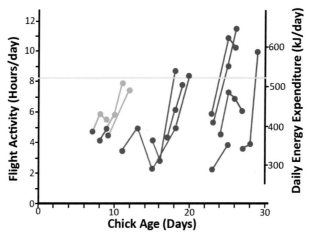

Figure 48. Effect of food deprivation to chicks on the male Kestrel's daily flight activity. Days on which sated younger chicks were replaced with hungry chicks are indicated by green circles. Days when food was removed from older chicks are indicated by blue circles. Red circles are control days (*i.e.* days on which there was no brood manipulation). Lines link days at an individual nest. The yellow horizontal line represents the daily maximal metabolisable energy intake from the work of Kirkwood (1983). Redrawn from Masman *et al.* (1989).

Food Consumption and Energy Balance

remained hungry, forcing the males to supply more food. The effect, of course, was to increase the hunting time of the males dramatically – Fig. 48.

The flight times of Fig. 48 are higher than those of Fig. 47, which at first glance is contrary to the conclusions which may be drawn from Fig. 47, but the Dutch team found that the Kestrels were increasing their use of hanging flight to reduce their energy output (!). So energy efficient is hanging flight that the male could extend his daily flight time to 11 hours, but limit his weight loss to only a further 20g (though this does represent a 10% loss on a weight which was already down by 10% before the experiment started). By contrast, the female was less able to utilise hanging flight and her weight continued to fall almost linearly. The experiment was terminated after 11 days as other data had already indicated that this was the maximum period over which the adult birds could sustain increased workload without exhaustion – Fig. 49.

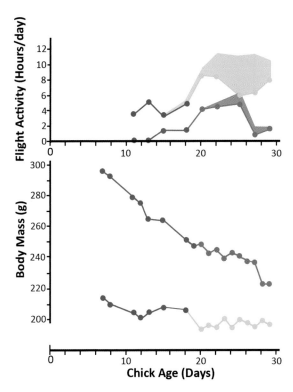

Figure 49. Flight times and weight changes in adult male and female Kestrels provisioning a brood of 5 nestlings. The blue circles are data points for the male, the red circles for the female, each on control days. The turquoise and purple circles are male and female on food deprivation days. The shaded areas are times when the falcons used hanging flight-hunting. Redrawn from Masman *et al.* (1989).

Masman *et al.* (1989) noted that their results re-opened the question of why Kestrels do not follow songbirds in foraging throughout daylight hours and so, perhaps, raising larger broods. The Dutch therefore looked at the daily energy expenditure of other birds to see how it varied. They found it was essentially linear with body weight, if both energy expenditure and weight were plotted on a log scale – Fig. 50.

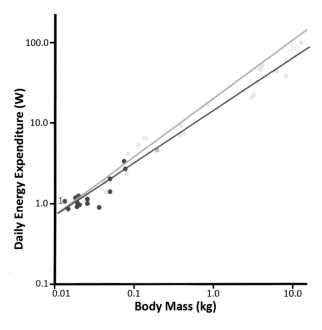

Figure 50. Daily energy expenditure by adult birds rearing chicks as a function of adult body weight. The red line is the parental care effort measured by various researchers for a number of species. Blue circles represent passerines. Number 1, the lightest passerine represented, is the Sand Martin at 12.6g. Yellow circles are non-passerines. Number 30, the heaviest non-passerine represented, is the King Penguin (*Aptenodytes patagonica*) at 13kg. The turquoise circle represents a 192g male Kestrel. A full list of the species shown on the figure is given in Masman *et al.* (1989). The green line is the derived relationship from the work of Kirkwood (1983). Redrawn from Masman *et al.* (1989).

Fig. 50 indicates that for bird species with weight differences of 10g to 10kg, parents work at 3-4xBMR when raising their young, and that larger species work further from their maximum sustainable energy intake (as defined by Kirkwood 1983) than do smaller species. As larger species have, on average, longer lifespans, reducing their breeding workload maximises their chances of breeding again. While Masman *et al.* are cautious about the conclusions of their work, it therefore seems that the foraging time for male Kestrels is defined by reproduction, the balance of brood number and food availability. With a defined supply of rodents, male Kestrels can feed a maximal number

Food Consumption and Energy Balance

of chicks. Beyond that number the hours spent hunting take a toll on the male which, ultimately, would be translated into a reduced chance of survival and, therefore, reduced likelihood of breeding again. This limitation is considered again Chapter 7 where the rate of prey delivery to chicks during their development from hatch to fledge is discussed.

One final, but equally fascinating aspect, of the work of the Groningen University group was their compilation of the daily time budget of male and female Kestrels for each month of the year or phase of their breeding/wintering cycle, and the variation of the activities with environmental conditions. This is shown in Figs. 51 and 52 overleaf. What is clear from these figures, and is succinctly set down in Masman *et al.* (1988a), is that in summer Kestrels maximise their daily energy intake (within limits set by their digestive systems) in order to maximise their reproductive output, while in winter they minimise their energy expenditure (rather than their foraging time) in order to minimise their daily energy requirements.

In some ways it is easier for the chicks – admire the view and wait for the next meal. But, of course, for both food consumption and energy balance the hard months are coming soon.

The Common Kestrel

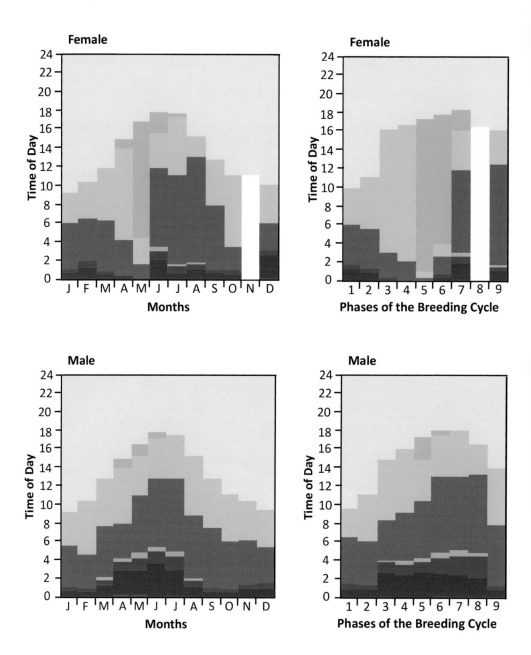

Figure 51. Time budget of male and female Kestrels relative to time of year (left histograms) and phases of the breeding cycle (right histograms). 'Perch' is defined as sitting above the ground on a perch which allows an unrestricted view of the local area. 'Sit' is defined as sitting on the ground, in a sheltered spot or in a nestbox. Phases of the breeding cycle are as given for Fig. 46. The colour key is given on the opposite page. Redrawn from Masman *et al.* (1988a).

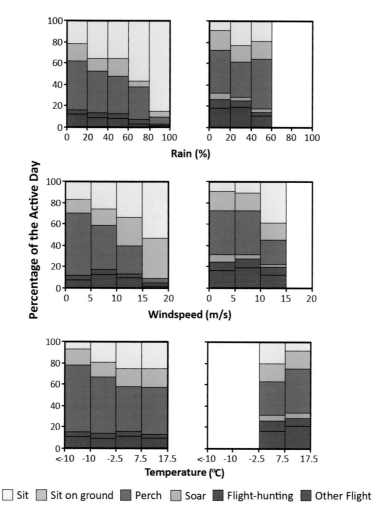

Figure 52. Time budget of adult male and female Kestrels (combined data) during the winter (left blocks), and for male Kestrels alone during Phases 6 and 7 of the breeding cycle (right blocks) relative to environmental conditions. Phases of the breeding cycle are as given for Fig. 46. The percentages were calculated relative to the active day, i.e. night-time resting is ignored. In this case 'sitting' is sub-divided into 'sit on ground' and 'sit' (in sheltered spot/nestbox). The colour key is given below the data blocks. Redrawn from Masman et al. (1988a).

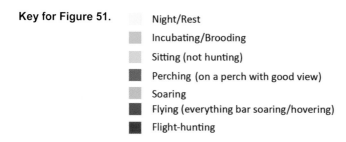

Key for Figure 51.

- Night/Rest
- Incubating/Brooding
- Sitting (not hunting)
- Perching (on a perch with good view)
- Soaring
- Flying (everything bar soaring/hovering)
- Flight-hunting

6 BREEDING (PART 1)

Age of First Breeding

There is considerable evidence for all four UK breeding falcons that birds rarely breed in their first year of life, with males significantly less likely to do so than females. In his study areas, Village (1990) noted that in general adult birds paired with other adults, with first-year birds tending to pair with other first-years if they were to breed. Cross-pairing by age did occur, but this was almost exclusively adult males with first-year females. Males seem much more likely to take the risk that a first-year female will be a dependable egg incubator and chick raiser, than females are that first-year males will be adequate hunters. Village observed 358 Kestrel pairs and found that 71% were adult–adult, 9% were first-year pairings, 17% were adult male–first-year female, with only 3% of adult females mating with first-year males. The number of birds apparently fit and able to breed, but not breeding, were – 25% (adult males), 21% (adult females), 67% (first-year males) and 34% (first-year females). Village (1985) noted that there was evidence that arrival time at breeding sites influenced these breeding rates. In general, first-year birds arrive later than adults which would explain the dominance of adult-adult breeding. However, this cannot explain the fact that adult male–first-year female pairs were six times more likely than first-year male–adult female pairs[1]. It seems that adult females have a marked preference for adult males, presumably working on the principle

---

[1] In 2017 the use of colour rings was introduced in southern Scotland for ease of identifying individual birds in subsequent years as an aid to studies on site and mate fidelity. That year a Kestrel pair laid seven eggs in a nestbox, hatching and raising all seven to fledge. In 2018 one of the fledglings, a male, occupied a nearby territory and paired with an adult female. The female laid seven eggs, all of which hatched. The pair then raised all seven chicks to fledge (G. Anderson and K. Burgoyne, pers. comm.).

that adult birds, being likely to have been through the breeding cycle before, know the ropes, as it were, and so will be dependable sources of food. The fact that they have survived through another winter is also an indication of fitness.

The numbers raise the questions of why first-year bird do not, in general, breed, and why there is a clear difference between the numbers of males and females that do. The differential mortality hypothesis suggests that reproduction is exhausting and may affect the life span of both sexes. Males provision the female and the brood, which means having to spend many hours flying and requires hunting skills: the former is hard work and may take its toll on the bird, while the latter needs experience which might explain the year spent acquiring skills before mating. Females must acquire the bodily reserves for egg laying and, later in the breeding cycle, must aid provisioning of a growing brood: again, both will take their toll on the bird, though neither would necessarily prevent a female from mating in her first year as lack of hunting skills should be alleviated by a good partner during the later stages of brood raising. After surviving their first winter birds of both sexes will likely be in a poorer condition than older birds because of their lack of experience. For young males that attempt to breed this potential reduction in body condition will be exacerbated by the need to feed themselves as well their mate and brood. In principle, first year females will regain body condition faster as they will be fed by their male partner during courtship and egg laying.

In a study of urban-breeding Merlins in Saskatoon, Canada Espie *et al.* (2000) attempted to explore differential breeding by considering the brood size of birds of differing ages, the hatch dates of the chicks, and the total number of fledglings produced by birds over their lifetime (the Lifetime Reproductive Success – LRS). Espie *et al.* also studied the number of young produced by individual birds which survived and returned to the study area. The results indicated that brood size was positively correlated with the age of parent birds, though the increase was more pronounced for males than for females. However, what was interesting was that the increase was apparent only for birds younger than the mean age of the population: for birds older than the mean age the brood size declined. In other words, birds which survived to breed again had larger broods until they reached the mean species age, but then had smaller broods as the reduced workload of rearing chicks increased the likelihood of surviving to breed again. The results also indicated that the hatch date of chicks was earlier for older parents, both males and females, and, as would be expected, the LRS was positively correlated with age for both sexes.

Considering their results, Espie and co-workers concluded that there was support for the differential mortality hypothesis for both sexes, though support was much more pronounced for females. Overall, as would be expected, the longer a Merlin lives the more chicks it will successfully raise, *i.e.* the higher the LRS. The researchers found that a female produced a mean of 9.2±6.2 fledglings in her lifetime, a male producing 7.4±5.9. The reduction in male

LRS is a result of losing a year to gain experience, *i.e.* male Merlins are trading loss of first year breeding for long term reproductive success. As a digression, the LRS for the female Merlins was higher than that calculated in a study of Merlins (of a different sub-species) in northern Sweden where Wiklund (1995) measured 6.4±4.6, with a range of 2 to 24. While the high standard deviations in both cases mean that there is considerable overlap in the two LRS distributions, the higher value for the urban-dwelling Merlins probably relates to the reduced nest predation of that population. It is possible that the LRS for urban-breeding Kestrels might show a similar increase relative to their stick-nesting rural cousins. It is also possible that the greater protection offered by nestboxes might increase the LRS for box-nesting Kestrels.

An alternative theory for delayed breeding posits that older males arrive earlier at breeding sites than first-year males and so establish territories which are richer in food resources, and it is certainly the case that adult-adult pairs tend to breed earlier than pairs in which one bird is a first-year, and that earlier breeding usually results in a higher number of fledglings. By contrast, later breeders, which often means pairs which are either both first-years or include one first-year, have a lower LRS: Warkentin *et al.* (1992), also studying urban Merlins in Saskatoon, noted that adult-adult pairs fledged, on average, 4.2±1.2 young, while pairs with at least one first-year bird, bred later and produced 2.6±1.6 fledglings. This theory again suggests that first-year birds, particularly first-year males, need to gather experience before they can mate. Further support for this theory comes from the numerous observations of third birds at nest sites. Often these are males which appear to be helping with provisioning the brood, and sometimes the female, and may even accompany the parent male on hunting trips, suggesting the juvenile is seeking to gain experience. However, it is worth noting that some observations of third birds imply the situation may be more complex. James and Oliphant (1986) report particularly interesting behaviour in the urban-dwelling Merlins of Saskatoon where there were several observations of a third bird aiding nest defence and food provisioning of the young. In one case a first-year male was taking food from an adult male, some of which he consumed, but some he passed on to the female at the nest. The younger bird accompanied the adult male on hunting trips, but was never seen hunting himself. The younger male also defended the nest, attacking both American Crows (*Corvus brachyrnchos*) and Black-billed Magpies, when the adult male was absent. Similar behaviour has been seen in British Merlins (Rebecca *et al.,* 1988), and it is likely that it also occurs in Kestrels. Frere (1886) records an instance where a gamekeeper shot an adult male Kestrel which was harassing breeding game Pheasants (*Phasianus colchicus*) only for a juvenile male to replace him within two hours: the gamekeeper also shot the replacement. Such replacements have been noted by later observers, and Village (1980) records the rapid (hours rather than days) replacement of an adult male, which he had removed from a breeding pair as part of an experiment, by a first-year male. As Village notes, the clear

Juvenile Kestrel. Many 'third birds' at nest sites are believed to be first-year males gaining breeding experience.

implication of such replacements is that areas in which Kestrels breed always contain birds which can breed, but which do not hold territories and so do not breed unless an opportunity presents itself.

Further evidence that might support the second hypothesis comes from a study of Kestrels in The Netherlands. Daan and Dijkstra (1988) examined the survival and probability of breeding of birds hatched from eggs laid at various times during the breeding season. This work, which considers the relevance of breeding decision on Kestrel survival is discussed in Chapter 11.

Territory

Village (1990) differentiates between a Kestrel's territory and its home range. Territory is the area around the falcon's nest from which it excludes conspecifics and competitors. Home range is the area over which the falcon hunts. The home range can be larger than the territory, but never smaller. In his Scottish study area, Village observed that non-breeding ranges were initially smaller than in summer, but that as the population declined, birds moved away from their breeding territories and range sizes increased. He also noted that ranges were defended, so that winter ranges became territories, probably due to the reduction in available prey numbers, voles ceasing to breed at the approach of winter: with a diminishing vole population, a bird's interest in increasing its home (*i.e.* hunting) range, and in maintaining it against competitors increases. In his English study area, Village found that the winter situation was different, reflecting the relative stability of the Kestrel population with reduced migration and breeding pairs often staying together and maintaining much the same range.

The change in population in Village's less productive Scottish study area also produced a noticeable change in ranges during spring and summer (Village, 1990). Ranges became larger as the population density of the birds increased: faced with the need to feed chicks the Kestrels expanded their hunting range, ranges being partially shared with adjacent breeding pairs. Initially, birds holding winter territories into which newcomers were seeking to advance attempted to hold on to their territories, displays and conflicts being seen, but ultimately pressure from incoming pairs forced territory sizes to shrink while, at the same time, home ranges increased with significant range overlaps occurring. Within the areas of overlap, hunting male Kestrels would occasionally have aerial jousts, but these were invariably short-lived and inconclusive, the birds returning to hunting when they were concluded. It is possible that males adjust their hunting times to avoid being in the same area as another, but as already noted, since the males hunt when they know their prey is most likely to be active, the more likely pattern is mutual tolerance in a situation of prey abundance after a brief display which concludes that neither bird is willing to be driven away. In times of prey shortage, mutual tolerance becomes a less attractive option, and in a study of Kestrels (and Long-eared Owls (*Asio otus*)) in Finland, Korpimäki (1987) noted that 'competition theory', the idea that the overlap of diet in neighbouring pairs of a species will reduce, apparently holds. In both species, neighbouring pairs took fewer *Microtus* voles and more alternative prey than did non-neighbours. However, what was notable was that the average number of fledglings produced by neighbouring pairs of Kestrels was lower (by about one chick) than in non-neighbouring pairs in both peak and low vole population years: the effect of overlapping ranges always reduces the efficiency of Kestrel breeding. This Finnish study is mentioned again when *Breeding Density* is considered below.

In his English study areas, Village (1990) noted that although the influx of newcomers was reduced in comparison to his Scottish area, so that territory sizes remained much the same, a similar increase in range size and bird behaviour was observed. Within their territory, males behaved very differently than they did within their ranges, the intrusion of other males not being tolerated as territory holders protected their females. In an experiment in which wild, breeding Kestrels were presented with an adult, caged Kestrel Wiklund and Village (1992) noted that the male of the breeding pair always responded aggressively to a male intruder, but a third of males displayed to an intruding female. Females were aggressive to female intruders before laying had commenced and after the eggs had hatched, but were less so between these periods. It might be expected to be in the interests of the female to keep her mate away from other females at all times, but Wiklund and Village consider the relative lack of female aggression during the egg laying period is explained by the female's increased body weight at this time, her reduced flying ability increasing the risk of sustaining damage in combat. Once the eggs had hatched, female aggressiveness to a live, caged female intruder increased with brood size. This, again, seems a natural response to the need to keep her mate away from other females during a period when he is sole provider of food, larger broods requiring more intensive hunting. By contrast, females solicited male intruders during the courtship period, though were more aggressive towards them once eggs had hatched (Figs. 53 and 54).

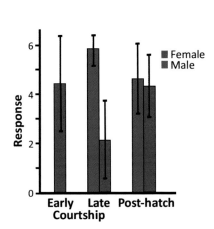
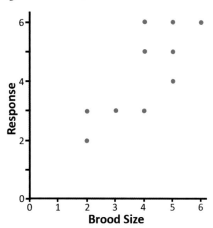

Figure 53. Response of male and female Kestrels to live decoys of the same sex at times during the early stages of breeding. Early courtship is defined as more than two weeks prior to laying and involved females only. Response varies from least to most intense. The black sticks are ± standard deviation on the mean value. Redrawn from Wiklund and Village (1992).

Figure 54. Relationship between brood size and aggression of the female Kestrel with chicks of 5-14 days old towards a live female 20m from the nest. Response varies from least to most intense as for Fig. 53. Redrawn from Wiklund and Village (1992).

Shrubb (1993a) notes that attacks by a territory-defending bird involving actual contact are rare, but when seen may not only involve the locking of talons in the air, but even fights on the ground, such fights occasionally ending in injury or death. More usually, a perched bird attacked by a territory owner will fly off, while the attacker of a flying intruder will position itself beneath the offender, fan its tail and force the offender upwards and away. Often during this manoeuvre the territory holder will perform the 'shivering' flight (see *Displays* below). Shrubb also noted a distinct seasonality to territorial displays, these peaking in early spring when the imperative was to secure a breeding territory, mate and nest site, during June when the need to feed growing chicks requires control of a hunting range, and in autumn when the dispersal of juveniles increases the pressure on available hunting land.

The overall pattern of behaviour noted above – smaller summer territories, with larger home ranges, followed by winter territories which correspond to home ranges, so that territories increase in size in winter, while ranges decrease has been noted elsewhere, for instance by Cavé (1967) in The Netherlands, and is likely replicated across the species' range, with winter territory defence being driven by the need to protect food supplies, and summer territory defence by the need to protect females from the attentions of other males.

The fact that territories and home ranges differ in summer, and that winter territories are dependent on prey availability makes defining the size of each difficult, but in general it seems that the summer territory of a Kestrel pair is 1-2km^2, while the home range is usually 2 to 3 times as large (but may be up to 5 times larger). In winter, territories/ranges may increase to 2-3km^2 in areas of stable population, but may be larger (up to 5km^2) in areas in which the population decreases (Shrubb, 1993a and Village, 1990). In all cases the variation in range size is dependent on vole population, Village (1990) noting a clear inverse correlation between size and vole density in his study areas.

Mate Selection
Village (1990) suggests mate selection is not random, noting instances where pairs seen early in the breeding season were later observed not to have bred. In each case the male had remained on his territory and bred with another female. Village conjectured that in some cases the first female had died, but noted a small number of cases in which the first female left and bred with another male. In these cases it appears that the first female was ousted by the second, implying that some females select mates other than by chance.

In a study on Kestrels in Finland, Palokangas *et al.* (1994) trapped wild falcons and noted female preference for males in controlled aviary conditions. The Finnish team found that females preferred males with bright plumage (as measured in visible light). In the wild the team noted that males with brighter plumage spent more time hunting than males with dull plumage and that females mated with bright males produced more fledglings than those mated with dull males.

This clearly suggested that male coloration was related to 'fitness'. All the Kestrels captured for the aviary experiment were released back to the area where they had been trapped after the testing had been concluded.

Another Finnish team, studying the parasite load of male and female Kestrels (Korpimäki *et al.*, 1995), found that brightly-coloured males carried a higher load of a particular blood parasite (*Haemoproteus tinnunculi*). As male blood parasite load has a negative impact on female laying date and clutch size – see Chapter 11 – this would seem to invalidate the finding of Palokangas *et al.* (1994), but Korpimäki and co-workers noted that blood parasite load was not related to male brightness. They also pointed out that use of the visible spectrum as a determinant of mate selection might not be the whole story because, as noted in Chapter 4, Kestrels do not necessarily see colour as humans do.

Young Kestrels at the point of fledging. As well as thumping its sibling during wing exercising, the active fledgling is showing its prominent, wide sub-terminal band, the significance of which is explored in the text.

This latter suggestion was confirmed when Zampiga *et al.* (2008) extended the work of Palokangas *et al.* (1994) by looking at the potential influence of UV signalling by male Kestrels. The researchers allowed captive females to view two males, one in front of, and one behind, a filter which blocked UV light. The females preferred the male with the UV reflecting plumage, as measured by the time they spent closer to the male without the filter. These studies suggest that plumage may also be an indicator of male quality when viewed in the UV part of the spectrum. More recently, in an experiment with Kestrels in Switzerland, Piault *et al.* (2012) suggested that the situation may be more complex than the overall brightness or UV reflectance of male plumage implies. In an experiment involving 20 Kestrel pairs breeding in nestboxes, the Swiss team measured the width of the sub-terminal black tail band in trapped adult Kestrels and in a control group of nestlings (Fig. 55a). The researchers then removed nestlings from some Kestrel broods and added them to others, creating two cohorts of chicks with enlarged or reduced brood size and, of course, manipulating the condition of the nestlings. The team then measured the width of the sub-terminal band of the chicks at fledging. They found that first-hatched chicks developed larger bands than their siblings in all broods, and that fledglings from reduced size broods developed larger bands than those from enlarged broods (Fig. 55b). The width of the sub-terminal band would therefore allow Kestrels another opportunity to determine the condition of a potential mate.

Any non-random plumage ornamentation comes at a cost for the individual bird, and so all these experiments suggest that only higher 'quality' birds

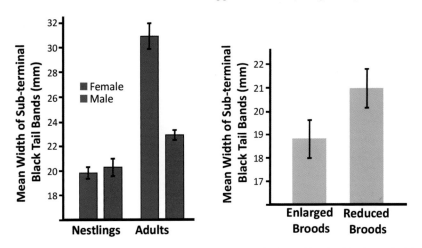

Figure 55a (*left*). Mean value (and ± standard error indicated by the black symbols) of the width of sub-terminal black tail bands of nestling and adult Kestrels.

Figure 55b (*right*). Mean nest value (and ± standard error indicated by the black symbols) of the width of sub-terminal black tail bands of nestling Kestrels in a brood size manipulation experiment. Both figures based on Piault *et al.* (2012).

would be involved. While the experiments were small-sample, they do suggest that plumage coloration is a factor in mate selection. But other observations suggest that additional factors may also be involved. In a study in central Finland, Hakkarainen *et al.* (1996) trapped Kestrels (12 males and 14 females) and kept them in aviaries where light as well as diet could be controlled. Body mass, wing and tarsus length of the male Kestrels were measured. In early spring the 'daylight' was increased suddenly from 7 to 20 hours day, and the diet of the females was increased to accelerate their sexual activity. The females were then shown the male Kestrels in a controlled manner that allowed an index of preference or avoidance to be registered. The males were also tested for hunting success by the straightforward, if rather gruesome, method of releasing rodents into their enclosures and measuring the success rate of attacks on them. The Finnish team found that the female Kestrels preferred males with a smaller body mass and smaller tarsus length (Fig. 56), but did not seem to express any preference by wing length.

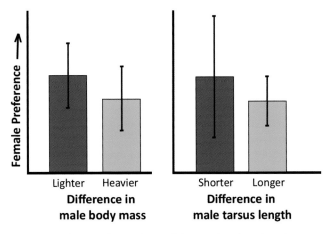

Figure 56. Female preference in relation to lighter rather than heavier competing males, and shorter rather than longer tarsi in competing males. The black symbols indicate ± standard deviation. Despite the size overlap, as indicated by the standard deviations, the differences in each case are statistically significant. Redrawn from Hakkarainen *et al.* (1996).

While Fig. 56 indicates a female preference, the preference was weak and apparent only when the differences in body mass and tarsus length were (relatively) large between males. Furthermore, hunting success did not accord precisely to the weak female preference as smaller, shorter-winged males were more successful than heavier, longer-winged males, though tarsus length seemed unrelated to hunting success. Hakkarainen *et al.* therefore concluded that there was a bias, albeit weak, of female Kestrels towards smaller males because it implies superior hunting skills and, hence, improves the likelihood of breeding success, noting that in years with low vole numbers, a male with better hunting skills would be a preferable mate. The study also provides some support for the idea that reverse sexual size dimorphism in falcons (see Chapter 2) is related to breeding protocol, smaller males benefiting from their smaller body size.

One final study on mate selection is also worthy of note. Working with American Kestrels, Duncan and Bird (1989) found that females did not discriminate against male siblings in their choice of mate, but did prefer males who displayed more actively. The authors suggest that high mortality in, and the limited number of nest sites available to, the species may mean that incest is rare in the wild and so may not influence mate selection. Since the Common Kestrel is subject to similar conditions, this lack of an apparently evolutionary disadvantageous strategy may also be seen in the UK population: it is known that incestuous relationships exist in other UK falcons. However, as the females in the Duncan and Bird study were captive it might be argued that they had little choice in mates and so might have exhibited behaviour which was rarer in the wild.

Displays

The most frequently described display involves the Kestrel pair soaring and circling at a considerable height, often with fanned tails. During these flights the male (much more rarely the female) will often exhibit a rocking (*e.g.* Village, 1990; Shrubb, 1993a) or rolling (*e.g.* Glutz von Blotzheim *et al.*, 1971) flight, moving horizontally with jerking wing beats while occasionally rocking (or rolling) from side to side to flash his underwings. Village (1990) believes this is most often seen when there is intrusion into the male's territory, suggesting it is as much a territorial as a courtship display: Shrubb (1993a) disagrees somewhat, seeing it much more as a mating display. Walpole-Bond (1938) and Glutz von Blotzheim *et al.* (1971) also mention a display in which the male makes mock attacks on the female, occasionally brushing her with his wings as she evades his approach. Shrubb (1993a) also notes this display, stating that occasionally a male will stoop so close to a perched female that she flinches away, and that if performed in flight the female will roll to point her talons at the approaching male. Village (1990) does not mention this display, and I have also not seen it.

Another courtship display is the V-flight, the male forming V-shaped wings as he points out a suitable nesting site to the female (Fig. 57). Village (1990) notes that the V-flight is invariably performed at speed, but others have suggested a more leisurely approach to the nest site by the male. The difference between the observations appears to derive from the starting position of the two birds – if the pair are high soaring, then the approach is rapid, but if they are at a lower altitude then it is slower. Females also perform the V-flight, sometimes as a prelude to potential nest inspection, on other occasions as an invitation to copulate. Shrubb (1993a) also notes that while the male V-flight

Figure 57. Delightful illustrations from Tinbergen (1940) showing (*above*) female (to the left) and male (right) V-flights, and (*right*) a juvenile using winnowing flight to solicit food.

seems primarily employed during mate attraction or nest selection, the female V-flight is more closely associated with territorial defence, the female often performing the flight while the male is seeing off an intruder.

The final common display is a 'winnowing' (Village 1990) or 'shivering' (Shrubb 1993a) flight, delightfully called the *zitterflug* – trembling flight – by Tinbergen (1940), in which either the male or female flies close to the nest, travelling slowly with wings vibrating quickly. Winnowing is usually a prelude to, or seen following, copulation, but may also be used to see off intruders. In flight, juvenile birds also winnow when soliciting food from adults (Fig. 57).

Shrubb (1993a) also mentions two further displays, each of which he saw rarely and only performed by female Kestrels. The first was a slow flight with exaggerated wing beats, similar to that seen in hawks and buzzards, the other a parachuting flight with legs dangling, again in the manner of buzzards. Shrubb also considers that perching may be a territorial display. He notes that during the winter Kestrels will often sit in prominent positions in poor weather, when seeking shelter would seem a better alternative, and considers that such behaviour must therefore have an ulterior motive, territorial defence being the obvious candidate. Although most courtship displays are aerial a Kestrel pair may sometimes indulge in perched displays, such a bowing and bill nibbling: the female may also beg for food.

Courtship
Courtship may start early for pairs which overwintered at breeding sites, in which case the courtship period may be lengthy. For birds arriving late at breeding grounds, courtship must be curtailed if egg laying is not to be

delayed: in his work at his Scottish study area Village (1990) noted a linear relationship between the length of courtship, measured as pair formation to first egg laying (Fig. 58), and the date of pairing.

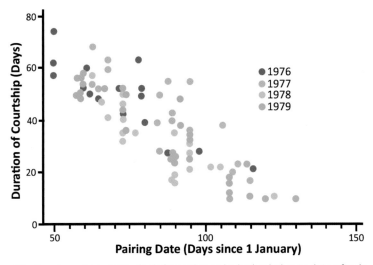

Figure 58 Duration of courtship (pair formation to laying) relative to date of pair formation in four successive years in southern Scotland. Redrawn from Village (1990).

The male feeds his mate during courtship (see Chapter 5 *Footnote 1* for thoughts on the correctness of this term). The female stops hunting: she will not resume until the nestlings are about 10 days old. That the female's hunting cessation allows a decrease in energy expenditure as a prelude to egg laying, which is intuitively obvious, is borne out by a study by Drent and Daan (1980) who also noted a correlation between the start of courtship feeding and the laying of the first egg. Drent and Daan simulated an increase in courtship feeding of the female by supplying an additional 100g of mice on every other day from February to mid-April to five Kestrels pairs of a total of 20 pairs in a study area in The Netherlands and noted that this influenced both the laying date of the first egg and clutch size (Fig. 59). Clearly, those falcon pairs given supplementary food laid earlier, and laid larger clutches than unfed control pairs. What is also clear from the control data is that clutch size reduces as the laying date of the first egg advances. The advance of laying date is considered again in the following Chapter when the timing of egg laying is dealt with in more detail.

After ceasing to hunt, the female is reducing her energy expenditure while the male is increasing his, catering not only for the food requirements of the female, but also in policing his territory.

Early in courtship, food passes may be aerial, the female rolling to one side to grab the offered prey from the male's talons. However, most food passes,

Breeding Part 1

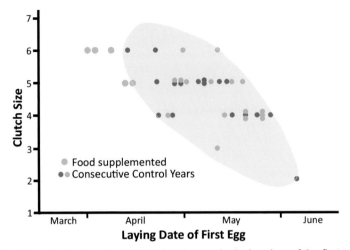

Figure 59. The effect of supplementary feeding on the laying date of the first egg and clutch size of Kestrels. Each symbol represents a single nest. The shaded area encompasses the distribution of first egg date and clutch size seen in the wild. Redrawn from Drent and Daan (1980).

particularly once the female has stopped hunting as a prelude to egg laying, are at a perch. To signal his intention of passing food, the male will take a favourite perch and give the '*clip*' call. This usually summons the female immediately, though several, increasingly insistent, calls may be needed. When the female arrives, the male may offer her prey held in his bill, or she may take it from his foot. The female usually gives the '*clip*' call, reinforcing the pair bond. Once incubation has started, food passes are invariably at the nest site, the male calling as he approaches, the female leaving the eggs and taking the prey away to consume it. In the study of nesting pairs in a barn in southern England (see Chapter 8), in 2017 the female always took the food at the nestbox and flew off to consume it. However, in 2018 the female invariably left the nestbox to take the food from the male outside the barn. In 2019 the female took food at the nestbox and departed. From this, and other behavioural changes noted in 2020, it is considered likely that the females differed in each of the four study years, though it is possible that the female in 2017 was the same as in 2019. In each year of the study the female consumed the prey away from the nestbox. However, other observers have noted the female eating at the nest site.

Pair Formation
It is considered that Kestrels are seasonally monogamous, though pairs are sometimes maintained from one year to the next: see the following Chapter for further data on mate fidelity. However, polygynous males have been recorded. Studying raptor species in Finland Korpimäki (1988) set down a hypothesis that linked raptor polygyny with abundant food resources and

THE COMMON KESTREL

Courtship feeding does not always go smoothly. In the photograph (*above*) the female snatches a bird from her mate. In the photograph (*opposite, above*) the female calmly accepting food from the male, but in the photograph (*opposite, below*) the female's impatience with her mate, perhaps having waited too long, causes a confrontation. All photographs *Torsten Prohl*.

Breeding Part 1

nomadic breeding dispersal. If his hypothesis was correct, Korpimäki suggested polygyny would be more common in rodent-eating birds than in bird-eating species; that polygyny would be more common in good rodent years; that polygyny would be more common in northern rodent-eating species as rodent population cyclicity was both more common and more pronounced in northern latitudes; and that polygyny would be more common in nomadic species with weak pair bonds and territoriality. Korpimäki also suggested that if his hypothesis was correct, he would be able to influence the likelihood of polygyny by supplementary feeding. By searching for data on observations of males feeding females at two or more nests, or being trapped near several nests, as well as his own observations, Korpimäki was able to support all the predictions of his hypothesis apart from that on supplementary feeding as he did not carry this out. As we shall below, Korpimäki was later to confirm polygyny by DNA analysis. From his data Korpimäki concluded that 4-10% of male Kestrels were polygamous in good rodent years.

From his own studies, and from studies in The Netherlands, Village (1990) suggests a figure of 1-2% of polygamous pairings in areas which do not see the exaggerated cyclical rodent peaks that occur in Fennoscandia. In either case, the low incidence is consistent with the view that it represents abnormal seasonal mating behaviour, arising only when conditions offer the opportunity. Shrubb (1993a) observed one such opportunity, noticing a female soliciting the male of an established pair with clear mating display flights over the pair's nest site, and at a potential nest site some distance away. The male drove the second female off, as, occasionally, did his mate, but she persisted. Ultimately the male stopped harassing the second female, though his mate still continued to do so. Shrubb then found the first female dead at the nest, sitting on four now chilled eggs: she had apparently been poisoned. Shrubb later discovered the second female incubating her own clutch and from the timing it was clear that the male had mated with her while his first mate was alive and incubating.

Glutz von Blotzheim *et al.* (1971) claimed that polygyny also occurred in colonial nesting situations, but without DNA analysis this is much more difficult to prove and so is conjectural. However, it is clearly the case that opportunities for male infidelity would be increased at such sites. Polyandry is also claimed by Glutz von Blotzheim and co-workers, and by Packham (1985a). While in most cases noted in these and earlier references, the behaviour could only be inferred by observation of two males feeding a single female, Packham observed mating in one instance, two adult males copulating with the same female at half-hour intervals. The two males did not approach each other closer than 30m.

A more definitive study of Kestrel bigamy was made by Korpimäki *et al.* (1996) who carried out DNA analysis of Kestrel egg clutches in Finland during a three-year vole cycle in 1990-1992. 1990 was a poor vole year, rodent numbers increasing in 1991, then declining again in 1992. Korpimäki *et al.* found no evidence of the paternity of any clutch being shared by more than one male in 1990 or 1992, but in 1991 7% of 27 broods (5% of 112 chicks) showed evidence of extra-pair copulation by the female (Fig. 60).

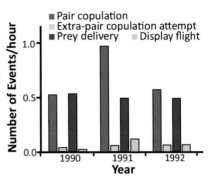

Figure 60. Mean number of within-pair copulations, male display flights, prey deliveries and extra-pair copulation attempts during a three-year vole cycle. Note that symbols for standard deviation have been removed from histogram entries for clarity. Redrawn from Korpimäki *et al.* (1996).

Fig. 60 indicates that prey deliveries remained constant across the three years, but within-pair copulations were higher in 1991 the difference being statistically significant. Display flight and extra-pair copulations were higher in 1991 (the good vole year): the differences were not statistically significant, but the sample sizes were small. However, the DNA analysis shows that extra-pair copulations were higher. The higher number of within-pair copulations suggests that at least one bird of the pair was attempting to curtail infidelity. We consider this again in *Copulation* below.

The timing of pair formation depends, in part, on the population of a given area. In areas where the winter population of birds is relatively stable, some pairs may stay together through the winter and will remain paired for the new spring. Where males and females have not remained paired, courtship behaviour may begin during the late winter, any remaining unpaired birds then pairing with arriving migrants. Once the surplus of wintering birds has been removed, remaining migrants pair with other migrants. In more northerly areas of the UK where few birds overwinter, the bulk of the population arrives in early spring, the early arrivals tending to pair up with the few resident birds, taking advantage of formed territories and the understanding of feeding areas to acquire potentially more competent mates.

In a study in The Netherlands, Meijer and Schwabl (1988) noted that courtship, the time between the first sign of pair formation and the laying of the first egg, lasted up to 50 days for birds that overwintered at, or were early arrivals at, a breeding territory, but could be as short as 13 days for pairs that formed late.

Breeding Density

There are difficulties in measuring the breeding density of any species. Some birds may not find partners and so will not breed, other pairs may form and mate, but fail to breed, Village (1990) considering that non-breeders may contribute up to 30% of the Kestrel population in any area. While that means a high proportion of pairs are breeding, assessing their density requires identifying viable nests. Kestrels breeding in nestboxes and on buildings are relatively easy to spot, but those using old stick nests, rock cavities, tree holes *etc.* scattered across the countryside are more difficult to find. Village (1990) noted that the problem of finding nests means that the larger the area being searched the more likely it is that nests will be missed, so that many quoted breeding densities, which rely on relatively small areas, may over-estimate the true overall density. Village considers that choosing areas of approximately 80-200km^2 gives the best estimate of the 'true' breeding density. In support of his claim, Village notes that in the study of Parr (1967) a breeding density of 1797 pairs/100km^2 was established for one 7km^2 area, and 203 pairs/100km^2 for another 10km^2 site, while in an area of 14,000km^2 of Finland Kuusela (1979) found a density of only 3 pairs/100km^2. On the basis that a study area of 80-200km^2 smooths out such anomalies, Village (1990) cites densities of 12 pairs/100km^2 for arable farmland, rising to 20 pairs/100km^2 for mixed farmland, and 30 pairs/100km^2 for grassland. However, Village notes that these values are subject to a high degree of change depending on prey population and weather conditions. Studies in the Pentland Hills show a similar dependency (Graham Anderson/Keith Burgoyne, pers. comm.) – see Table 5.

Year	2013	2014	2015	2016	2017	2018	2019
Breeding Density (Nests/100km^2)	1.4	3.2	6.0	4.9	5.3	9.5	4.9

Table 5. Kestrel breeding density in an area of 285km^2 of the Pentland Hills, Scotland.

Table 5 illustrates not only the dependence of breeding Kestrels on prey and weather, but on the availability of nest sites. The principal difference between the data for 2013 and 2014 was the installation of a further 7 nestboxes which allowed more Kestrel pairs to breed. There was a further increase in the number of falcon pairs breeding in subsequent years even though the number of nestboxes did not increase. The change in those years was an increase in occupancy, probably due to Kestrel chicks raised on the Pentlands returning to their natal area. In 2017 there was a further increase in breeding pairs, but that year the weather was unusually bad with a prolonged period of continuous heavy rain and high winds at the start of the breeding season causing several pairs to fail. In 2018 there was a further increase in nestbox numbers, this coinciding with a year of peak vole numbers. The number of pairs breeding increased significantly, and the mean clutch also increased to 5.5 eggs per clutch (149 eggs in 27 clutches, with 2 clutches of 4 eggs, 11 of 5 eggs, 12 of 6 eggs and 2 of 7 eggs). The following year the number of available nest sites was the same, but the vole population crashed and both breeding density and mean clutch size declined (to 5.1, with no clutches of 7 eggs).

Village's density for mixed farmland agrees reasonably well with data from Germany where Kostrzewa and Kostrzewa (1991) noted densities of 9-17 territorial pairs/100km^2 depending on conditions the previous winter and on vole numbers (Fig. 61).

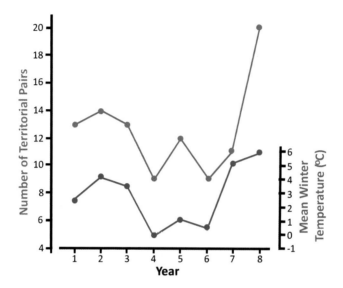

Figure 61. Variation of the number of territorial Kestrel pairs and mean winter temperature through an eight-year period in the 1980s in a mixed area (72.5% farmland, 16.0% urban and 11.5% woodland) on the west bank of the Rhine in northwest Germany. During the period of study there were higher numbers of voles in Years 3, 5, 7 and 8, and a lower number in Year 2. Redrawn from Kostrzewa and Kostrzewa (1991).

However, the number of egg-laying pairs in the study of Kostrzewa and Kostrzewa was 6-15/100km^2 which implies that non-breeding pairs amounted to 11-33% of the total (*cf.* the 30% of Village (1990) quoted above). The German data suggests that if the winter has been harsh not only are there fewer territorial pairs, but a greater proportion of them will fail to breed. The effect of harsh winters on Kestrel survival is considered in Chapter 11 where the work of Kostrzewa and Kostrzewa is mentioned again: harsh winters not only affect falcon mortality, but their ability to breed.

The German data agrees with what would intuitively be thought – the number of breeding Kestrel pairs in an area would depend on winter survival, and on the food supply in spring. This was confirmed by Village (1980, 1982) in studies of British Kestrels. Village found that the number of Kestrel breeding pairs varied with vole numbers and that there were more first-year birds in the population when vole numbers were high. The effect of winter survival (of both adult and juvenile birds) was to reduce the size of the home range and increase the breeding density (Fig. 62).

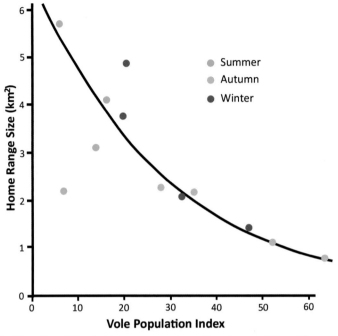

Figure 62. Variation of Kestrel home range size with vole population. The range sizes are the mean of five taken in each of summer, autumn and winter. Vole population numbers are from the mean number caught in 48 traps over five nights. The line is a regression fitted to all the data. Redrawn from Village (1982).

Kostrzewa and Kostrzewa (1990) noted a similar reduction in home ranges near Cologne, Germany, where, as noted above, Kestrel density in spring was positively correlated with temperature in the previous winter. They also identified a positive correlation between the temperature in May and June and the number of young fledged by a Kestrel pair, successful fledglings per falcon pair rising from 3.5 if the mean May-June temperature was 13°C, to 5.1 if it was 16°C (Fig. 63). Kostrzewa and Kostrzewa also showed a correlation between breeding density and spring rainfall, higher rainfall leading to reduced breeding.

Village (1990) noted that the number of breeding pairs in an area was always less than the theoretical number that could be accommodated. Village also estimated that 6-13% of breeding Kestrel pairs failed to produce eggs. Adding this to the number of non-breeding birds it is clear that as a species, Kestrels are not as successful at breeding as might be assumed, raising the obvious question of why some Kestrels and Kestrel pairs do not breed.

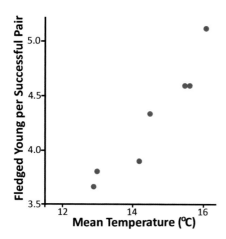

Figure 63. Variation of the number of chicks successfully reared to fledge by Kestrel pairs in north-west Germany with ambient temperature in May and June. A regression of the data by the study authors was significant at the 99.9% level. Redrawn from Kostrzewa and Kostrzewa (1990).

The relationship between food supply and breeding density is intuitively obvious, and is occasionally seen in spectacular fashion, particularly in northern areas where peak lemming years are accompanied by an increase, for example, in the breeding density of Snowy Owls (*Nyctea scandiaca*) – *e.g.* Potapov and Sale (2012). In the case of Snowy Owls, not only does the breeding density increase, but clutch size may also rise dramatically. Female owls can lay from one to a dozen eggs, raising nestling numbers in accord with the lemming population which can achieve staggeringly high densities, much greater than is seen in vole populations. As we shall see below, Kestrels are more limited in clutch size (seven seems to be the maximum, though larger clutches have been claimed – see Chapter 7 *Clutch Size* – but they can take advantage of increased vole numbers by both increasing breeding density and raising more nestlings to fledge. In Britain, and in many other areas throughout the Kestrel's range, where the population is essentially sedentary and rodent cyclicity is relatively subdued, large increases in breeding density are not seen. But where the Kestrel population is largely migratory, sharp increases may occur, *e.g.* in Finland in the studies of Korpimäki (1984b), Korpimäki (1985c), Korpimäki (1994) and Norrdahl and Korpimäki (1996) where an increased vole population saw additional falcon pairs rapidly arriving and breeding. The study of Korpimäki (1984b) shows both the reduction in breeding pairs of Kestrels during a crash in the population of Microtus species, and the subsequent recovery in breeding pairs when the population recovered, together with population increases for Bank Voles and Sorex species (Fig. 64 overleaf).

Those populations in Britain which are partially migratory show a similar pattern on the few instances where the vole population reaches epidemic levels. Adair (1891 and 1893) reported a sharp increase in the populations of both Kestrels and Short-eared Owls during a vole 'plague' which afflicted the Scottish Borders, and the subsequent decline in the population of both when vole numbers collapsed. Mitchell *et al.* (1975) report a similar, though apparently

The Common Kestrel

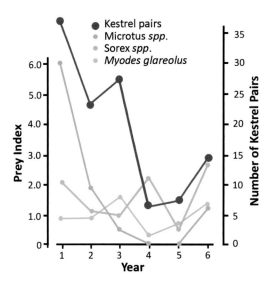

Figure 64. Variation of Kestrel pairs with abundance of rodent species over a six-year period in western Finland. Prey Index is given by the number caught in 100 traps during a single night. The correlation between Kestrel numbers and Microtus numbers was >95% significant. Redrawn from Korpimäki (1984b). Please note that in his report Korpimäki refers to the Bank Vole as *Clethrionomys glareolus*. The name has now been changed and the new form has been used in the Figure.

more subdued, vole peak in 1971-72 and a consequent increase in the breeding numbers of the same two species, as well as in clutch size (the latter more pronounced in the owls). In his study in western Finland Korpimäki (1985c) reported that there was no time lag between the population changes of vole predators (Long-eared Owl, Short-eared Owl, Boreal (Tengmalm's) Owl, Hen Harrier and Kestrel) and their microtine prey, noting that the ability of the predators to move into an area of high prey density aided population stability by limiting rodent number peaks. Korpimäki also noted the high degree of mobility of the predators which aided a rapid response, and that even at times of rodent scarcity there were invariably small numbers of predators.

The obvious question raised by the observed Kestrel breeding density increase of Korpimäki (1985c) and the increases seen during Scottish vole 'plagues', is how those predators that were not already resident in the area were able to recognise the increase in rodent population density. In their study of Snowy Owls, Potapov and Sale (2012) postulated that the owls move in loose congregations which they termed 'boids', owls staying within 5-7km of each other so that visual signalling remained possible. A boid would be a congregation of several hundred owls spread out over an area with a long axis of 300-350km. One owl finding a high lemming density would, by stopping and preparing to breed, alert and attract close members of the boid and hence a build-up of predator numbers would occur, later arrivals having to establish territories at the fringes of the peak lemming area, or to continue to search the tundra. The 'boid hypothesis' was set down in early 2011: later that year a peak lemming year in Scandinavia saw an influx of Snowy Owls after many years during which breeding had been spasmodic or entirely absent, the influx being in accord with the hypothesis. It is interesting to speculate whether something similar was happening in the early 1890s, with overwintering Kestrels, discovering the increase in vole density and preparing to breed, visually alerted other falcons. While the 'boid hypothesis' provides a potential explanation of the observed phenomena, proof would require the continuous tracking of large numbers of individual birds over many seasons and is unlikely to be feasible in the immediate future.

Riddle (2011) plots the mean clutch size of Kestrel pairs with time in his Ayrshire study area over the period 1986-2009 and notes the position of good vole years (Fig. 65), clearly indicating a correlation. Interestingly, Riddle notes that Kestrel numbers peaked in the years of high vole population, then declined, whereas the population of Barn Owls peaked the year after a high vole population, and then declined.

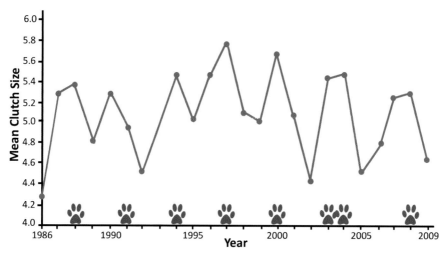

Figure 65. Variation of Kestrel clutch size during the period 1986-2009 in south-west Scotland. The paw prints indicate years with good vole numbers. Redrawn from Riddle (2011).

However, the relationship between breeding Kestrels and the availability of prey is not quite what it seems, as food resources are not the only factor influencing Kestrel breeding. As mentioned above, Kostrzewa and Kostrzewa (1990) note the influence of rainfall and temperature, confirming earlier work with Kestrels breeding on Dutch polders. There, Cavé (1967) had also noted that breeding density was negatively correlated with rainfall, particularly in spring, and positively correlated with temperature. Rainfall impeded hunting and reduced the above-ground activity of the voles which were the main prey of the Kestrels. This effect was most apparent in spring, as later in the year there were more hours of daylight, an advantage to both predator and prey. In spring a reduced food intake affected the ability of the female

Kestrels to breed. Cavé conjectured that temperature had an effect because higher temperatures reduced the energy expenditure in thermoregulation. Counter-intuitively the breeding density appeared to be independent of vole population density, but here, too, several factors need to be considered. If voles are scarce, adult Kestrels can switch prey – in the case of the Cavé study to young Starlings – but there was also evidence that male chicks weighed less than in years of more abundant voles, apparently because they were losing out in competition to their female siblings (which did not show a weight reduction). The rate of brood desertion was also higher. Such factors would mean that provided the vole population density varied within a 'normal' range, the Kestrel breeding density was (relatively) insensitive to it, steep rises or falls in bird breeding being seen only when the vole population change was more pronounced. Data collected by Village (1990) over a four-year period in his Scottish study area supports this, though the sample size was too small for a statistically significant result. What Village discovered was that in his farmland study areas, where the Kestrel breeding density was lower because the vole habitat was reduced, winter temperature had a much larger effect. As lower temperatures meant an increased likelihood of snow which adversely affected the birds' ability to catch voles, there was a significant reduction in breeding density as mean temperature decreased.

Breeding density is also dependent on nest site availability, the increased availability of nestboxes in areas with poor natural sites over recent years having been a significant factor in increasing Kestrel numbers in those areas. In a study area of, chiefly, young upland conifer plantations in southern Scotland, Village (1983b) noted that some areas had an excess of usable nests (corvid stick nests), but that these were grouped together in small woods. In general, only one pair of Kestrels nested in each wood. This meant that even in good vole years additional falcon pairs might be prevented from nesting in the same wood by the behaviour of a territorial pair so that available stick nests remained unused. The placing of nestboxes in suitable positions therefore meant that Kestrel pairs that would otherwise not have bred were able to breed successfully (see Fig. 66). Village noted that in good vole years nesting pairs might be as close as 250m. Village (1990) notes a case in which two nests were 40m apart and that the male of one of the pairs was the two-year old son of the male at the second nest. However, he notes that he found no evidence that close positioning of nests implied some relationship between member(s) of the pairs.

Close nesting in urban Kestrels has been found in several studies. Salvati (2001) studied two sites in Rome. In an inner city area he noted up to 3.9 pairs/km^2 (390 pairs/100km^2) with nest separations as low as 22m, and 0.6 pairs/km^2 in a suburban area. In the inner city area the falcons were nesting semi-colonially in scaffolding holes in ruins and medieval towers of the old city. Salvati notes that the mild climate of spring-time Rome is clearly an aid to the Kestrels, as was the fact that their diet was chiefly avian.

Breeding Part 1

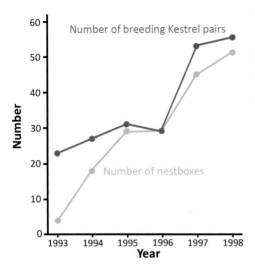

Figure 66. Variation of the number of Kestrel pairs breeding, and the number of nestboxes installed in central Spain during a six-year period. The Spanish data confirmed the finding of Village that the erection of nestboxes in an area of southern Scotland increased the Kestrel breeding density.

The figure was redrawn from Fargallo *et al.*, 2001. See further comments on the study of Fargallo *et al.*, 2001 later in the text.

The suggestion of colonial, or semi-colonial nesting, is not confined to cities, Hagen (1952) recording an area of birch woodland and associated rock outcrops in Norway which held more than 20 Kestrel pairs. Piechocki (1982) records a colony of 15 pairs in an old rookery in Germany the season after the Rooks (*Corvus frugilegus*) had been culled. Piechocki also records 7-9 pairs nesting on a gasometer near Leipzig. The gasometer had about 100 identical cavities, Piechocki noting that some female Kestrels were laying single eggs in several of these having apparently forgotten where they had laid previously. Johansen (1957) notes more than 10 nests in a small wood on the Russian steppe, while Peter and Zaumseil (1982) noted a colony of up to 28 pairs of Kestrels inhabiting a 400m long motorway bridge near Jena, Germany. Bustamante (1994) found 12-15 pairs of Kestrels in a colony with several pairs of Lesser Kestrels in an abandoned sandstone quarry near Seville, in south-west Spain, that was 20m high and about 200m wide. The minimum distance between neighbouring nests was only 6m. Bustamante notes that he saw no signs of aggression between fledgling Kestrels, but there were several instances between the fledglings and their Lesser cousins. Occasionally Kestrel fledglings solicited food from Lesser adults, but these were never successful.

Further information on Kestrel colonies comes from Japan where Hyuga (1955) observed a colony of 17 pairs (*F. t. interstinctus*) on a natural cliff face. During the late 19th century vole plague in the Scottish Borders mentioned above, Adair (1891 and 1893) recorded six nests in one wood: this colony seems to have arisen as a result of prey abundance, and there is some evidence to suggest that other colonies may similarly relate to available food resources: in his study in Rome, Salvati (2001) found that the reproductive success of a pair was low if the breeding density was high, reflecting the idea that overlap of hunting ranges limited access to food when the number of hunting Kestrels rose, a finding in agreement with a Finnish study (Korpimäki 1987) which found that dietary breadth (*i.e.* number of prey species taken) of near-neighbour Kestrels was wider than that of non-neighbouring pairs. Korpimäki also noted that neighbouring pairs staggered their hunting times to avoid conflict, this and the need to expand their diet resulting in near-neighbours producing fewer fledglings than non-neighbours.

One further report on the colonial nesting of Kestrels comes from southern Russia, close to Rostov, an area of steppe between the Black and Caspian seas. There, Ermolaev (2016) noted that Rooks had nested in trees planted as shelter belts in the steppe, occasionally creating extensive rookeries which might cover more than two hectares. Within these rookeries there were also Kestrel colonies, nests being occupied throughout the rookery. What Ermolaev found was that the breeding success of Kestrels at the centre of a rookery was one-third lower than that of pairs at the rookery's periphery. Given the potential for predation that might seem reasonable, but Ermolaev found that success was correlated not only with position within the rookery, but height above the ground (the higher the nest the higher the success) and position within the tree (greater success for nests on side branches). Ermolaev also looked at the age profile of female Kestrels relative to position of nests, using egg diameter as the means of assessing age. On that basis, the age range of nests at the periphery of rookeries was wider than for nests at the centre, which is surprising as it would be intuitively assumed that older, more experienced females would take nests that were likely to be in 'better' locations. However, there was considerable overlap in the limits of egg diameter so caution must be exercised when considering age range across the rookery.

Nest sites
As noted in Chapter 1, the 'True Falcons' do not build nests, either using the nests of other birds, making a scrape on the ledge of a cliff or on the ground, or using a rock crevice, tree hole or a human-assisted site such as a building or nestbox.

Mester (1980) and Piechocki (1982) both considered that it was the male Kestrel that chose the nest site. Village (1990) watched a male Kestrel systematically searching the canopy of an area of woodland clearly looking for old nests and added that males make a great deal of noise at nests sites which could be interpreted as attempts to persuade the female that a good site has been located. However, Village does not endorse the idea that males always chose the nest. Shrubb (1993a) was also less confident that males were wholly responsible, suggesting both sexes were involved, having seen females examining nest sites as well as being shown potential sites by the male. In my own experience males which overwinter at sites with an available nest resource, *e.g.* at a farm which has a nestbox in a barn, will often inspect the box, and even dissuade other species (owls, pigeons) from using it before showing it to

In each of the years studied in Chapter 8 the male Kestrel inspected the nestbox first, the first female visit being a few days later.

a potential mate. Overall, while available evidence is not conclusive, it seems that males are at the very least involved in nest selection, but the final decision likely rests with the female.

Kestrels are extremely eclectic in their choice of nest sites, a fact which, added to their dietary flexibility, has been a factor in their success as a species, particularly after mankind began to drastically modify the environment. Prior to humans offering convenient options, Kestrels were restricted to the use of the stick nests of other species and naturally occurring features, the fact that such a range of sites has been used indicates that the birds were long capable of adapting to circumstances. Ever since man began to construct buildings and other structures which offered convenient ledges and some protection from mammalian predators, these have been used, with Kestrels later taking advantage of nestboxes to become one of the more frequent users. In his study of 'natural' Kestrel nesting habits across Europe (*i.e.* ignoring nestboxes), Village (1990) notes wide variations in nesting choice. In his Scottish study area stick nests in trees and rock ledges were the most used sites, while in his English lowland site tree holes were the most abundant form. In lowland sites

Kestrel chicks in a (probably corvid) stick nest.

throughout Britain, buildings offered a reasonable alternative to stick nests and cliff sites, but ground nesting was minimal. Village (1990) quotes data from lowland and upland Germany which show similar usage. Village also quotes work by Korpimäki (1983) in lowland Finland where 100% of nesting was in stick nests prior to the introduction of nestboxes.

Stick nests are most often those of corvids, particularly Carrion Crows (*Corvus corone*). In an analysis of BTO nest records over a 50-year period (1937-1987) Shrubb (1993b) found that 83% of stick nests were Carrion Crow. Of the rest, Magpie constituted 11%, Sparrowhawk and Common Buzzard contributing 2% each: the remaining nests had been built by Raven (*Corvus corax*), Wood Pigeon (*Columba palumbus*), Jay, Rook and Heron (*Ardea cinerea*). Old squirrel dreys may also be used, the Kestrels scraping a hollow in the top. In Shrubb's study two Grey Squirrel (*Sciurus carloinensis*) dreys were reported in a total of 556 stick nests. Shrubb also analysed the trees in which the stick nests were found, noting 45% in pine species, though many types of tree, both coniferous and deciduous, had been utilised by the original nest builders. In general Kestrels avoid stick nests close to the ground and prefer trees which stand with others rather than alone.

The stick nests used by Kestrels are usually one year old as many do not survive the winter to be used again. However, some are reused, extreme examples being known. In Russia's Tuva Republic (close to Mongolia's north-western border) a Magpie nest was used in consecutive years in 2004 and 2005,

Kestrel chicks in a crevice in a cliff face. *Mike Price*.

Breeding Part 1

Above   Kestrel chicks on a cliff ledge.

Below   Kestrel chicks on a ledge situated on sloping, vegetated ground.

Kestrel chicks in a huge hollow tree. *Dave Anderson.*

then used again in 2010 when a male chick, hatched in the nest in 2004, was the male of the pair (Karyakin and Nikolenko 2010). If a suitable remnant lining exists in a stick nest, this will be used as a substrate, the female Kestrel laying directly on to it, but if nest cup has collapsed, a scrape in the remaining material may be made before laying. Female Kestrels will occasionally remove old nest lining material or, more unusually, may add material if the nest is messy. Kestrels are known to have used nests constructed in the same season if the builders have abandoned the site. Historically there have been instances of larger falcons evicting corvids from recently constructed nests: indeed, this behaviour was once thought to be have been relatively common, though it is now thought to be rare. There are no instances in the literature of Kestrels evicting nest builders, though as noted above, eviction of potential users from nestboxes are known.

Kestrels also use tree holes, choosing anything that will accommodate the female, but avoiding deep holes. In most cases such holes are natural, but holes excavated by Green Woodpeckers (*Picus viridis*) where the surrounding tree has begun to rot, enlarging the hole, will also be used. There is no evidence that Kestrels prefer one tree species to another, choice being based solely on the availability of a suitable hole. Females reject holes which do not have a suitable base for the eggs. In holes that are chosen the female rarely makes any scrape, accepting the hole base as she finds it. Interestingly, Kestrels will share a hollow tree with other species: in a trawl of BTO nest records Shrubb (1993a) noted 13 instances of sharing with Barn Owls, seven with Jackdaws (*Corvus monedula*), six with Stock Doves, four with Shelduck (*Tadorna tadorna*), three with Little Owls (*Athene noctua*), two with Tawny Owls and one with a Black Redstart (*Phoenicurus ochruros*). While in most cases the entrance to the tree

Kestrel chicks in a very confined tree hole. *Ant Messenger.*

was not shared, there are known cases (one with a Barn Owl, another with the Black Redstart) where a common entrance was used. Cases of Barn Owls and Kestrels sharing nestboxes are also known (see Chapter 10). Given the fact that the larger owls are predators of Kestrels or eggs, while others are potential prey species of Kestrels, these associations are remarkable.

An equally tight tree hole to that above. There were five fledglings in the hole, which often meant that one or more were sat in the entrance so that when an adult arrived with prey it was unable to perch to deliver it, circling in front of the tree several times before the youngsters worked out they need to move back to make room. The tree was also dead and so had few branches. With the hole at about 5m this meant that several chicks landed on the ground on first flight and were unable to take-off. Replacing them was difficult and it is not known whether any were lost before achieving flight.

Above Ground-nesting Kestrel on the Orkney Islands. *Gordon Riddle*.

Left Kestrel ground nest in Germany. *Ralf Wassmann*. Herr Wassmann writes that the bare patch chosen by the female falcon for her two eggs was a 'hearth', and it does indeed appear that a small fire had been lit in the field. While the Orkney falcons have little choice but ground nesting, finding an example in a country such as Germany which is rich in buildings, trees *etc.* is rare.

In the absence of trees, Kestrels will nest on the ground, though this is rare. Balfour (1955), in a famous paper on Kestrels breeding in the Orkneys, where the falcons' diet included the Orkney Vole (*Microtus arvalis orcadensis*), noted the use of tunnels in heather and rabbit burrows. When Riddle (2011) visited the area, he found that Kestrels were still using ground nests, though by then the annual number had dropped from 19 (at the time of Balfour study), to just four. There are other examples of ground nesting in sand dunes adjacent to the North Sea, and a more curious one reported in Germany by Wassmann (1993) who found a female Kestrel sat on two eggs in an area of bare ground in rough pasture.

It is usually assumed that Kestrels avoid ground nesting because of the possibility of predation, but as Merlins and harriers regularly nest on the ground this would only be true if there was some reason why the nests of Kestrels were more likely to be discovered, which seems unlikely. More likely is that the areas frequented by the other two raptors are less suitable as hunting grounds for Kestrels, despite their dietary flexibility. That said, personal experience is that Kestrels and Merlins share a habitat in Scotland's Pentland Hills where there have been several examples of Kestrels in nestboxes within a few tens of metres of nesting Merlins, though in all these cases the Merlins were in trees (stick nests or baskets set up as potential nesting sites).

In his analysis of BTO nest records Shrubb (1993b) noted 27% of nests on rock ledges, 20% in tree holes, 17% in stick nests, 16% on buildings and 15% in nestboxes. The remaining 5% included 3% where the nest site was

Kestrel fledglings on the sill of a church window. The equipment to the left is a CCTV system allowing people to watch the falcons.

not adequately described, and 2% which comprised tree forks, ground nesting (mostly in the Orkneys) but including rabbit burrows, and ledges on steep banks. Shrubb noted marked regional variations in the percentages for each site type, attributing these, at least in part, to the likelihood of local predation. Anecdotal evidence suggests that in the 25 years since Shrubb's survey the number of building and nestboxes has risen, while the use of stick nests has declined.

The Kestrel's enthusiasm for nesting on buildings is responsible for its German name (*Turmfalke* – Tower Falcon). The birds will take any structure on offer – old barns, high rise housing or office blocks, church towers, bridges *etc*. Pike (1981) tells the story of a Kestrel pair nesting in a box section (measuring 142mm wide, 355mm high, 370mm deep) of a, normally inaccessible, girder in a Birmingham brewery. During repair work the girder was inspected and two eggs were found. The workman carefully removed these as he needed to paint the box section. After painting he returned the eggs. A few days later he noted that the female had laid three more eggs, and then incubated the clutch despite the frequent, noisy, comings-and-goings of bricklayers and painters. All five chicks hatched and fledged successfully, aided, in part, by a management decision to minimise disturbance once the young were learning to fly. More recently (2018) a Kestrel pair was found nesting in a steel girder supporting the carriageway of the Forth Road Bridge in Scotland despite the 80,000 vehicles which make use of the bridge each day. Great care was taken to avoid disturbance of the pair who were nicknamed Mr. and Mrs. Younger after the former owners of the company making Kestrel Lager. Repairs to the girder were postponed until after the brood had hatched.

The Common Kestrel

Above   Kestrel eggs in an open-fronted nestbox.

Below   Male Kestrel leaving a close-fronted nestbox after delivering prey.

Kestrel chicks in an old shopping basket.

Kestrels have readily taken to the use of purpose built nestboxes, becoming the most frequent users of boxes in many areas. Village (1990) notes that in the case of boxes attached to high-rise flats, human occupants are treated to grandstand seats during the brood raising phase, but may find that opening their windows causes the Kestrels, who have clearly lost their fear of humans, to aggressively defend their territories. Kestrels also make full use of boxes in rural areas: once having erected a box in a tree close to my home in the hope of encouraging an owl to take up residence, I was surprised to find the first tenants were a pair of Kestrels who successfully raised three chicks.

Many boxes are reused, even if the adults change. In work in eastern China Mingju E *et al.* (2019) found that Kestrels preferred to use boxes which had been used in previous years rather than taking boxes which were clean, or to which raptor materials, *e.g.* the feathers of Ural Owls (a local predator of the falcons) had been added. The Kestrels were obviously picking up information on nest-site quality from previous breeding (by both Kestrels and other, non-predatory, species). This finding echoed previous work on the falcons by Sumasgutner *et al.* (2014b) in Finland. Both studies will be mentioned again in Chapters 7 and 11.

In a fascinating study of urban breeding Kestrels in Israel, Charter *et al.* (2007) investigated whether the 'urbanisation' of the falcon had benefitted the species and, as an input to that conversation, whether choice of nest site influenced breeding success. Earlier studies (*e.g.* Pikula *et al.*, 1984, Salvati 2002) had claimed higher reproductive rates for urban Kestrels, but Charter *et al.* noted that there was a need to separate the effect of the environment from that of the nest site, as the site was itself an important factor in breeding success. Charter and co-workers noted that Kestrels nesting in rock cavities or in closed nest sites on buildings show higher breeding success than those

Kestrel chicks in an old woven basket.

in more open sites such as stick nests and open nestboxes, and that it is the reduction in predation in closed sites which accounts for the increased success. To illustrate this the researchers found that many urban Kestrels in Israeli cities were nesting in flower pots (!) on window sills and that some houseowners, not noticing their avian lodgers, were watering the flowers (!!). Other owners were deliberately removing eggs, while some falcons which merely nested on the windowsills were predated by crows. Charter *et al.* found lower reproductive success overall (measured as young fledged per laying pair) in cities than in towns and villages, but considered this was largely attributable to higher prey availability in the latter, *i.e.* that the greater security of many nests in urban areas was not enough to compensate for the relative lack of prey.

The suggestion of Charter *et al.* (2007) that nest site choice was a factor in breeding success confirmed earlier work (Fargallo *et al.*, 2001) in which the success of Kestrels in Spain was measured for a range of different nest sites, both nestboxes and natural sites (the latter including stick nests in trees and on pylons, and cavities in buildings). Fargallo and co-workers noted that the population of Kestrels increased after the introduction of nestboxes (see Fig. 66): the increase was due almost entirely to the provision of nestboxes, the number of falcon pairs nesting in trees, in holes and on pylons remaining more or less constant. Kestrel pairs in the nestboxes began laying earlier and had higher breeding success in terms of the number of fledglings raised. Fargallo *et al.* considered the latter effect was due to lower predation in nestboxes, but

Breeding Part 1

An extraordinary Kestrel site, the eggs having been laid in an excavated area at the base of an Osprey nest. The nest failed. *Mark Rafferty*.

cautioned against assuming earlier laying was correlated with nestbox usage as it might also have been due to site quality. However, while these studies imply that urban living is advantageous to Kestrels, a study in Vienna (Sumasgutner *et al.*, 2014a) suggested a very different story. In their study the Austrian team found that the predation risk could be significant in urban environments, with corvids and Beech Martens[2] taking eggs and nestlings. Sumasgutner *et al.* also noted a more compelling issue – the reduction in rodent availability to an individual pair of falcons due to the increase in sealed soil (*i.e.* land covered in buildings, concrete or tarmac and therefore largely unavailable to rodents) and competition if the Kestrel breeding density increased. Forced to shift to hunting small birds, the Austrians noted that as Kestrels were less efficient at hunting avian prey and birds were less profitable, in dietary terms, than rodents, the overall effect was an increase in nest desertions, lower hatch rates and smaller fledged broods, particularly in the most Kestrel-dense city areas.

---

[2] The Beech Marten (*Martes foina*) is a mustelid of similar size to the Pine Marten (*Martes martes*). It is sometimes called the House Marten as it has become essentially commensal with humans in continental Europe, frequently occupying attics (where it causes damage to electricity cables) and being responsible for damage to dozens of cars daily because of its enthusiasm for sleeping under car bonnets to take advantage of warm engines. Largely terrestrial, but an efficient climber, the martens explore crevices in buildings for food, occasionally finding Kestrel nests. Beech Martens are confined to the warmer parts of Europe, their absence from the UK and the rest of northern Europe meaning they are not a danger to northern Kestrels.

It seems that whether the breeding success of urban Kestrels is higher or lower than their rural cousins is city dependent, varying with the availability of rodents and the effect of predation. But one equally interesting aspect of urban Kestrel life is the suggestion of Rejt *et al.* (2004) that city falcons may be in the process of becoming both genetically and ethologically separate from the rural community. In a study in Warsaw, Rejt *et al.* found that the urban birds had lower genetic variability, perhaps due to isolation and subsequent inbreeding. As the urban falcons showed a marked increase in nest-site fidelity, the dietary and nest site preference differences between them and the Kestrel population of the surrounding rural area meant the researchers considered this could eventually result in the creation of an isolated sub-population[3]. However, as Rejt *et al.* noted, further work was required before definite conclusions can be reached. That later study has indeed been made, by Riegert *et al.* (2010) in the Czech Republic, and did not support the idea of genetic separation between urban and nearby rural Kestrel populations, although it did confirm the idea of more inbreeding in urban populations. The Czech study also failed to confirm any ethological difference, though it noted that urban female Kestrels tended to be heavier than their rural cousins.

Copulation
Copulation starts significantly before egg laying, van Boekel (2019) notes copulation on 2, 26 and 29 September, while Riddle (2011) noted several examples of autumnal copulation, adding that this was mostly in fine weather when the day length equated to spring. Dickson (1987) recorded it in November, while Masman *et al.* (1988a) record it as occurring in January) and believed it could therefore be considered to contribute to pair bonding.

The female solicits copulation by making the '*clip*' call. She bends forward, fans her wings and raises her tail, moving it to one side when the male mounts her. The male beats his wings rapidly and fans his tail to maintain position and gives the *kik-kik-kik* alarm call, the female responding with '*clip*' calls. Copulation takes a few seconds only, the pair then separating and spending time preening. From data collected in the study of Chapter 8, the average timespan of copulation in 2019 was 11.5s (range 4.3-15.3s) with time of day ranging from 06.50 to 18.34, while in 2020 the mean duration was 6.7s (range 5.2-8.8s), and occurred at times between 06.02 and 17.20, though most were recorded between 06.00-07.00. Copulations increase in frequency as egg laying is approached, reaching a peak during egg laying when the birds copulate 7-8 times daily (Masman *et al.*, 1988b, Korpimäki *et al.*, 1996, the latter giving an average time between copulations during the day of 42 minutes, consistent with the 45 minutes quoted by Riddle 2011), more frequently in

---

[3] In an intriguing study by researchers from the Estonia, the UK's Exeter University and the USA (Sepp *et al.*, 2018) the suggestion is made that urban-living birds have a slower pace of life than their rural cousins and can therefore invest more time in self-maintenance and, hence, have longer lifespans. While the authors conducted valid research, it is difficult to avoid the amusing conclusion that chilled-out urban Kestrels might indeed become a distinct population.

the early morning than at other times. Copulation usually occurs close to the nest, sometimes within it.

As noted in Fig. 60 (p183) within-pair copulation rates increase in years when rodent numbers are high (Korpimäki *et al.*, 1996). However, the DNA results of the Finnish team suggest that when conditions allow, both male and female Kestrels may explore the option of extra-pair copulation to enhance the chances of passing on their genes. Korpimäki *et al.* suggest that an increased copulation rate was a means of avoiding extra-pair copulation. The Finnish team also note that in good vole years males spent more time guarding their mates, therefore protecting their investment in the brood. This, and the extra copulations, mean the male has less time for infidelity. For female Kestrels, other than chasing away females potentially soliciting their mates, there is little that they can do to prevent their male partner from mating with other females, particularly once incubation has begun and policing other females has become increasingly difficult. Their own behaviour in terms of extra-pair mating may also be regulated by the need to maintain male presence, both for themselves and their brood, such care being potentially jeopardised by infidelity.

Mating Kestrels. *Torsten Prohl.*

7 BREEDING PART 2

A female Kestrel intending to breed and having already decided upon a mate and nest site, is faced with two critical decisions – when to start laying and how many eggs to lay. Research has shown that each of these decisions has a significant impact on the likelihood that her eggs will result in surviving offspring. Environmental factors as well as those associated with the female herself are clearly drivers of the decisions, time of year being the factor which most obviously springs to mind. In the early part of this Chapter I explore this dilemma, before moving on to discussing breeding.

In one of the array of impressive studies at Groningen University, Masman *et al.* (1986b) showed that without the constraints imposed by the physiology of males and females, a pair of animals could breed at any time when the availability of food allowed the female to successfully overcome the energy expenditure of reproduction and the pair could provide enough food to raise the resulting offspring. In practice, most animals are 'programmed' to breed at a specific time, the time being defined by the food availability as food is not usually available consistently. Masman *et al.* calculated the daily energy expenditure of male Kestrels during the year (Fig. 67) and found that from August to mid-January the shorter northern day meant the male did not have the time to catch the food

Breeding Part 2

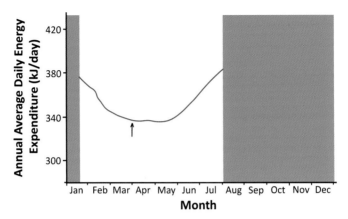

Figure 67. Annual average daily energy expenditure of a male Kestrel through the breeding season. The arrow indicates the average start of courtship. The blue shaded areas are parts of the year where reproduction is precluded by day length. Redrawn from Masman *et al.* (1986b).

required to feed a female and brood as well as feeding himself. In principle Kestrels could breed in the autumn when the vole population is comparable to that of spring, but the male is then constrained by hunting time (*i.e.* by daylight hours). To prove the point, Karyakin (2005) discovered three juvenile Kestrels during the first week of December at a site in south-western Russia and speculated that they were from an autumnal brood, breeding having been prompted by a very high vole population density in the area. Karyakin does not record whether the juveniles survived the winter, but as he refers to them as juveniles rather than nestlings, they had clearly fledged.

Prey availability

Prey availability is obviously a significant driver for the Kestrels to breed. In years with high vole abundance, Kestrels breed early and lay large clutches: in years of low vole density the falcons breed later and produce smaller clutches (Cavé, 1968; Daan and Dijkstra 1988). Village (1990) noted that in his study area in Scotland Kestrels laid earlier than their English cousins despite being 300km further north. Village hypothesised that the Scottish study area was better vole habitat so that, all other conditions being equal, the northern birds had better chances of catching prey.

The influence of food availability is highlighted in studies in The Netherlands by Cavé (1968) and Dijkstra *et al.* (1982). Cavé took two groups of female Kestrels and supplemented the food of one group. The ovaries of birds were then examined post-mortem at regular intervals: the better fed birds exhibited a more advance state of development than those with poorer rations. The work of another Dutch team in which supplementary feeding during courtship advanced the laying date and clutch size of Kestrels has already been mentioned (Chapter 6 *Fig. 59*). In a further study in The Netherlands over a four-year period (Dijkstra *et al.*, 1982), dead rodents were placed in the nestboxes of some wild Kestrel pairs, but not in the boxes of others: again the supplementary-fed pairs bred earlier than the unfed pairs, the period by which laying was advanced being as much as three weeks in poor vole years, but less in good vole years when the food difference between the fed and unfed birds was less stark. By observing the control (unfed) groups across the four-year period, Dijkstra *et al.* were able to show that, on average, in good vole years falcon pairs laid 14 days earlier and had clutches with one more egg, than in poor vole years.

The increase in female body weight in each of these Dutch studies indicates that the accumulation of bodily reserves is important in the timing of laying. However, another Dutch team (Meijer *et al.*, 1988) while confirming this finding, did not confirm that supplementary feeding had an effect on clutch size, the differences in their case being non-significant. In a separate study in Finland on the provision of supplementary food, Wiehn and Korpimäki (1997) observed that in pairs with an artificially fed female, the bird reduced her hunting rate and prey delivery rate, while the male did not. Nevertheless, in all cases of supplementary feeding, whether in good or poor vole years, the number of fledglings surviving in a given brood increased.

While these studies have not always shown an increase in clutch size, they have shown an advance in laying date, implying that the reproductive output of Kestrels is food-limited whatever the abundance of prey. That the date of laying of the first egg by any female Kestrel is dependent on food supply seems entirely reasonable. However, the difference seen in laying date between individual females in a given study has caused considerable debate and has resulted in the development of two hypotheses. In an ideal situation a Kestrel pair would time their breeding so that the peak food requirement of their nestlings corresponded to the time of maximum food availability. The first hypothesis, the constraint hypothesis, therefore suggests that the optimal laying date is the same for all pairs in a population, but that not all females are able to take advantage of it, either because of differences in their condition or in the territory of the pair. Because the female's condition is defined by resource availability, they may be ready to breed only after the optimal laying date has passed, resources representing a constraint on their breeding. The second hypothesis, individual optimisation, contends that laying date is adaptive and depends on parental quality so that each individual female begins egg laying at her own optimal laying date. Numerous papers have been produced that deal with these hypotheses and the various experiments which have been carried out to test them. One of the most interesting for its mix of mathematics and experiment (the latter by reducing or enlarging clutch sizes by taking chick(s) from one nest and placing them in another) was another by a team at Groningen University (Daan *et al.*, 1990b). The Dutch manipulated brood size 10 days after hatching. What they found is summarised in Fig. 68.

Breeding Part 2

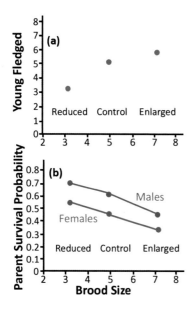

Figure 68. Effect of brood reduction and enlargement on the number of chicks fledged (upper graph) and survival of the parent birds to the following year (lower graph). Redrawn from Daan *et al.* (1990b).

Although Fig. 68a notes that the number of chicks fledged in enlarged broods increased, the survival of an individual chick declined in those nests, *i.e.* the number of chick deaths in enlarged nests was higher than average. Fig. 68b notes that the effect on parent birds was equally pronounced. Parents with reduced broods were more likely to survive to breed again, their workload having declined, but the survival probability of parents (both males and females) with enlarged broods was reduced, *i.e.* parental effort raised more chicks, but at the expense of a shortened life. The Dutch team then defined 'reproductive value' for both parents (Vp) and chicks (Vc) and showed that for the chicks 'value' declined with laying date – as a chick you are more likely to survive to breed if your egg is laid early – but rose steadily with parental effort at first until the increased risk of high clutch sizes caused it to decline again (Fig. 69). For parents the situation is different as their likelihood to breed again declines with the effort of chick raising (Fig. 69).

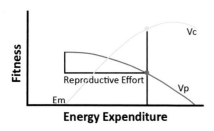

Figure 69. General scheme of clutch size optimisation theories. With increasing energy investment in the current season's reproduction, the value of the clutch (Vc) increases. But the rate of increase declines due to competition for food and increased risk at higher clutch sizes. The residual reproductive value of the parent birds (Vp) declines with increasing energy expenditure. The dots indicate the optimal position. The 'Reproductive Effort' is the energy requirement beyond Em, the energy requirement of non-reproduction (*i.e.* the energy requirement of self-maintenance). Redrawn from Daan *et al.* (1990b).

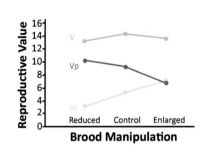

Figure 70. The reproductive value of the brood (Vc) and parent birds (Vp) and their sum (V). The scale of reproductive value is defined by the mathematical calculations of Daan et al. (1990b), but is essentially arbitrary in terms of the numeric value. Redrawn from Daan et al. (1990b).

While the solution for parents seems obvious – do not breed and live longer – that is hopeless in terms of passing on their genes. The Dutch therefore added Vc and Vp to give an overall reproductive value (Fig. 70) and found that evolution had already organised Kestrels to provide the optimal solution.

In later work on Kestrels in Spain Aparicio (1994a, 1998), though taking issue with the Dutch interpretation of some of their data, noted that his own work manipulating Kestrel broods agreed that the quality of the Kestrel pair, in terms of the hunting ability of the male and the initial body weight of the female, *i.e.* individual optimisation, were critical in determining both laying date and clutch size. While that view conformed to the general consensus among experts, a complication was recently added by the suggestion of Tomás (2015) – for all avian species – that it is hatching date rather than laying date which is critical as greater information on food availability is more clearly available as egg development proceeds and the female can adjust the timing of hatching by varying the intensity of incubation. Tomás' idea has considerable merit, but for falcons the variation in observed incubation periods suggests that the female's ability to hold back hatching is severely limited. At this point it is also worth noting the study of Martínez-Padilla *et al.* (2017) who found, in their study of survival and the Lifetime Reproductive Success (LRS - see Chapter 6 for further information) of Kestrels in Spain, that while the laying order of eggs had little effect, hatching order did. The Spanish team found that hatching order predicted the 'fitness' of a chick which, in turn enhanced both the probability of its survival (measured as likelihood of the bird returning to the natal area the following year) and the LRS for both sexes.

The debate between the theories will doubtless continue, but what is clear from the numerous studies of breeding Kestrels is that clutch size declines as the date of the first egg advances through the year, *i.e.* the later the first egg is laid, the smaller the clutch will be.

Daylight hours

While the window for breeding opens, as far as the male Kestrel is concerned, during the early days of spring, another mechanism is required to ensure that female Kestrels are also ready to reproduce. From the work of Bird *et al.* (1980) it is the start of the longer days of spring that triggers the female

Prey exchange from male to female Kestrel during incubation on an electricity pole close to the nest site.

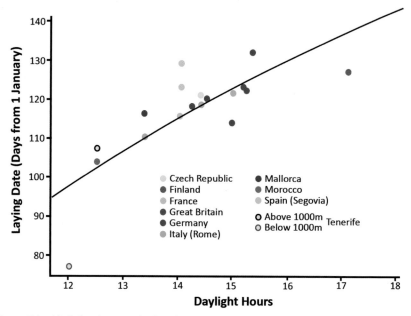

Figure 71. Variation in mean laying date with daylight hours. The later data point for Great Britain is from southern England, the earlier data points are from southern Scotland. References for all data points are given in Carrillo and González-Dávila (2010). Redrawn from Carrillo and González-Dávila (2010).

to prepare for breeding. Bird and co-workers experimented with captive American Kestrels, exposing them to artificially shortened days during October, then to artificially longer days in November. This change induced 6 of 10 pairs to breed in January at a time when, in the wild, the probability of raising young was vanishingly small. Bird *et al.* noted 18 of 27 eggs were laid when the ambient temperature was below 0°C, but the fertility, likelihood of hatching and fledging success were comparable with eggs laid at normal times. A similar result was mentioned in Chapter 3 for Kestrels (Meijer, 1989) when the effect of photo-period on moult was considered. By artificially altering the day length from 1 December (to 17.5 hours for one group of Kestrels, 13 hours for another), Meijer noted that egg laying was advanced to 29 December (17.5hr group) and 15 February (13hr group). In all cases moulting was synchronised with laying.

In a study in Spain, Carrillo and González-Dávila (2010), collected data on the laying dates of Kestrels breeding from Tenerife to Finland, and plotted these dates against day length (Fig. 71). From their study, the Spanish concluded that day length, and both winter and spring weather, affect laying date. As the days lengthen into spring, and the temperature rises, the male Kestrel's basal energy expenditure reduces, and he can hunt for longer periods. He is therefore able to fend for his female and, later, his brood, while the lengthening days ready the female for breeding.

A female Kestrel peers out of an old shopping basket in which she has nested, having delivered prey to her brood.

Winter and spring weather

As noted in the previous chapter, winter weather influences both Kestrel survival during the winter and the falcons' arrival time at breeding sites. Spring weather also affects a Kestrel's hunting efficiency, and these factors will influence the onset of laying on an annual basis. In a study of urban Kestrels in Vienna, Kreidrits *et al.* (2016) noted that dry winters increased the population of mammalian prey, while warm, dry spring weather meant courtship and egg laying were earlier.

Field data

Experimental and observational data have shown that environmental conditions and the condition of the female after winter's hardship influence the timing of the first egg, while further evidence confirms that the availability of food is the proximate dictator of Kestrel egg laying, Meijer *et al.* (1990) noting that a female's first egg was laid 2-3 weeks after she had begun to be fed by her mate.

I conclude this examination of female first-egg date selection by looking at field data on actual timings. That timing is remarkably fluid, the Kestrels of northern Europe breeding in April and May, with those of southern Europe, north Africa and the Indian sub-continent commencing a month or so earlier. In equatorial Africa Kestrels breed in October and November (the dry season), while the Rock Kestrel of South Africa breeds in September-November, the southern spring. But even this apparently clear pattern is overwritten by curiosities: Kestrels in southern India breed earlier than those on nearby Sri Lanka, while Kestrels in the Horn of Africa breed during the rainy season. Table 6 indicates the mean laying dates from various countries of northern Europe.

Country	Latitude (°N)	Laying Dates			Number of Nests	Source
		Good Vole Year	Poor Vole Year	All Years		
Finland	61			13 May[2]	136	Korpimäki, 1986b
	63			7 May[2]	131	Korpimäki, 1986b
Sweden	57			1 May[3]	37	Wallin et al., 1987
Scotland	55	28 April	11 May	3 May	127	Village, 1990
	55	18 April	8 May	29 April[2]	142	Riddle, 1987
England	53	4 May	17 May	12 May	263	Village, 1990
The Netherlands	53	16 April	24 May	27 April	705	Meijer et al., 1988
Germany	48-55			29 April	1197	Kostrzewa and Kostrzewa, 1997
Czech Republic						
-Urban	49			26 April	44	Pikula et al., 1984
-Rural	49			2 May	252	Pikula et al., 1984
Bohemia	49	19 April	4 May	27 April	238	Plesník and Dusík, 1994
France	47	22 April	11 May	3 May	82	Bonin and Strenna, 1986

Table 6. Variation of mean laying dates in northern Europe.

Notes:
1. All quoted dates are the mean for the number of nests in the study. Riddle (2011) quotes an earliest date of 17 March (in 2000, a year with an early spring, benign weather and a good vole population) while Santing (2010) gives 16 March for the first egg in 2007, an equally good year for weather and vole numbers in The Netherlands.
2. These dates are median, i.e. the date on which half the females had started laying.
3. Calculated by assuming that the first egg was laid 36 days prior to hatching.

Note that the data sets in Table 6 were used as they represent a sample which includes the date of the first egg laid with differing vole populations in habitats which are comparable to Britain. A more comprehensive list of mean laying dates with geographical position, hours of daylight and clutch size is given in Carrillo and González-Dávila (2010) who tabulate data from 20 or so studies across Europe and north Africa, from the Canary Islands to Finland (28°N-65°N), and for altitudes from sea level to 1300m: see Fig. 71 for the effect of latitude on laying date. See, also, Zellweger-Fischer *et al.* (2011) who provide updated data for Switzerland.

Intuitively, a correlation between laying data and latitude would have been assumed, but the data of Table 6 suggests this is not the case. In Finland, where the Kestrels begin their breeding cycle before the last snows have melted, increased day length compensates, at least in part, for lower temperatures, bringing to mind the data of Bird *et al.* (1980) who noted no change in fertility and breeding success when American Kestrels were manipulated to lay clutches in sub-zero temperatures. In his studies Village (1990) noted that laying in Scotland began earlier than that in England despite the Scottish study area being 300km further north. The histograms of first-egg dates from Village's studies are also interesting (Fig. 72). Whereas the histogram of English dates is symmetrical, as might be anticipated, that of the Scottish Kestrels is skewed towards the earlier first-egg date. Village explored this, and found a correlation between mean first-egg date and the laying of the first egg, but no correlation between the mean and either the last laying or the duration of the laying season. In other words, the end of the laying season was (more or less) constant in all years: birds may start to lay early if conditions allow, but the early laying of some birds lengthens the laying season rather than moving the entire season forwards in time.

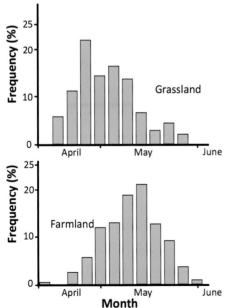

Figure 72. Percentage of first eggs laid as a function of time of year for Kestrels breeding on southern Scottish grassland and east-central English farmland. Redrawn from Village (1990).

Female Kestrels delivering prey to hungry (and noisy) broods in nestboxes.

The Eggs

From the work of Cavé (1967) in The Netherlands, egg development in the ovary begins during the autumn and proceeds through winter and spring. Cavé assumed that winter conditions – weather and food availability – would influence egg development and attempted to test this experimentally. Adjusting the temperature at which his captive Kestrels were held did not prove successful at mirroring environmental conditions, but by giving some birds more food, and others less, Cavé was able to show that this did indeed affect egg development. The result was later confirmed by another Dutch team, Meijer and Schwabl (1988). Final development of the egg starts about nine days before laying (Meijer *et al.*, 1989) with seven days of rapid follicle growth followed by two days during which the egg membrane, albumen and shell are formed. From the work of Meijer *et al.* the energy cost of producing an egg is 99.7kJ, with the yolk requiring 65kJ and the albumen 34.7kJ. The rapidity of egg development means that the female will, at the point of lay of the first egg, hold the entire clutch, each egg at a slightly different stage. For a four-egg clutch, the total mass of the eggs is about 84g, the Dutch team noting the change in body mass of a female during the courtship, laying and early incubation phase of breeding of one bird during two successive years (Fig. 73).

Egg Laying

To accumulate the energy reserves required for egg production the female reduces, then ceases, the time spent hunting, eventually relying entirely on food supplied by her mate. In Britain, where the majority of Kestrels do not migrate, this usually means the female stops hunting 2-4 weeks before the first egg is laid, but in Scandinavia where the population is entirely migratory females cannot afford the luxury of such a period of hunting abstinence, needing to replenish energy reserves drained by migration as well as preparing for egg production: in southern Finland, Korpimäki (1985b) noted that female Kestrels were fed by their mates for only one week prior to laying. During the period when the male is provisioning his mate, Shrubb (1993a) noted that males fed their mates every 2-3 hours, suggesting 4-6 meals daily. For further details of the feeding rate of females and chicks during courtship, incubation and chick rearing, see *Chick Growth* below. Usually the feeding follows a set pattern, occurring at regular times during the day, the male's approach being stealthy, especially during incubation, to minimise the detection of the nest site by potential predators. The female spends an increasing amount of time on or near the nest as egg development begins.

Egg Laying Period

After the first egg has been laid the female usually produces eggs every other day: Village (1990) estimates a two-day interval, deciding not to risk extreme disturbance of the female by seeking a more accurate number. In Germany Hasenclever *et al.* (1989) calculated a mean of 2.03 days (48.72 hours) from a

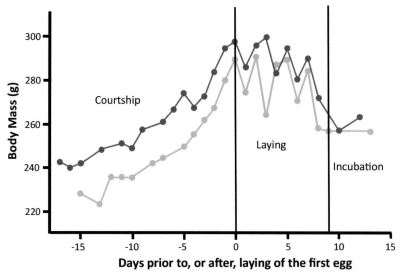

Figure 73. Variation in body mass of a female Kestrel breeding in captivity. The two graphs are for the same female which laid four eggs in each of two successive years. Redrawn from Meijer *et al.* (1989).

total of 57 eggs, but with an extraordinary spread from as short as 24 hours to as long as five days. The single egg laid five days after the previous one was the last of a clutch.

Aparicio (1994b) calculated an average of 2.12 days (50.88 hours) from 66 eggs in Spain. However, the Aparicio study included some pairs which had been given supplementary food in order to assess the effect on laying interval, the 2.12 days being an average of extra food pairs (mean laying interval 2.04 days = 48.96 hours) and a control group which were not fed (mean interval 2.50 days = 60.00 hours). Aparicio's study sought to understand the effect of female energy (*i.e.* food) intake on clutch size and concluded that food intake was more important than the start of laying on determining the clutch. However, as we have seen above, this conclusion is not as straightforward as it might at first appear.

In the work described in Chapter 8 with continuous camera observation of a Kestrel pair in southern England, the average time interval in a clutch of three eggs in 2017 was 46.16±3.65hrs. However, the laying of the fourth egg was much better defined, at 45.55±1.06hrs. In the same study, but in 2019, the mean interval of a five-egg clutch was 46.62±0.30hrs. In 2019 the mean interval for the first four eggs was 45.84±0.34hrs, while the final egg was laid 48.93±0.16hrs after the fourth. Chapter 8 notes that in 2020 the situation was very different.

Eggs

Kestrel eggs are elliptical and measure 34-47mm along the major axis and 27-39mm across the minor axis (Village, 1990 and references therein, and

Boileau and Hoede (2009)). Village's larger egg sizes are data from a small set of eggs in Germany and Poland, whereas the study of British eggs indicated 35-45 x 29-35 with a sample size of 306. The relatively large differences in egg dimensions can be due to physical size differences between females, or differences in their condition when they start laying, but may also be seen in the eggs of an individual clutch. Three hypotheses have been set down to explain within-clutch differences. The first posits that it represents an attempt to stop sibling rivalry and so reduce chick mortality, with larger eggs being laid later so that asynchronous hatching produces smaller, early chicks and larger, later chicks. The second hypothesis suggests that as larger eggs may take longer to incubate, smaller eggs should be laid later to aid synchronous hatching. The third idea is that the female bird controls egg size, aware that limitations in her own condition requires fewer resources to be invested in later eggs as they are less likely to produce chicks which will fledge. However, the three hypotheses require all other factors to be equal. In reality, they are not, as synchronous hatching is more productive (in terms of raising chicks to fledge) when prey availability is high during chick rearing, with asynchronous hatching likely to be more productive when prey resources are poor. The hypotheses also fails to account for female condition. A female in poor condition may be unable to invest similar bodily resources in all her eggs so that later eggs will be smaller, while a female in first class condition may choose to invest greater resources in early eggs. To attempt to shed light on the conundrum, Aparicio (1999) studied intra-clutch differences between Kestrels in central Spain. Aparicio had a control group of 10 falcon pairs and a second group of 11 randomly-selected pairs in which he supplemented the food of the female. The results of his study are shown in Figs. 74a and b.

The noted differences in egg volume in Figs. 74a and b, while statistically significant, were within the normal scatter of egg sizes, and the effect on nestling survival could not be distinguished from the effect of hatch date. Aparacio could not, therefore, distinguish between the hypotheses. He did, however, note that the duration of incubation increased from 27 days for an egg of volume 18ml to 29 days for one of 25ml.

When first laid Kestrel eggs weigh about 20g: at hatching this weight has reduced by about 15% due to water vapour loss during chick growth (Fig. 75 overleaf). In a study of six eggs by Meijer *et al.* (1989) the mean weight was 21.5±1.8g, this weight comprising shell (2.1±0.2g), albumen 14.8±1.5g and yolk (4.6±0.4g). For a female of average weight (230g) a clutch of 4 eggs therefore represents about 35% of body weight, though as the eggs are formed on a conveyor belt this is not a reasonable calculation in terms of the bird's input to the process. As noted above, Meijer *et al.* (1989) also calculated the caloric value of each egg as 99.70kJ. Apportioned to the production process of a clutch of four eggs this represents an energy requirement of 72kJ/day. As the daily energy intake of a female was calculated as 388kJ, egg formation therefore requires the female to devote 19% of her daily energy input to egg

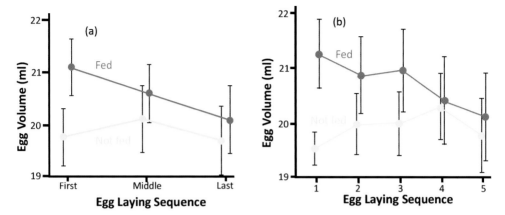

Figure 74. Variation of average egg volume with laying order for Kestrel females given supplementary food and other females that were not fed. In Fig. 74a all clutches, numbering 3 to 6 are included. In Fig. 74b only clutches with 5 eggs are included. The black symbols represent one standard deviation (SD) on the mean value of volume. Although there is overlap of the SDs, the differences for all clutches (Fig. 74a) between the fed and unfed volumes are significant. Redrawn from Aparicio (1999).

production throughout the egg formation and laying phase. To prepare for this, Meijer and co-workers found that female Kestrels increased their body weight by about 70g (from about 230g to about 300g for an average bird). Of this weight gain 64% was accumulation of body reserves, the remainder being the development of oviducts and eggs.

Kestrel eggs are white/pale buff, with heavy speckling, usually of red-brown, but varying to darker brown. Village (1990) notes that there is considerable variability in the speckling with some eggs being almost plain white with only a few brown spots, while others are almost a uniform dark chocolate brown. There is also variability within a clutch, and in general the last egg is the least coloured, both ground colour and speckling. Village (1990) is almost certainly correct in suggesting that this results from depletion of protoporphyrin, the pigment which creates the speckling.

Clutch size

The usual clutch size is 4-6. Single eggs and clutches of seven have been recorded, though these represent less than 1% of all clutches (see Chapter 6 *Footnote 1* (p166) for a remarkable occurrence of 7-egg clutches in southern Scotland. Clutches of 2 are also rare and may result from partial predation rather than being full clutches. Records of clutches greater than seven exist, but Village (1990) considers these probably arise from two females laying in the same nest. However, Korpimäki (1984b) observed an eight-egg clutch in his studies of Kestrel population variations with mammal numbers in Finland, and Riddle (2011) records a similar clutch in southern Scotland in 1997 when good weather and a peak in the local vole population coincided.

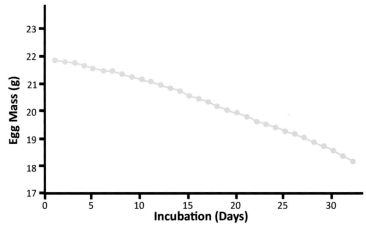

Figure 75. Weight loss of a Kestrel egg during incubation. Redrawn from Hasenclever (1999).

Heukelen and Heukelen (2011) photographed a nest (on a haystack within a barn in The Netherlands) with a clutch of 10 eggs which they considered had been laid by the same female rather than involving egg dumping. As evidence they claim a second female was never seen in the vicinity, and the large quantity of unconsumed prey suggested a bumper vole year: at least nine of the eggs hatched and eight chicks fledged. Considering the difficulty female Kestrels occasionally have in covering smaller broods, the female had performed an outstanding job of keeping the total brood warm: the male's hunting achievement was equally noteworthy. See overleaf for photographs of large clutches.

As we have already seen (Chapter 6 *Fig. 65*) Kestrel clutch size varies with available prey. Collecting clutch size data in Wales, Shrubb (2003) noted that sizes showed considerable annual variation: examining 83 nests over the period 1997-2002 Shrubb observed a mean clutch of 3.00 in 2001 compared

In the nestbox of the study detailed in Chapter 8 one of the cameras captures an image of the first egg of a five-egg clutch, the photo of the damp, shiny egg allowing the time of laying to be accurately measured.

Breeding Part 2

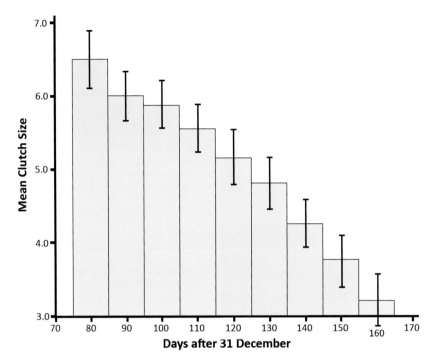

Figure 76. Variation of clutch size with the date of laying of the first egg. Black symbols are standard deviation on the mean. Drawn from data in Beukeboom *et al.* (1988).

to a mean of 5.00 in 1999. Given the relatively subdued cyclicity in rodent numbers in Britain (in comparison to Scandinavia) the change from 3 to 5 is considerable. As noted in Chapter 6, within a given year clutch size also varies with the laying date of the first egg (see Chapter 6 *Fig. 59*, and Fig. 76). Carrillo and González-Dávila (2010) investigated the effect of latitude on clutch size, but used evapotranspiration (the sum of water evaporation and plant transpiration) as a proxy for resource availability. Their data show that northern Kestrels lay larger clutches than more southerly birds, but this could result from both inherently greater resources (*e.g.* lemmings in Scandinavia) and longer day length (more hunting time). The data shows minimal differences in clutch sizes for Kestrels living in mainland Europe, but a significant difference for falcons breeding in Mallorca, Tenerife and Israel, where evapotranspiration is higher, reducing prey availability.

Dijkstra *et al.* (1990a) manipulated the brood size of Kestrel pairs in The Netherlands, their results indicating that Kestrels can raise larger clutches to fledge than are seen in the wild in all but exceptional circumstances. Using nests with, on average, 5 eggs, two extra eggs were placed in some, while two eggs were removed from others. The parent birds fed and raised the manipulated broods, but both the male and the female hunting times were

Top left Seven-egg clutch photographed on Scotland's Pentland Hills where such clutches seem to occur more often than elsewhere.

Top right Eight-egg clutches have been seen in Finland and, this photograph, south-west Scotland. *Gordon Riddle.*

Left Ten-egg clutch photographed in haste with a camera-phone in south-west Netherlands in 2011. *Cock van Heukelen.*

affected. The male flight time increased by 26% (total energy expenditure rising by 18%) when the brood was enlarged, and fell by 18% (total energy expenditure falling by 6%) when the brood was reduced. In the enlarged broods, hatchling mortality was about 20%, compared to <2% in reduced and control broods, but once the brood size had been reduced by nestling deaths, the offspring were successfully raised, an average of one extra chick being raised in comparison to control broods. These findings confirmed earlier work by another Dutch group (Masman *et al.*, 1989) which was mentioned in Chapter 5 when considering the energy costs of breeding for male Kestrels (see Chapter 5 *Fig. 44*).

Dijkstra *et al.* (1990a) and Deerenberg *et al.* (1995) also investigated the parental cost of provisioning a large brood and suggested that the mortality of parent birds was significantly influenced by brood size. However, these studies and that of Masman *et al.* (1989) defined mortality as the failure of adult birds to return to breed in the following year, which is an indirect indicator, capable of being influenced by other factors. By contrast, Daan *et al.* (1996) were able to show that parent bird mortality really had been increased by recapturing

Breeding Part 2

birds in subsequent years and by recovering and examining birds which had died subsequent to brood manipulation experiments. Studying 200 Kestrel pairs, Daan *et al.* found that the survival of both male and female Kestrels was influenced by brood size (Fig. 77).

From Fig. 77 it appears that for female birds survival to breed decreases with enlarged clutch size. This would certainly influence the female's decision on clutch size, though in the Dutch experiment females were 'choosing' clutches which were reducing their likelihood to breed again. For the male the situation appears to be different, with a 'cliff-edge' in terms of clutch size, before survival is influenced. In each case the data suggests that the seasonal imperative to breed and to produce offspring takes precedence over the possibility of producing fewer offspring this year and being in good condition to breed again next year. Daan *et al.* found that most parental deaths occurred in winter implying that exhaustion due to breeding is not an immediate danger, *i.e.* the late summer/autumn death rate of adults does not increase, but influences the bird's chances of getting through another winter.

The Dutch finding was confirmed by Tolonen and Korpimäki (1994) in Finland, but a later study of brood manipulation in Finland (Korpimäki and Rita 1996) added a further complication. In their study the Finnish researchers found no increased mortality in broods which had been artificially increased. However, they did find that the parent birds did not raise all the chicks to fledge even if the vole population was high and they would have been able to do so. This apparent contradiction was explained by Korpimäki and Rita as being due to the difference in the environment of the Dutch and Finnish birds. In The Netherlands the vole population was essentially stable, so for the parent birds the attempt to raise more chicks enhanced the possibility of their gene survival even if their own lives might be shortened (the so-called intra-individual strategy where the trade-off is between current parental effort and future (adult) survival). In Finland, the cyclic fluctuations in vole population meant that the best strategy for the parent birds was to maximise their own survival (the so-called inter-generational strategy, where the trade-off links current parental effort with the number and size of offspring, which

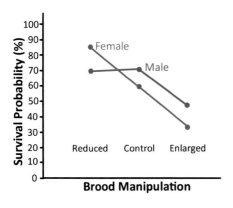

Figure 77. Probability of survival of male and female Kestrels with reduced, control and enlarged broods. Drawn from data in Daan *et al.* (1996).

The male Kestrel admires the completed five-egg clutch in the study of Chapter 8, 2019.

thus affects offspring reproductive value). The Finnish experiment therefore appears to explain why Dutch female Kestrel appear to take a decision which reduces their survival chances.

In further experimental work by the Dutch Groningen University group Beukeboom *et al.* (1988) not only manipulated clutch sizes, but installed thermistors in the studied nests to observe female incubation behaviour. They found that females with clutches of three or four eggs began purposeful incubation earlier than those with clutches of six, the start of persistent incubation apparently being predicated on clutch size (Fig. 78). As clutch size in general decreases with the date of laying of the first egg (*i.e.* early layers produce larger clutches than late layers – see Fig. 76), this means that females laying late in the season begin incubation sooner than those laying early in the season. The decline is almost certainly controlled by hormonal changes, these defining an egg-laying period so that all females have a laying period in which the start point is defined by the laying of the first egg and the end point is

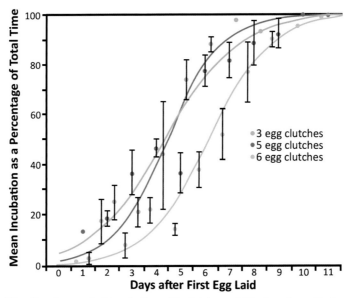

Figure 78. Development of incubation with clutches of differing sizes. Redrawn from Beukeboom *et al.* (1988).

Breeding Part 2

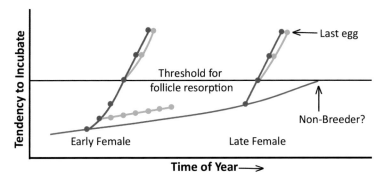

Figure 79. A model for incubation behaviour and clutch size control. The laying of six eggs by an early-laying female and of four eggs by a late-laying female are shown, based on an experiment in which eggs were removed from five early and three late females in The Netherlands. Eggs laid are indicated in red. Eggs laid, but removed are indicated in green. In the early laying females removal of eggs suppresses the development of incubation so that removal of four eggs, after the laying of two, stops incubation and permits the female to lay four more eggs. If two eggs are removed at a later time, these are also replaced. In late-laying females removal of eggs after two have been laid does not promote a larger clutch as once the female has reached the threshold for follicle resorption the clutch size is fixed. Redrawn from Meijer *et al.* (1988), but extended by the present author beyond the data of that study by suggesting that a continuation of the lower red line defines a point at which, presumably, females can no longer lay.

defined by hormone levels. This suggestion is consistent with the finding of Village (1990) that the end of the laying period is the same for all females irrespective of the first egg date.

In a continuation of the work of Beukeboom and co-workers, Meijer *et al.* (1990) hypothesised that control of clutch size was associated with 50% incubation, *i.e.* when the female spends 50% of her time incubating. Those females which lay early in the season have a low tendency to incubate and may lay four eggs before incubation reaches 50% of their time: a late laying female will reach 50% incubation after only two eggs. From experiments with egg removal and additions, the Dutch group suggest that the decision of the female to stop laying and to resorb developing follicles occurs approximately four days prior to the laying of the last egg. In each case this means that only two further eggs will be laid after the decision is made, resulting in a preponderance of six-egg clutches for the early females and clutches of four for later females (Fig. 79).

Second clutches

Normally Kestrels lay a single clutch, but evidence from a montane population of Kestrels in central Spain (Fargallo *et al.*, 1996) suggests that in certain circumstances second clutches may be laid. Fargallo and co-workers noted that three of 11 pairs of Kestrels began second clutches after the chicks of the first clutch had fledged. In each case the three pairs had laid their first clutches earlier than any pair with a single clutch. The second clutches were of one, three and four eggs, and all were incubated, but only the clutch with three eggs hatched: all three of the chicks fledged successfully. Fargallo *et al.* believe that two factors influenced the Kestrels' decision to attempt to breed again. Firstly, the benign weather, the southerly position of the site (40°N) extended the breeding season (in relation to northern-dwelling Kestrels). Secondly, there was high prey availability.

Santing (2010) records the laying of a second clutch of three eggs in the same nestbox in The Netherlands in which a Kestrel pair had raised four fledglings from six eggs. In 2007, a good vole year, the first clutch was started on 16 March (which was 8 days earlier than any other Kestrel pair in the neighbourhood). After the chicks had fledged, a second clutch was started on 28 May. However, while the timing of the second clutch was consistent with a double clutch, Santing was unable to positively confirm that the same pair was responsible.

Although double clutches are rare, females may lay a second clutch if disturbed during laying or incubation, the first eggs being taken by predators or chilling. Village (1990) noted that in 60% of cases of egg chilling the repeat clutch was laid in the same nest, the female then incubating all the eggs though, of course, only the second clutch would hatch. However, Village notes, the female only does this if there are few eggs from the first clutch remaining as she needs to ensure she can cover the entire number during incubation. Females may also abandon clutches if there is a food shortage or if her mate dies. In such cases she may lay a second clutch if better food resources become available or a new mate is found.

In the case of death of a partner, the remaining bird (of either sex) may take a new partner with the newly-formed pair taking the clutch or brood, or they may start a new clutch. The fact that there are 'spare' birds available for pairing requires that there are non-breeding birds in the population, but as already noted, this is usually the case. Village (1990) experimented by removing one bird of a pair and watching the response. In three of four nests in his Scottish study area the removed bird was replaced, while at 25 nests in his English study area, 11 birds were replaced. Of the 25 English removals, 13 were male, 12 were female, but there were more female replacements (8 of 11, 72%) than male (3 of 11, 27%): in all three cases the replacement males failed to breed, but four of the replacement females bred successfully, in addition to one who took over the clutch of her predecessor. The implication of the failure to breed of all three replacement males was that they were poor hunters: two of the three were first-year males, which might suggest this was the case, as an ability to survive your first winter does not necessarily mean you are up the task of supporting a female and a brood. In the English study area, one removed male bred again with his original partner when he was released. Some other removed/released birds bred the following year, occasionally with their original mate. However, nine removed birds were not seen again after release. The experiments suggest that many non-breeding Kestrels are in that position because they lack a territory, or a mate, or both.

Incubation

The female's brood patch develops prior to laying, feathers of the belly and lower chest being lost to expose an area of richly vascularised skin, blood heat then being more easily transferred to the eggs. In the study of Beukeboom *et al.* (1988) mentioned above, the researchers concluded that the eggs both stimulated incubation and clutch fixing by follicle suppression, though whether the stimulus was visual or tactile (*i.e.* contact between the brood patch and an area/volume of eggs) or a combination of the two could not be differentiated.

The female periodically turns the eggs during incubation. She also occasionally gives a soft call hoping to initiate a response from her mate as reassurance that he is close by, as he often sits within hearing distance when resting from hunting. As a contrast to their American cousins, male Eurasian Kestrels do not develop brood patches or incubate. Some males spend very little time with the eggs while the female is away, but others cover the clutch while the female feeds, preens or bathes. Packham (1985b) observed a male which not only showed a brood patch, but took turns to incubate, and gives references for others who have also seen males incubating. However, this report of a male brood patch appears to be unique, suggesting anomalous behaviour.

During the breeding of Kestrels in the four years 2017-2020 (Chapter 8) the 2017 male spent long periods covering the eggs, though there were no observations of him having a brood patch. On one occasion he spent 192 minutes with the eggs, and over a four-day period he spent a total of 14hrs 38mins (15% of the total time) covering the eggs. It is difficult to believe that the male was not incubating given such a time. Once the eggs had hatched, the male spent almost no time with the chicks and did not brood them. For further information on males covering the eggs see Chapter 8.

Females avoid defecating at the nest: they usually break from incubating for no more than 20-30 minutes during which time they not only defecate, but preen and stretch. However some females are much less committed to the eggs as will be seen in Chapter 8. But while avoiding nest site defecation, females are often less fastidious when it comes to casting pellets, these occasionally accumulating around the eggs during incubation and forming part of the substrate for the hatchlings. Incubating females may sit tight if approached (to the point of allowing the observer to catch them: Village (1990) claims 10% of his catches were made this way) or may fly off immediately when approach is detected. As a panicky escape may result in eggs or chicks being thrown out of the nest any approach should therefore be made with the utmost caution.

Because incubation does not begin with the first egg and some hatchings are asynchronous, defining the incubation period precisely is difficult. Village

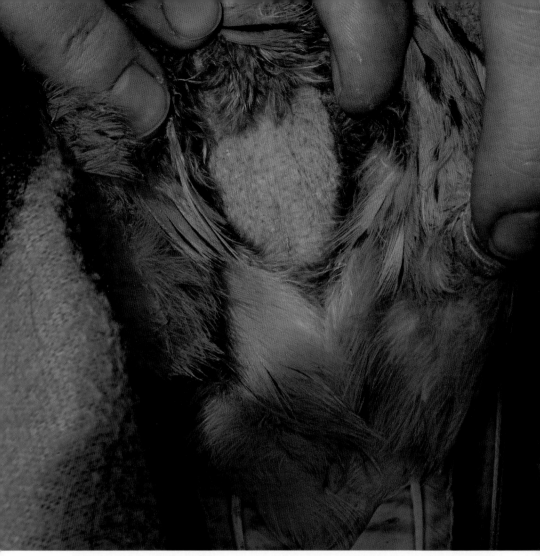

Brood patch of a female Kestrel. The bird was an imprinted falconry bird that had bred annually for several years. Though used to being handled, the bird was wrapped for the photograph, the procedure taking only a few seconds. The bird was unharmed and seemed unfazed by the activity. She had laid a five-egg clutch, all the eggs hatching. The chicks were raised normally by the female and the captive male Kestrel.

(1990) notes periods from 26 to 34 days, with a mean of 30.8 days. The data from the breeding considered in Chapter 8 is consistent with Village's data, a more precise incubation period not being possible because of the difficulty of knowing when incubation started.

With delayed incubation and synchronous hatching all or most eggs hatch on the same day (in Chapter 8, in 2017 4 eggs hatched by 18.04 on 24 April, the fifth by 15.44 on 25th; in 2019, 2 eggs hatched between 10.53 and 16.41 on 29 May, the other three by 14.10 on 30 May; in 2020 the first egg hatched at 16.02 on 31 May, and all four had hatched by 17.00 on 1 June). Asynchronous hatching can take 4-5 days to complete. The advantages and disadvantages of synchronous and asynchronous hatching in birds have long been debated, with evidence being inconclusive. The favoured theory is that

Female (*left*) incubating (and sleeping) and male (*right*) covering the eggs. From the study of Chapter 8.

the choice is an adaptive behaviour in Kestrels arising from the cyclicity of the vole population, synchronous hatching being 'chosen' in good prey years, asynchrony in poor prey years as it maximises the chance of the older chicks surviving, reasoning which has been termed the 'brood reduction hypothesis'. The obvious area to attempt studying the hypothesis is Scandinavia where vole cyclicity is maximal. Wiebe *et al.* (1998a) carried out such a study in Finland over 12 years, a period that included several instances of high and low vole populations as the local population varied on a three-year cycle. The study failed to support the brood reduction hypothesis. For clutches of equal size, Wiebe *et al.* found that hatching was more synchronous in low vole years, asynchronous clutches dominating when vole numbers were increasing. Within a pair, asynchronous clutches were correlated with the age of the female, but not the male. The Finnish team suggest that this paradoxical result might be influenced by the stability of local vole populations. In a low phase vole numbers are low, but stable, while in increasing and decreasing phases, small areas might see fluctuating numbers. Low, but stable numbers might encourage a Kestrel pair to select a small, synchronous brood. Fluctuating numbers, hinting at an unstable supply, might then suggest a larger, but asynchronous, clutch. Overall, Wiebe *et al.* felt that the differences revealed by their study were not sufficiently clear cut, and that other factors needed to be investigated before a conclusion could be reached. A second study, also in Finland, by Wiehn *et al.* (2000), was carried out in 1996, when vole numbers were high, and 1997 when they were lower. Both the synchrony and chick feeding were manipulated, the former by moving hatchlings, the latter by supplementing the food at certain nests. This study agreed with Wiebe *et al.* (1998a), finding that synchronous hatching was more profitable (in terms of number of fledglings successfully raised) than asynchronous hatching whether food was scarce or abundant, and that this result was seen even if chicks were given supplementary food (Fig. 80). However, what was also clear was that parental feeding workload was higher in synchronous nests, Wiehn *et al.* noting that differences in parental care may be as important as any adaptive behaviour.

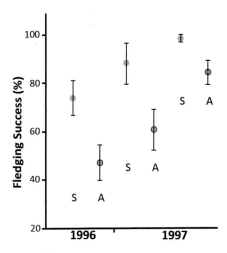

Figure 80. Fledging success in two years. Each pair of dots represent synchronous (S) and asynchronous (A) hatchings as control (orange and green dots) and manipulated (black ringed dots) broods. The orange dots are broods which were manipulated by the adding or removal of chicks to create synchronous or asynchronous broods. The green dots are broods which were similarly manipulated, but supplemental food was given to all chicks. The symbols represent ± standard error on the mean: the number of nests involved in each study was approximately the same (varying from 10-14). In all cases synchronous hatching proved more profitable. Redrawn from Wiehn *et al.* (2000).

In a study in Spain, Martínez-Padilla and Viñuela (2011) also looked at the brood reduction hypothesis, by examining the 'quality' of chicks in synchronous and asynchronous broods in which a nestling died. They found the only 'quality' difference between the chicks was that in asynchronous broods the chicks were smaller, but their immune response was higher than for chicks in synchronous broods. This suggested that more resources were put into immune response than into growth, *i.e.* the chicks were trading size for an enhanced ability to counteract disease. Martínez-Padilla and Viñuela noted that brood reduction was more likely to occur in asynchronous broods and that asynchrony was more likely later in the breeding season. While the Spanish study can be interpreted as support for the brood reduction hypothesis, as with the Finnish studies, other factors may also be at work.

But if, for the moment, we assume that the female Kestrel wishes to achieve hatching synchrony, how does she do so? This was explored in Finland by Wiebe *et al.* (1998b) who studied the incubation behaviour of 17 wild Kestrels to assess its impact on hatching times. They found that in general hatching occurred in the order eggs were laid (*i.e.* asynchronous hatching) and that both this, and the total time span of hatching, corresponded well with incubation behaviour. The Finnish team identified three incubation modes (Fig. 81). In 'rising' incubation the female steadily increased the time spent incubating as the clutch grew, though the starting time for commencing, and the intensity of, incubation could vary with egg number. In 'steady' incubation the female would start incubating, then hold a steady rate before a final increase. Finally,

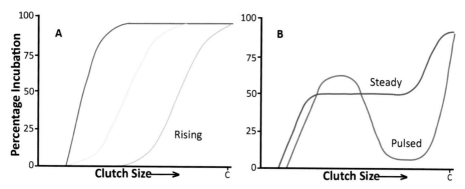

Figure 81. Possible incubation schedules. The curves show the percentage of the day spent incubating as the clutch size increases to its final size (C). Fig. A is typical 'rising' incubation with differing start times and rates of rise. Fig. B are atypical patterns where the incubation rate does not rise steadily with increasing time and clutch size. Redrawn from Wiebe et al. (1998b).

there was 'pulsed' incubation where the rate of incubation rose, then fell before rising again to the maximum rate.

The effect of these variations in incubation rate was to increase the time span of asynchronous hatching (Fig. 82) with the total incubation time before the last egg was laid being linearly related to the hatch time span.

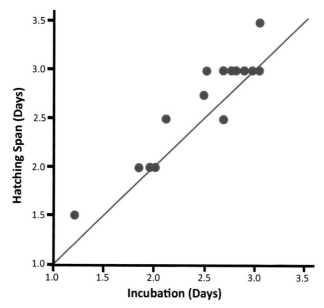

Figure 82. Relationship between total incubation time prior to the laying of the last egg and the hatching span. Data based on 15 clutches in Finland. The line represents perfect correspondence between incubation and hatching times. Redrawn from Wiebe et al. (1998b).

But while the Finnish study showed that the female's incubation behaviour influenced the time span of hatching, it was still not definite that the bird was in complete control of the incubation pattern. Could the time the female spent incubating be constrained by other factors? In a second report, (Wiebe *et al.,* 2000), the Finnish researchers noted what the females were doing when they were not incubating. The time spent away from the nest was not correlated with either male delivery rates or prey abundance: indeed, only one-third of the females hunted while not incubating. Neither was time away correlated with ambient temperature: the females were deliberately choosing not to incubate.

Interestingly, although the females were nesting in boxes, and could therefore have taken advantage of the shelter afforded by the box by staying in it (sitting beside, rather than covering, the eggs), the birds chose to leave the box. This could, perhaps, imply that the sight of the clutch was a stimulant to incubation and so was being avoided, though in the study explored in Chapter 8, during the early days of incubation one female spent long periods of time off the eggs, but sat at the box edge (facing away from the eggs), preening or stretching, or merely sitting: it is possible that this apparently unusual behaviour may have influenced subsequent events.

It also seems possible that females can choose incubation patterns that are not related to asynchronous hatching: in Britain (C. Dobbs pers. comm.) one female Kestrel stopped incubating during a period of very cold weather with snow, covering the eggs long enough to avoid chilling, but delaying hatching so that the male would be able to hunt when conditions improved. Delayed incubation was also seen by Riedstra and Dijkstra (2017) in The Netherlands. Laying of a five-egg clutch began on 1-2 May, so hatching was assumed to be 28 May-1 June, but on 22 June the chicks were only 7 days old. Incubation had been delayed by 14 days, again presumably because of a cold weather spell, perhaps in combination with a poor rodent supply.

Before leaving the debate about the differences between hatching synchrony/asynchrony, one further study needs due consideration. All who have handled Kestrel chicks will be aware that they carry an ectoparasite burden. Indeed, some ringers believe that the reason chicks move away from each other and the nest as they near fledging is an attempt to avoid parasites – though it is worth noting that the consensus holds that the behaviour is an attempt to be first in the queue when the next prey delivery is made. In a study in Switzerland, Roulin *et al.* (2003) looked at the parasite burden of chicks from asynchronously hatched broods of several species, including Kestrels. The aim was to test the captivatingly-named Tasty Chick Hypothesis (TCH) which suggests that chicks hatched later in asynchronous broods are more attractive to ectoparasites. Chick hatch rank was identified in the nests of several species, and the chicks were checked for immune response and parasites. The results showed that immune response was greater in earlier hatched chicks of Barn Owls, but not in Great Tits: the burden of the tick *Ixodes ricinus* was random across hatchlings; and the louse fly *Crataerina melbae* infested older rather than younger hatchlings in the Alpine Swift (*Apus melba*). These results do not support TCH as a global hypothesis, but the burden of the fly *Carnus haemapterus* was greater in late-hatched chicks in both Barn Owls and Kestrels, which might imply that Kestrels do indeed produce tasty chicks.

Chick growth
It has long been known that immediately prior to hatching chicks communicate with the parent bird by bill tapping and also by producing soft noises (*e.g.* Brua, 2002). In fascinating work on eggs of the Australasian Pig-nosed Turtle (*Carettochelys insculpta*) Doody *et al.* (2012) found that vibrations within developing embryos were being transmitted to adjacent siblings and, therefore, throughout the clutch, and that these vibrations were stimulating synchronous hatching. Noguera and Velando (2019) extended this work to the eggs of the Yellow-legged Gull (*Larus michahellis*). They found that information was also being passed by vibrations between embryos. Eggs were removed from clutches in the wild and placed in incubators within a laboratory. Some eggs of a clutch were then exposed to the calls of potential predators, while others were not. The unexposed eggs were then replaced in the clutch. A later examination of the eggs showed that embryonic changes and the levels of stress hormones in all eggs were similar: vibrations within eggs were being used to pass information between siblings. Noguera and Velando note that in the wild developing embryos are subject to the alarm calls of adults and that such information passing between eggs may assist embryos at different stages of development and that this would explain why within just a few hours of hatching all chicks will crouch on hearing an alarm call.

Chicks can be heard pipping from the egg a day or so before hatching, though it is not known if this is to alert the female or a further attempt to synchronise hatching. Hatching is slow and exhausting, the chick chipping away at the shell with its egg tooth for up to 24 hours before being completely free: some chicks are so exhausted by the hatching process that they do not survive. The egg tooth disappears a week or so after hatching. On hatching, the chicks are semi-altricial and nidicolous, and weigh 14-18g. They are covered in a first, white, down which dries quickly once they are brooded by the female. The chicks are blind, the eyes opening fully only by the third or fourth day after hatching. However, the chicks can lift their heads to call for food and are voracious, to such an extent that from a very early age they will readily take any item of food they can reach irrespective of size, and attempt to swallow it. The female must therefore be very careful to provide, bill to bill, pieces which can be easily swallowed. Village (1990) notes that the chick's enthusiasm for taking anything that comes within reach occasionally results in prey items, *e.g.* legs, sticking out of their bills, and that the female has sometimes to retrieve these if they are not, ultimately, swallowed. The wonder of it is that more chicks do not die from choking. Since most chicks which die do so of starvation, competition between siblings does appear to involve survival of the fittest, but, as Shrubb (1993a) observed, some females carefully ensure that all chicks receive food. Shrubb suggests that this behaviour is likely to be more common when food is plentiful, competition and starvation resulting when food becomes scarce. The female is responsible for feeding the chicks when they are not able to feed themselves, but there are instances where the male has been known to raise chicks when the female has died. Tinbergen (1940) records one such example, with the male dismembering prey for 13-day old chicks and successfully raising them, while Riddle (2011) records a male raising five 10-day old chicks to fledge, with some assistance from human helpers, when the female died following a Tawny Owl attack.

The ravenous hunger which drives them to take any available food item means the chicks grow quickly, weight gain following a sigmoid (S-shape) curve until the young birds lose weight after fledging (Fig. 83). This S-shape is seen in many species and is independent of whether the chicks are altricial or precocial at hatch. The growth pattern was explored in two seminal papers by the American ornithologist Robert Ricklefs (1968, 1973). The 1968 paper covered 105 species, these including two falcons (Merlin and American Kestrel), and developed a mathematical formulation for the curve.

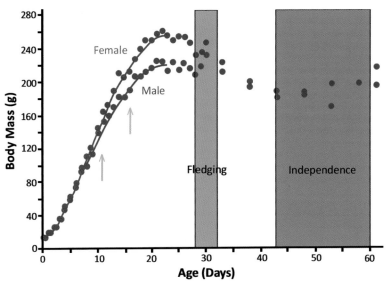

Figure 83. Mean mass, from hatch through fledgling to independence, for male and female nestlings from wild Kestrel nests in The Netherlands. Juveniles were trapped to measure masses at ages beyond fledging. The final masses beyond independence are 'standard' for adult Kestrels of a given maximum nestling mass. The green arrow indicates the point of peak growth, the orange arrow indicates the upper point of maximum curvature of the regression line, *i.e.* the point at which the growth rate starts to slow as the chicks approach the maximum mass they will achieve. Redrawn from Dijkstra *et al.* (1990a) with arrows from Steen *et al.* (2012).

More recently, Steen *et al.* (2012) have developed a mathematical model for European Kestrels which allows the point at which weight gain peaks (10 days) and the point at which the maximum growth period ends (15 days) to be identified. These points are identified on Fig. 83.

Chick weight doubles within two days of hatching, and rises steadily so that at about 15 days the chick is at 80-85% of adult weight. Weight gain then slows, the chick reaching 90% of adult weight at about 3 weeks of age: the parent birds adjust food delivery to suit chick demand. Steen *et al.* (2012) used video cameras to observe ten Kestrel nests in south-eastern Norway over a period of about 16 days per nest at varying chick ages. The monitoring times for the nests overlapped, but no single nest was monitored throughout the period from hatching to fledging. Overall a prodigious 3595 prey items were delivered (60.2% voles, 19.4% unidentified small mammals, 9.8% shrews, 4.0% birds, 2.7% lizards, 3.9% unidentified and 0.3% prey fragments: no insect deliveries were recorded). Steen *et al.* then calculated the mass of prey delivered to each chick as they grew (Fig. 84).

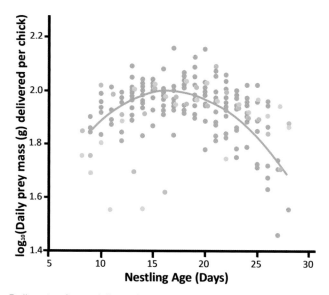

Figure 84. Daily rate of prey delivered to each nestling at 10 nest sites in south-eastern Norway. The regression line was computed from a statistical model of the data. Redrawn from Steen *et al.* (2012). Superimposed on the Norwegian data are data obtained from the study of breeding Kestrels set down in Chapter 8 for 2017 (Green circles) and 2019 (Turquoise circles). Note that, if needed, these circles are offset slightly to the right to avoid confusion with the Steen *et al.* data points. The green and turquoise circles show a similar pattern to the regression line, but with a greater spread than the Norwegian data, largely due to the lower points which represent days on which the adult Kestrels could not hunt. It seems the Norwegian Kestrels either had fewer bad weather days, or were better at coping with bad weather or, and perhaps more likely, had a more stable rodent population than the UK.

The Norwegian researchers also plotted prey mass delivery against chick growth to produce a fascinating association between the two (Fig. 85). Steen and co-workers found that maximum prey delivery peaked at a chick age of 15-17 days. However, they were unable to conclusively distinguish whether the decrease in prey supply was then defined by a lessening in demand by the chicks or a seasonal decline in prey availability. What the Norwegian team did notice was that prey size declined with chick age which they had not expected. They suggest this might be a result of the parent birds dismembering larger prey at first because of the small gape size of younger chicks, and then providing smaller prey when the chicks are first able to feed unassisted, smaller prey being easier to dismember.

In his study of Kestrels in The Netherlands, Tinbergen (1940) observed prey deliveries in Kestrel pairs throughout the breeding cycle (Fig. 86). Chapter 8 presents data covering a similar range for British Kestrels.

Riddle (2001) records 10 voles being delivered to a nest in six hours. I photographed six prey deliveries (four rodents, one lizard and one bird) to a brood of five 14-day old chicks during a period of 2hrs. 8mins. The female Kestrel stood guard for some, but not all of the time, so it is not clear what the total hunting time of the pair was, but there were five prey captures in the total time, *i.e.* one every 25 minutes including time to and from the hunt site and, on occasions, prey passes from male to female. This implies either a high hunting success rate or frequent hunting attempts with a lower success rate. Riddle (2001) records 28 voles being delivered to a brood in one day, a number which was replicated during the studies of Kestrel breeding in a barn in southern England in 2017 (see Chapter 8): in that case there were 26 voles,

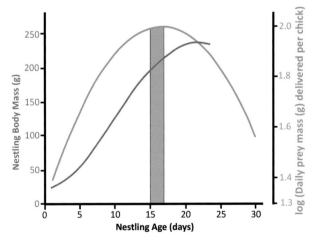

Figure 85. Growth curve of Kestrel chicks in comparison to the regression line of Fig. 83 for an average chick. The shaded area represents the period of maximum metabolisable energy intake from the work of Kirkwood (1981): while growth rate has slowed, the energy intake peaks as the chicks attain weights above those they will have as adults. Redrawn from Steen *et al.* (2012).

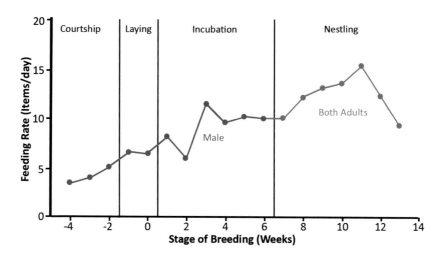

Figure 86. Feeding rate of the male to the female during courtship, egg laying and incubation and by the parent birds to the nestlings during the period to fledging. The horizontal axis is in weeks starting from '0', the point at which the clutch is complete. The data is from Tinbergen (1940) and is based on numerous Kestrel pairs in the wild. For further information on feeding rates see figures in Chapter 8.

one mouse and one bird. In 2019, 29 prey items were delivered on one day (27 voles and 2 birds). At a Kestrel nestbox in 2020, on a still day with ground mist there were no prey deliveries for 4 hours, then 4 deliveries 16 minutes, three by the female and one by the male, though one of the female's items had been taken from the male in the nest tree. At the same site on a later visit the male made 3 deliveries in 9 minutes, though it was not possible to say if any had been from cache.

The overall sex ratio of the chicks is 1:1, but in The Netherlands, Dijkstra *et al.* (1990b) noted significant variations in the ratio dependent on both the sequence of eggs and the date of laying of the clutch. Males dominated in early clutches, being about 58% of nestlings, but in later clutches the number of males declined to about 32%. In early clutches, eggs which would develop into male birds tended to be laid earlier in the clutch (eggs 1-3), while in later clutches 'male' eggs were laid later (eggs 4-7) (Fig. 87 overleaf).

That the sex ratio varies is not surprising as it has been observed in many other birds, with either males or females dominating in the later clutches of different species. But a more rigorous study of Kestrels in Finland (Korpimäki *et al.*, 2000) showed that with Kestrels the situation was more complicated than the Dutch team had suggested. Using blood sampling and analysis, Korpimäki *et al.* found that the sex ratio was dependent on food supply, with more male eggs being laid when food was scarce. Moreover, the sex ratio was also dependent on the condition of the parent birds (defined by weight .v. size), the proportion of males to females in a clutch declining as the condition of either parent declined. The probability of a yearling male breeding in the

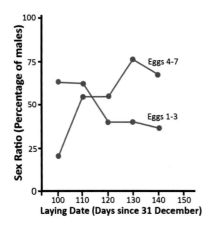

Figure 87. Variation of sex ratio (as percentage of males) with position of egg in the laying order and time of laying of first egg. Redrawn from Dijkstra *et al.* (1990b).

spring of the following year declines with birth date, whereas that of a yearling female does not, so if all other factors are equal, by producing male chicks early and females late, a female Kestrel is maximising the potential of passing her genes to the third generation. This effect was modelled by Daan *et al.* (1996) who concluded that the observed preponderance of male chicks born early and female chicks born late was in line with the evolutionary hypothesis that early breeding parents are usually in the best condition and territory, and will therefore produce an excess of the gender which profits most from an early birth date.

In general, the chicks are brooded more or less continuously by the female for about 10 days. The reason for this was established in a study of hand-reared Kestrel chicks by Kirkwood (1981) who noted that the chicks were largely inactive and produced little heat during first days after hatching, almost all food intake being converted into growth: weight gain also dominates plumage development, this being relatively slow which, of course, exacerbates heat loss unless the chicks are brooded. This lack of heat production means the chicks are unable to regulate their body temperature until they are about 10 days old[1] – second down only appears when the chick is about 6 days and seems essential for thermoregulation. In the absence of female brooding the chicks cool rapidly, cooling being accentuated by wind and rain.

The female usually feeds the meat of prey items to the chicks, contenting herself with the entrails and skin that remains, and so loses weight unless the male brings a surplus of food. After the 10-day period the females spends progressively less time brooding, preferring to stand guard from a nearby perch, though she may brood for long periods if the nest is exposed and the weather is wet or cold. Once she has reduced her brooding time the female starts to hunt, both for herself and the chicks, and the male starts to deliver prey directly to the nest.

In a study in Czechoslovakia (as it then was) Pikula *et al.* (1984) noted that prey was delivered at fairly regular intervals during the day, though with a slackening towards evening. In another study by the Dutch researchers at Groningen University (Masman *et al.*, 1988a), nestboxes were organised so

---

[1] Kirkwood (1981) noted that for laboratory-raised Kestrel chicks thermoregulation depended on ambient temperature and body weight. At 25°C it was achieved at a body weight of about 80g, *i.e.* age 8-9 days.

This Kestrel chick, aged about 14 days, still has its egg tooth which is usually lost by 11 days of age.

that prey deliveries could be observed. The researchers then varied the number of nestlings in the nest. Adding more chicks had little effect on male prey deliveries when the chicks were less than 10 days old, but once they were older and the male was delivering prey directly to the nest, the delivery rate increased, presumably because the male was able to judge the hunger of the chicks. When the chicks in a brood were given extra food, the male reduced prey deliveries, reinforcing the idea that he could gauge chick hunger and alter his hunting behaviour appropriately. Further information on prey deliveries will be found in Chapter 8.

Initially the chicks can do little but move their heads to accept food, but once they are mobile they grab at food to forestall attempts by siblings. However, it is not until Day 20 that the characteristic mantling of food is seen. During periods when they are not eating or competing for food the chicks exercise their wings and practice their killing skills with attacks on remnant prey items. They also stare from the nest site, seemingly taking in the local area as though composing a mental map: if that is the case then it is likely to be time well spent when the time to leave the nest and start exploratory hunting arrives. As the chicks age they become more aggressive to the parent birds bringing food, the parents no longer attempting any feeding, often spending minimal time at the nest apparently to avoid injury. In addition to the female's pellets which accumulate at the nest, the chicks also produce pellets as soon as they ingest fur and feathers as well as meat. Once they are mobile, the chicks move to the edge of the nest to excrete, but their aim is not always true, and with prey items occasionally being cached or left uneaten, over time the nest, particularly the edges, becomes rank.

Above First hatch of a clutch of five. The chick probably hatched within the last hour.

Below Dead hatchling removed from a nestbox. The chick probably died aged about 36 hours.

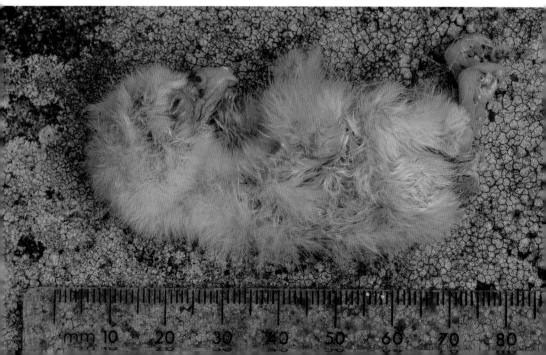

Above   Kestrel chicks in a corvid nest at 2 days of age.

Below   Kestrel chicks at 6 days of age.

Above   Kestrel chicks at 16 days of age.

Below   Kestrel chicks at 23 days of age.

Kestrel chicks at 28 days of age. With five near-fledge chicks the corvid nest is only just big enough (*above*), wing flapping causing disapproving looks (*below*).

During the early days of chick feeding the female Kestrel's feet can become coated with fur, feathers and the mashed up remains of other detritus in the nest. The coating is worse in wet weather, and particularly bad in nestboxes before the chicks are old enough to defecate through the box opening.

While most interest lies in the development of plumage on the chick, it is worth noting that growth is highly asymmetric, the legs and feet being near adult size before the flight feathers emerge. The initial white down of the chicks is replaced by a coarser grey down 6-7 days after hatching, this new down emerging between the first down plumules. At about the same time the sheaths of the primaries emerge, followed by those of the tail 2-3 days later. The primaries erupt from their sheaths on Day 13 by which time the chicks take food from the adults and eat it themselves (Village, 1990). Contour feathers start to grow at about 8 days of age, emerging from the follicles of the first down, pushing it out. Table 7 sets down the development of an 'average' Kestrel chick. Figs. 88a-d (overleaf) illustrate the development of feathers and the tarsometatarsus in chicks.

Fledging can occur with chicks as young as 28 days: with some clutches it may take about 35 days. In a study in Finland, Massemin *et al.* (2002) found that while early growth rates of late-hatched chicks in asynchronous broods were slower, by the age of 26 days these chicks had attained the weight

Breeding Part 2

Age of Chick	Development
0	Eyes closed or partly closed. Covered in white down. Wings used to support body in sitting position when gaping for food, though begging not necessarily directed at adult. Food call is whining whistle. Will make a barking cheep if uncomfortable, *e.g.* when too cold or hot.
2-3	Eyes fully open. No longer uses wings for support when feeding. Excretes by raising rear end.
4	Preens down on breast, back and wings. First pellet regurgitated.
5	Scratches head with foot.
6	Follows movement and so can beg to female.
8	Second down emerges. Grabs at prey. Moves to back of nest to excrete.
11	No longer brooded except in extreme circumstances. Preens tail stump. Egg tooth lost. Tail quills emerge.
12	Bobs head while watching movement, *e.g.* flies.
14	Cleans talons.
16	Stands up.
17	Tears at prey with talons. Can now stand and wing shake, but unsteadily.
20	Achieves full body weight. Reacts to call of adults before they arrive. Mantles prey.
21	Can feed without help, holding food between medial talons and tearing pieces off.
22	Stands on one leg while stretching the other leg and one wing. Co-ordinated wing flapping.
24-26	Preens as adults do. Down starts to shed as body feathers emerge, down loss aided by regular preening. A third down starts to grow: this reaches its maximum length (of about 12mm) when the bird is 6-7 weeks old. At this point the plumage has reached maturity pre-first moult.
29	Jump/fly 1-5m. Primary and tail feathers almost full-grown. Vigorous wing flapping often accompanied by excited *kee-kee-kee*.
31	First real flights, up to about 50m, but unable to take off from the ground.
35	Down almost gone. Start to leave the nest for periods of time.

Table 7. Constructed from data in Village (1990), supplemented with data collected in the field.

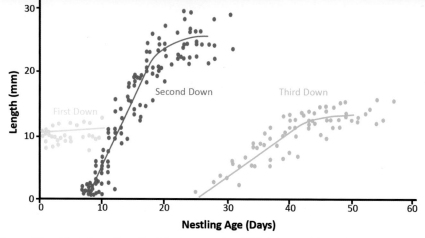

Figure 88a. Variation of length of first, second and third down with nestling age. Redrawn from Kirkwood (1980). Regression lines drawn by present author.

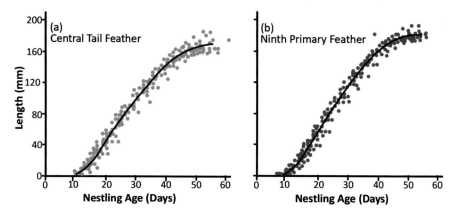

Figure 88b. Variation in length of (a) central tail feather and (b) ninth primary feather with nestling age. Redrawn from Kirkwood (1980). Regression lines drawn by present author.

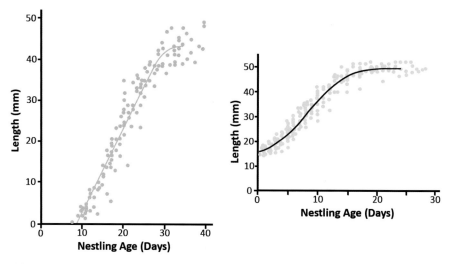

Figure 88c (*left*). Variation of length of dorsal contour feathers with nestling age. Redrawn from Kirkwood (1980). Regression lines drawn by present author.

Figure 88d (*right*). Variation of length of the tarsometatarsus with nestling age. Redrawn, including regression line, from Kirkwood (1980).

of their siblings. When they leave the nest, young Kestrels are heavier than adult birds, but lighter in weight than a maximum achieved a couple of days prior to fledging (see Fig. 83). In most observations the young stay reasonably close to the nest after first leaving it, still being dependent on their parents for much of their food. When delivering food to fledglings that have left the nest the adults call as they arrive, prompting reply calls that direct them to where a chick is sat. As soon as their flying techniques allow, the chicks will fly up to meet the returning adults, allowing them to gain an advantage over their siblings. The young bird flies to the adult, then rolls sideways to take the prey item from the adult's talons. The period over which the juveniles require food from the adults (or over which the adults continue to provide it as the two periods may not necessarily be the same) varies considerably. Masman (1988a) noted times from 14-32 days, consistent with a study by Boileau and Bretagnolle (2014) in western France who found 3 to 31 days with a mean of 18 days for 25 Kestrel fledglings. The French team noted that as dependency on the adults reduced, the young Kestrels moved further from the nest site (Fig. 89). The mean distance travelled by 22 young falcons at the end of dependency was 372m, but the standard deviation on this was large (±243m) as one bird had travelled 1000m – this is clearly shown by the single arm deviation plotted in Fig. 89: the minimum distance travelled was 50m. In a study of colonial breeding in Spain Bustamante (1994) noted that the

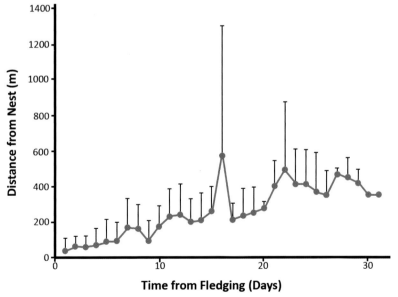

Figure 89. Average dispersal distance travelled from nest by juvenile Kestrels in relation to time after fledging in a non-colonial breeding area of western France. Vertical symbols are one standard deviation. Tracking of the birds was by observation of colour bands and from attached radio transmitters. Redrawn from Boileau and Bretagnolle (2014).

young infrequently returned to the nest area once they had left it, perching on ledges or the top of the colonial cliffs and sometimes gathering in groups of unrelated birds. In this case the mean dispersal distance (over a period of about 4 weeks) was only 37m (but with distances to 252m). The fledglings also stayed closer to unrelated young birds than they did to their siblings. It is not clear whether this behaviour is specific to colonial nesting.

Boileau and Bretagnolle (2014) also noted that most feeding was by the male falcon, females stopping their contribution to post-fledge feeding after only three days (Fig. 90). However, this finding is at variance with other studies which suggest continuing female involvement. Interestingly, Boileau and Bretagnolle noted that there was a positive correlation between the body condition (as measured by wing length and weight of the chicks at age 31 days) of the youngsters and the time of continued feeding – the male bird was preferentially feeding those youngsters which appeared to have a better chance of survival (Fig. 91).

Boileau and Bretagnolle noted that the decision by the male to continue feeding a fledgling did not appear to be influenced by the hatching order of the chicks. The positive correlation between post-fledge feeding and body condition seen by the French researchers recalls previous work on Kestrels in Spain. There, Vergara and Fargallo (2008) noted that sibling rivalry for food in the post-fledging period differed from that between nestlings, with male chicks tending to obtain larger prey items delivered by their parents, principally because larger items were left on the ground for retrieval and males had superior flying abilities. In addition, males with greyer rumps fared better,

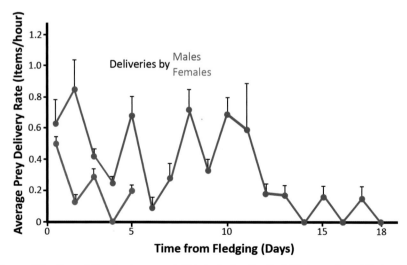

Figure 90. Prey deliveries by parental birds during the first 18 days of post-fledge dependency of juvenile Kestrels in western France. Vertical symbols are one standard deviation. Redrawn from Boileau and Bretagnolle (2014).

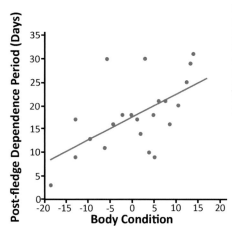

Figure 91. Variation of post-fledge dependence period on body condition (defined by considering wing length and body weight at age 31 days: negative numbers are 'poorer' condition, positive numbers are 'better' condition) in 22 juvenile Kestrels in western France. Redrawn from Boileau and Bretagnolle (2014).

in terms of larger prey, than male siblings with browner rumps. Vergara and Fargallo suggest that grey rumps are an indication of better body condition and that this enables the grey-rumped youngsters to out-compete brown-rumped siblings. This finding might also explain the result of Boileau and Bretagnolle (2014) as coloration would allow adult male falcons to more easily distinguish better quality offspring.

In a later study in Spain Vergara *et al.* (2010) again showed the importance of fledgling male coloration in the post-fledge period, but also noted that the 'quality' of the parent birds (quality being measured by female body mass and earlier laying date) was important in defining the length of the post-fledge feeding period. 'Better' parents feed post-fledge youngsters for longer, particularly in scarce prey years, though even then, 'better' juveniles were fed

Female Kestrels in nestboxes can stop brooding their chicks when they are 7-10 days old, but this basket nest on the Pentland Hills was exposed to rain and wind, and the female kept brooding until the chicks were 14-16 days old, though the job became increasingly difficult, the youngsters being both larger and more active.

preferentially. The work in Spain on Kestrel coloration was later extended to female Kestrels by López-Idiáquez *et al.* (2016). The Spanish team placed stuffed females close to (within 10m) of the nestboxes of breeding females. Some of these decoys had grey rumps, while others had brown rumps. What was clear was that breeding females reacted more aggressively towards brown-rumped decoys, implying that grey-rumped females were also indicating higher 'quality' and so were less likely to be attacked. López-Idiáquez *et al.* also noted that females with larger clutches reacted less often and less aggressively to decoys, a finding which would be expected: attacking another female is a high-risk strategy and the larger a female's clutch the more she has to lose if an attack results in injury.

From about 10 days after leaving the nest, juveniles start to catch insect prey on the ground, soon graduating to catching flying insects. Most observations suggest the juveniles start to flight-hunt about 20 days after leaving the nest, but Bustamante (1994) observing fledglings reared in a colonial nesting area in western Spain noted flight-hunting at a mean of only 9.7 days: early attempts were close to the ground and frequently unstable. The juveniles become proficient at the technique about four weeks after leaving the nest. Fledglings also play, with sticks or feathers, using these as proxy prey which are duly 'killed': fledglings will also 'pluck' prey items with their talons. Juveniles also interact with siblings (see Fig. 92), interactions including 'beaking' (a mutual grooming-like activity most often seen when one fledgling picks down from

Figure 92. Activities of juvenile Kestrels in relation to time after fledging in a non-colonial breeding area of western France. Vertical symbols are one standard deviation. Activities other than hunting and feeding, but involving siblings, were defined as 'socialising'. Redrawn from Boileau and Bretagnolle (2014).

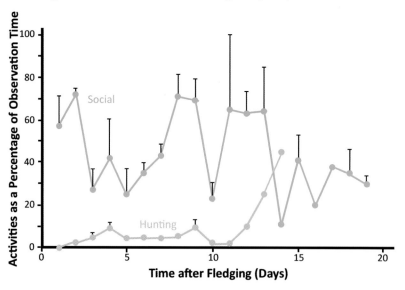

One near-fledgling Kestrel 'beaking' a sibling.

another's head), but more often involving mock attacks on a perched sibling or high-speed chases, such chases sometimes including more than two birds.

In another observed incident one boisterous youngster stooped repeatedly on another, perched, fledgling (it was not clear whether the perched bird was a sibling or from another nest), one mock stoop getting out of control and involving the youngster in a very precarious looking braking manoeuvre, after which the mock stoops ceased for the remainder of the observation - see photos on p262. Similar mock (or assumed mock) attacks by a juvenile on an adult female have also been observed (Oakley-Martin, 2008). Juveniles also occasionally interact socially with their parents. Shrubb (1993a) reports a delightful observation from Gwent, Wales in which the adult birds spent about 20 minutes dropping sticks and balls of sheep's wool for their youngsters to chase and catch (see Hewitt (2013) for more general thoughts on prey dropping).

At this corvid nest in southern Scotland five chicks had reached 28 days of age and would fledge within the next two days. The restricted space in the nest meant that when prey was delivered there was pandemonium, the frantic scrabbling towards the adult, with legs and talons extended, meaning the female did not actually land on the nest edge, merely throwing the food and flying off to avoid injury.

Breeding Success

Multiple factors are involved in determining whether the path adult Kestrels take from pair formation to seeing fledglings hunt for themselves is smooth, one of the more interesting being whether the falcon population of a given area is sedentary or migratory. In England (Village, 1990) and Germany (Kostrzewa and Kostrzewa, 1997) where the population is sedentary the number of territorial pairs which failed to lay eggs was high (13%), while in the partially migrant population of Scotland it was significantly lower (6%: Village, 1990; 7%: Riddle 1987) and in the completely migrant population of Finland it was lower still (3%: Korpimäki and Norrdahl, 1991). Examining what lies behind this increasing failure rate as populations become more sedentary requires consideration of several possible inputs.

The influence of winter weather on the number of breeding pairs in an area has already been mentioned in the previous Chapter where the studies of Kostrzewa and Kostrzewa (1990, 1991) in Germany were considered. Those studies noted that rainfall had a negative effect on Kestrel breeding, while fledging success was positively correlated with the temperature in May and June (see Chapter 6 *Fig. 63*). In the 1991 paper the German researchers also indicated that while the territorial density of Kestrels was independent of 'vole rank' – the vole population being ranked as either 'low', 'intermediate' or 'high' – the breeding density (*i.e.* the number of Kestrel pairs which laid eggs) was correlated with vole population. As would be expected, the researchers also found that the number of fledged chicks from a Kestrel pair was dependent on the vole population.

Various studies in Britain, Finland and The Netherlands also noted a relationship between the laying date of the first egg and the success of the pair in fledging young (*i.e.* of breeding success) and the number of fledged young (Fig. 93). As already noted above, the study of the timing of the first egg and subsequent clutch size has been much debated, but it is clear parental 'quality' is influential, the poor quality of either or both parents influencing the success of the pair in producing any chicks, and in nurturing those chicks that do hatch through to fledging. Overall, Village (1990) showed that 30% of Kestrel pairs failed after laying eggs and that 32% of successful pairs lost either eggs or chicks.

Village (1990) noted that complete breeding failure was more frequent during incubation than after the chicks had hatched, the most likely reason being nest desertion, this occurring in over 50% of nests in each of his English study areas, and in over 70% at his Scottish study area. Usually the female abandons the nest if she is receiving inadequate food from the male. Cavé (1967) confirmed what is intuitively obvious, that this is most likely due to the male being unable to catch sufficient prey in poor prey years, or being unable to hunt in poor weather. However, the fact that nest failure is positively correlated with time of laying indicates that parental quality is also a factor. Human interference and predation are also likely contributors, while historically, egg

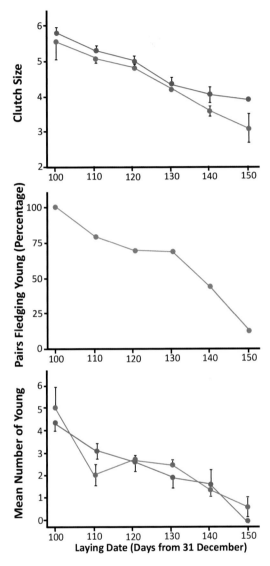

Figure 93. Decline in mean clutch size, mean number of chicks fledged per laying pair and percentage of pairs fledging young with laying date of the first egg. Blue circles are data from southern Scottish grassland, red circles are data from east-central England farmland, and purple circles are combined data from the two sites. In all three cases, data are for a 10-day laying period. Redrawn from Village (1990).

addling and breakages due to organochlorines were undoubtedly significant factors.

From his studies, Village (1990) considers that about 50% of Kestrel eggs laid failed to produce fledged young, with the bulk of failures (about 70%) occurring at the egg stage. Egg failures were mainly due to eggs not hatching (76%), though there were also losses from broken eggs (9%) and eggs falling from the nest (5%): in 10% of cases the cause of failure was not identified. Losses among chicks were 28% at hatch, 5% in falls from the nest and 55% where the probable cause was starvation: in 13% of cases no cause was identified. Village notes that most unknown chick losses were within 10 days of hatch, and that dead chicks were likely to have been eaten by siblings or parents, a rather sad and macabre occurrence. The statistics of Village are in reasonable accord with those of Shrubb (1993a) who studied the records of Kestrel breeding in Britain during the period 1950-1987, a total of 1350 eggs in 308 nests. Shrubb found 48% of eggs failed to produce fledged young, with 64% of the failures occurring at the egg stage. Shrubb's analysis also confirmed that the bulk of losses were due to non-hatching, either because they were infertile or had been deserted. However, Shrubb also found that the theft of Kestrel eggs by man was a

significant cause of failure, as was the taking of chicks. Shrubb quotes an analysis of Mather (1986) who noted that the number of hatchlings removed from the nest by would-be falconers had increased dramatically after the film *Kes* was released in 1969.

However, the data on fledglings/laid egg of Village (1990) and Shrubb (1993a) are not borne out in the study of 2013-2019 on Scotland's Pentland Hills by my colleagues Graham Anderson and Keith Burgoyne who found much higher breeding success in all those years (Table 8).

Year	2013	2014	2015	2016	2017	2018	2019
Fledglings/laid Egg	0.63	0.79	0.76	0.52	0.90	0.80	0.82

Table 8. Breeding success measured as number of chicks fledged per laid egg.

The curiosity of the data of Table 8 is the high values in 2017 and 2019 relative to 2018. 2017 was a year of prolonged poor weather, while 2019 was a year of poor vole numbers. However, the breeding density in both 2017 and 2019 was low so that, perhaps, only the fittest, most diligent Kestrel pairs bred and these were able to breed successfully. 2018 was one of the best summers for years in the area with continuous days of sunshine and warmth, the weather aiding a vole density peak. But while breeding conditions appeared ideal, the Kestrel breeding density was twice that of 2017 or 2019 so competition for nest sites and prey increased. This competition was not solely between Kestrel pairs, as some nestboxes were taken by Tawny Owls and some by squirrels. Table 8 also suggest that breeding success is higher on the Pentlands

The position of the radio-controlled camera meant that the female Kestrel delivering breakfast to her chicks is barely visible in the photograph (*opposite*), though breakfast, a Field Vole, can be seen. In the photograph (*above*) one hungry chick is anticipating he will be next to be fed.

than in other areas. My colleagues and I consider this is primarily due the installation of nestboxes. In 2017 those Kestrel pairs in stick nests lost chicks due to exposure, the continuous rain and chilling winds taking a heavy toll. By contrast chicks in nestboxes were protected from the elements, though, of course, the adult birds still had to contend with the difficulties of hunting in poor weather. Chicks in nestboxes were also protected, to a greater degree, from predation.

One interesting aspect of Village's (1990) study was that the productivity of the Kestrels – *i.e.* the number of chicks raised per unit area of countryside – was positively correlated with breeding density. This implies that when conditions are good – fine weather and high vole population – more Kestrel pairs can breed early, clutches are larger, and the survival rate of chicks is high. Of course, this relationship cannot continue indefinitely, but the Kestrel breeding density is relatively low and the rodent population cycles seen in Fennoscandia rarely occur. However, even at low breeding densities, an effect might be expected within nests if direct competition between siblings led to higher chick mortality, but Village did not see this and neither did Cavé (1967) in his study of brood survival rates in The Netherlands.

A final factor worth considering in the survival of chicks is the experience of the adult pair, and here the data of Village (1990) showed a clear effect both in terms of parental age, and whether the adults were newcomers to an area or had bred there previously. As might be expected older adults and pairs who had previously bred together in the same area, bred earlier than other pairs, first-year male and first-year female partnerships bred later, as did pairs new to an area. As earlier breeding leads to higher clutches and better chick survival

In a year of peak vole population Gordon Riddle took the photograph to the left of 33 dead voles and a couple of mice piled up around a brood of well-fed Kestrel chicks. The study of Chapter 8 did not see anything as excessive, but there were frequently dead rodents lying on the platform and within the nestbox. The photograph to the right, taken at about midnight one day when the chicks were near fledgling shows five dead rodents on the platform.

rates, the best a Kestrel chick can do is to ensure his parents are older birds, at home in the area and with each other. Bad news for the chick is a pair of first-year birds who have just moved in. Sadly, of course, Kestrel chicks find choosing their parents no easier than the rest of us.

Nest, Territory and Mate Fidelity

In considering fidelity in both birds and animals, Greenwood (1980) distinguished between *natal dispersal*, in which birds move from their nest site to a first breeding (or potential breeding) site, and *breeding dispersal* in which birds which have bred move to a different site the following year. In each case, Greenwood noted that while many species of animals are faithful to their natal area, in birds the tendency is for males to be more philopatric than females while the reverse is true for many mammals. Although Greenwood

Strenuous wing-flapping by older chicks can produce clouds of down mixed with dust created by the break-up of the keratin sheaths of emerging feathers.

Above A near-fledgling chick demonstrates the way to consume a Field Vole, watched by an envious sibling. Unless they are very hungry or feel they might be the victim of piracy, most adult Kestrels do not eat voles this way. It is possible that chicks are both keen to take in as much food as they can, but also that swallowing is a way of preventing siblings from attempting to share the prey.

Left This near-fledgling grabbed a Common Shrew from a sibling, and to make sure it would not be grabbed back started to swallow it. But having grabbed the rear end he swallowed it tail-first. The back legs have already disappeared, but the front legs seem about to poke the youngster's eyes out. In fact the chick did manage to swallow the shrew.

included only Merlin and Hobby in his analysis, the fact that in those two and most other avian species males are more philopatric than females implies that Kestrels will likely share the philopatry of their falcon cousins. That the degree of philopatry differs in the two sexes in either case reduces the likelihood of incest, which is clearly beneficial to the species. There is also evidence that female birds prefer philopatric males because occupation of a territory suggests both a willingness to defend it and its potential suitability for hunting, each a good sign for successful breeding. Knowledge of a territory also implies some understanding of local predators which is equally advantageous.

These issues were neatly illustrated in an experiment in central Finland (Hakkarainen and Korpimäki, 1996) with three owl species – Eagle Owl (*Bubo bubo*), Ural Owl (*Strix uralensis*) and Tengmalm's (Boreal) Owl. Of the three Tengmalm's is by far the smallest and is predated by the two larger owls. However, the larger owls cannot access the small tree holes used by breeding Tengmalm's to take nestlings. Tengmalm's Owls are also out-competed by Ural Owls for the small rodents on which both prey. Eagle Owls take

The Common Kestrel

much larger prey and so are not direct competitors for resources. By erecting nestboxes in Eagle and Ural owl territories Hakkarainen and Korpimäki were able to show that the breeding of Tengmalm's Owls was less affected in Eagle Owl territories than in territories of Ural Owls where the smaller owls suffered both predation and competition for prey. Most breeding attempts of Tengmalm's Owls near Ural Owl territories failed during courtship and those that succeeded saw clutches laid 11 days later, a delay which would inevitably have resulted in lower fledgling rates. Tengmalm's Owl pairs close to Ural Owls invariably consisted of younger males and females: experienced males were taking the best sites and experienced females were recognising the advantages they offered.

In his studies in the UK, Village (1990) identified factors which influenced the fidelity of Kestrels, both to nest area and to mate. His two study areas were very different, in Scotland most of the Kestrels were migratory, only 11% being resident, with 66% of birds being summer-only visitors: there were no winter visitors. In England 52% of the birds were resident, the area attracting fewer summer-only and more winter-only visitors than in Scotland. One consequence of the non-resident nature of the Scottish birds was that most individuals bred only once (70% of males, 80% of females) though Village did see a small number breeding in successive springs. In England, with its higher resident population, the number breeding in successive seasons was also higher, around 50% of males and 40% of females, with males breeding in up to six successive seasons, and females in up to seven (in a seven-year study).

Female Kestrel observing her chicks. The chicks are near-fledge and the female has not brought prey. Instead she stared at the chicks for some time, then left. It appeared she was telling them that if they wanted another meal they would need to fly. Photographs of more definite behaviour by a male is shown in the next Chapter, while the most extreme behaviour by a female is shown on the opposite page.

On this occasion the female arrived with a mouse. There were four near-fledged chicks, two on the box roof, one in the box and one on a branch off-camera to the left. All were excited to see her arrive and anxious to be given the prey. But she landed on a perch at the box and in full view of the four she began to eat the mouse. She ate about one-third before flying off. The lesson seemed clear – follow me if you want the rest of the prey. Three chicks fledged that day (one getting prey from the female high at the back of the nest tree an hour or so later). The chick in the box fledged the following day.

Playtime with a sibling almost ends in disaster. Perched high on a factory building a chick dived towards a sibling (*left*), but lost control as the speed increased (*right*), and only managed to avoid plummeting into the ground at the last moment.

In general, Village noted, those birds which successfully bred were more likely to return to the area, a feature which was equally likely for both sexes in Scotland, but more likely for males in England. It would therefore seem that the more likely a bird was to remain in a breeding area, the more likely it was to be breed there, suggesting territorial and, perhaps, nest fidelity, and, though less likely, mate fidelity. For the Scottish birds it was much less likely that either would return, and much less likely that they would mate with the same partner if they did. However, in England where many more birds were resident, the percentage of birds mating again with the same partner was not significantly different, suggesting that migration did not influence mate fidelity to any great degree.

More significant in determining mate choice was the success of the previous year's breeding. This was particularly the case for English birds, though as Village points out, this result may have been due to the much smaller data set for the Scottish area masking a similar effect. For his English birds, Village found that if a nest failed, 76% of males and 70% of females changed partner (though the difference in the percentages between the sexes was not significant, *i.e.* males and females were equally likely to change). If the nest was successful, 58% of males and 63% of females stayed with the same partner. From Village's work it can therefore be confidently concluded that if a Kestrel pair are successful in breeding, then provided that both birds meet again the following spring they are more likely than not to remain together.

Breeding Part 2

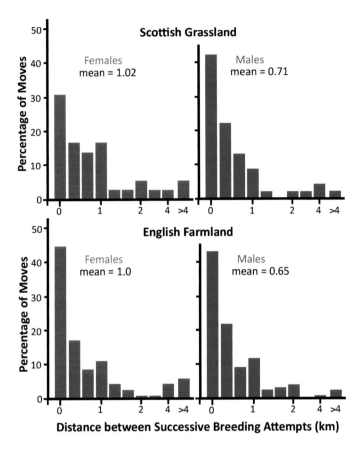

Figure 94. Distances moved between successive breeding attempts by female and male Kestrels in southern Scotland and east-central England. The distances were similar for both sexes, but in general females moved further than males. Redrawn from Village (1990).

In his study of Kestrels in Ayrshire in 1980-1992, Riddle (2011) noted that at one site, a female returned in three successive years, while another returned three times in four years. However, Riddle found that generally females returned no more than twice, and often did not return at all, data which is consistent with Riddle's view that the life expectancy of adult Kestrels was four years: his results suggest that Kestrel mortality may have a significant effect on partner choice.

Village (1990) also considered nest site fidelity in more detail (Fig. 94), finding that movements averaged 0.7km for males and 1.0km for females (combining data from both his Scottish and English sites, which showed similar numbers). These distances included birds which used the same nest site in successive years. If those data were excluded, movements rose to 1.1km

The ultimate in nest fidelity? Kestrels had used a ledge on a steep, part-vegetated coastal cliff in south-west Scotland for many years. When the local council installed plastic netting to prevent rock falls, the falcons continued to use the ledge, squeezing through the netting to reach the site and successfully raising chicks. *Gordon Riddle*.

for males and 1.6km for females. Older birds were less likely to move between breeding attempts, especially older females, where 47% moved after their first breeding season, but only 29% after subsequent ones. These percentages imply a high degree of fidelity not only to a nest site territory, but to a nest site. While for some pairs this is due to the number of nest sites being limited: personal observation suggests that fidelity to a nestbox, rock cleft and even to stick nests if they remain serviceable, is high. As noted in Chapter 6 (and will be mentioned again in Chapter 11) both Mingju E *et al.* (2019) in eastern China and Sumasgutner *et al.* (2014b) in Finland have noted that Kestrels appear to obtain information on the quality of a nestbox, in terms of breeding success, from the detritus left from the previous year, enhancing fidelity.

The nest site fidelity of urban Kestrels has already been noted (Chapter 6 *Nest Sites*). That this may be, at least in part, territorial fidelity was illustrated by a study of male Kestrels overwintering in a town in the Czech Republic (Riegert and Fuchs, 2011). Winter snow cover made rodent predation difficult, but some male Kestrels did not migrate, choosing to stay in the town, supplementing their diet with birds and insects, and roosting near the nest sites used during the previous breeding season. This allowed the males to be in a good position to defend their territories and so acquire mates early in the following breeding season. In Britain one urban nest site, a church window ledge, is known to have been used over a period of at least 40 years (C. Dobbs, pers. comm.). Clearly that time period, many Kestrel lifetimes, means numerous changes of birds, especially as there was no reliable data for successive breeding by the same pair, or for individual birds. In the study of Chapter 8 the same nestbox was used in four successive years, but the box it replaced had been used for six years.

The box top was the prized position for this brood. Getting there was easy if you could use a sibling's head as a step. Getting down again was also easy if the sibling was still available. The crop on the chick in the box suggests it was the one that received the last prey delivery.

Watched by its siblings, one chick's attempt to jump from perch to box ends in disaster (*above*). The chick landed on a slender branch about 3m down and 3m from the ground (*below left*). After a climb an alpinist would have been proud of (*below right*) the chick ascended the trunk to where it split in two, then the left trunk to a point above the box. From there it attempted to jump to the box roof, failed to cling on and fell again. But one outstretched foot caught the box edge and the chick regained safety (*photos opposite*).

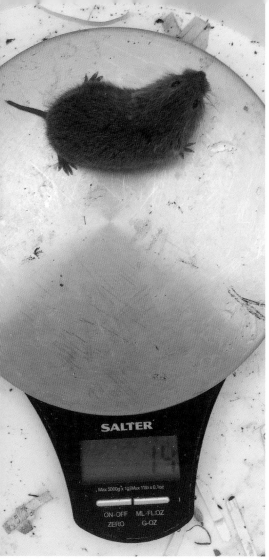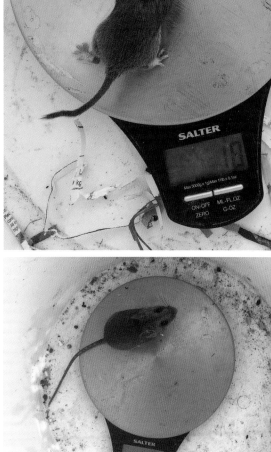

Trapped (and later released) prey species from across the years. A Field Vole *(top left)*, Bank Vole *(top right)* and Wood Mouse *(right)*.

and a five-egg clutch was completed on 25 April. Given that hatching was more or less synchronous, four chicks having hatched by the early evening (18.04) of 24 May and five by 15.44 on 25 May, it would appear that the female began incubation after the fourth egg had been laid, probably at some time on 24 April. That would suggest an incubation time of 30-31 days, consistent with the mean 30.8 day suggested by Village (1980).

To ensure minimal disturbance of the female during the early period of incubation, the camera system was not switched on until 2 May. Thereafter

The adult pair from 2018.

at the opposite end of the barn. The position of the computer was chosen to allow unobtrusive access to it. Elsewhere, stands for motion-activated trail-cameras, were mounted on barn support beams to provide back-up/additional information.

2017

During March 2017 a pair of Barn Owls showed an interest in the box, as did both feral pigeons and Wood Pigeons (though the latter were much more interested in the box top than the interior), but by early April a pair of Kestrels had taken possession. Neither male nor female falcon had been ringed. The male Kestrel was first observed presenting prey (a Wood Mouse) to the female at 12.41 on 6 April. The female inspected the box, and was presented with more prey, this time a vole, at 14.20. Over the following two weeks the male spent time each day at the box. The female was present less often, but did eventually start to roost close by. She also made attempts to dissuade the Wood Pigeons from their continuing interest in the box top.

Because of concerns that the new box might not be to the liking of the falcons it was decided not to install a nest camera. The interval between egg laying cannot therefore be established, but the first egg was laid on 17 April

8   Four Years in a barn in southern England

by L. Newberry, S. Newberry, N. Sale and R. Sale.

A short report including data from this Chapter for 2017-2019 can be found in Sale *et al.* (2020).

As an aid to studying the breeding behaviour of Kestrels a project began in 2017 in a barn in southern England. Several years previously an old wooden box had been placed high in the rear of a barn in the hope of attracting birds to breed. Barn Owls were expected, but it was Kestrels that took possession of the box and bred successfully. Over the next few years Kestrels returned regularly. In 2017 a new box was erected in place of the now somewhat time-worn original. The barn is large and has an open front. Bird access is therefore easy, but the position of the box, high and at the back of the barn, means it is well-sheltered from wind and rain. The new box had a landing platform separated from the breeding area by a small upstand designed to prevent active young chicks from falling. Two perches were added, as was a translucent 'roof' section to allow enough light into the box for photography of chick-rearing. An event-recording video system was installed about 2m from the box, suspended from the barn roof and connected to a computer system located

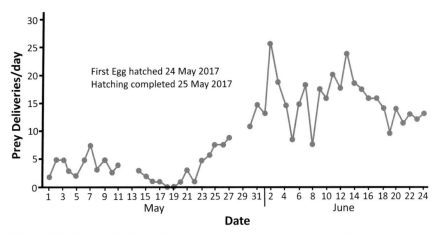

Figure 95. Prey deliveries by the male to the female during the last three weeks of incubation and by the two adults during chick rearing in 2017. The two episodes of power failure during which data could not be recorded can be clearly seen. The chick death occurred on the morning of 2 June.

surveillance was continuous apart from two unfortunate periods when the video system was disabled. Early in the incubation phase the power supply to the barn (in fact all local premises) failed. Data collection was therefore not complete during the period 12-13 May, and such data as was obtained on those days has been ignored. The power supply problem resulted in the local supply company deciding on an upgrade. During this, a further interruption occurred during 28-29 May. This was much more serious as it occurred soon after chick hatching: again data were lost on both days.

The complete breeding period of the Kestrels coincided with almost continuous good weather, with benign hunting conditions in terms of wind speed, temperature and rainfall, and associated visibility. Weather data were collected hourly throughout the breeding period from Middle Wallop Meteorological Station 11km from the barn.

Observations early in the breeding season indicated the primary hunting area of the Kestrels. In these areas live traps were set up to allow data to be collected on the species comprising the local mammal population, and their weight. The traps had food, water and bedding to minimise the trauma of the collected animals. The traps were visited regularly, the animals being weighed and released immediately. The barn video footage allowed the species of prey being brought to the box to be identified in most cases, and its size to be assessed. Combined with the weight data from the trapped mammals this allowed an estimate of the weight of prey delivered to the chicks.

Incubation
Fig. 95 shows the prey deliveries to the breeding box by the male during the observed incubation period (*i.e.* 2-24 May), and by the adult pair during

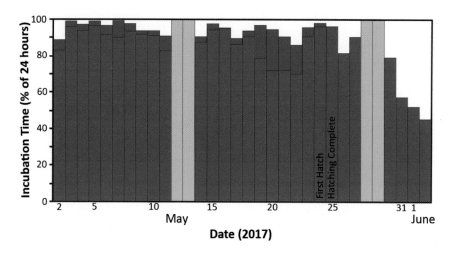

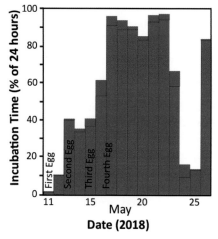

Figure 96a and b. Percentage of time the female (red rectangles) and male (blue rectangles) covered the eggs during the latter stages of incubation in 2017 (*above*), 2018 (*right*) and 2019 (*opposite page*). For further information see caption on opposite page.

the chick-rearing phase. Fig. 101 is a comparison of deliveries during 2017 and 2019. Fig. 96a shows percentage of time the female spent incubating the clutch and the time the male covered the eggs. Incubation times for 2018 and 2019 are shown in Figs. 96b and 96c for comparison.

During incubation the male delivered 60 prey items to the female in 20 days (3 rodents/day on average). However, during the same period there were numerous occasions when the female left the box, sometimes for extended periods and so could have been hunting for herself or exchanging prey outside the barn. On 4 May, a day when the male delivered three prey items, the female left the box seven times (three times with the delivered prey, and on four other occasions). On average that day, the female spent 16.3mins away from the box after taking delivered prey, and 12.5mins away from the box at other times, but on two of the four occasions when she left the box without a prey exchange, she was away for 17 and 20mins. On these occasions there

Four Years in a Barn

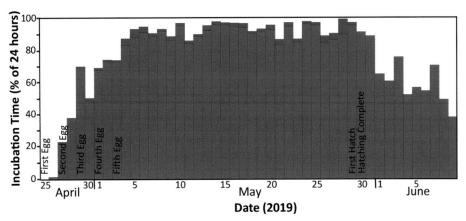

Figure 96c. Percentage of incubation time by female and male Kestrels during 2019. For 2017 and 2019 the time the female spent with her brood post-hatch has been included to the point at which she was no longer with the chicks overnight. Note that while in the early days after hatching the female was spending much of her time brooding, particularly during the night, the time with the hatchlings also includes periods of feeding. In both years the time spent brooding declined with chick age, but the time spent feeding increased. In both years the female also spent time with the chicks when she was neither brooding nor feeding, *e.g.* time at the box edge preening or time apparently doing nothing more than watching/guarding the brood. In 2017 the female spent several nights roosting on the perch outside the box after ceasing night brooding. In 2019 the female roosted outside the barn as soon as night brooding ceased. What is notable in 2019 was that the days when the female spent more time with the chicks than was expected by the declining trend (*i.e.* 3rd, 5th and 7th June) were days when prey deliveries were less than expected, the female apparently compensating for lower food intake by increased brooding.

In 2017 incubation is believed to have started on the fourth egg (24 April), but the cameras were not switched on until 2 May. The grey rectangles are days on which there were power cuts: partial data on some of those days has been ignored.

In 2017, the male was assiduous in covering the eggs, obviously so in the latter stages of incubation. The 2018 male was no less diligent in the early days of incubation. The female's incubation pattern was consistent with 'steady' (Wiebe *et al.*, 1998). Steady incubation sees the female increasing the incubation period as the first eggs are laid, reaching a plateau which is maintained until the clutch is complete. Incubation then rises quickly to a maximum. In 2019 the male covered the eggs less often in the early stages of incubation, though again he covered for longer periods as hatching approached. However, the female's behaviour differed from that of 2018, suggesting 'pulsed' incubation (Wiebe *et al.*, 1998), which sees the female incubating for longer as the initial eggs are laid, then incubating for a shorter period until the clutch is complete, before rising to a maximum.

was adequate time for prey exchanges, feeding, cleaning, preening *etc.* outside the barn. On each of four successive days 19-22 May the male delivered fewer prey items, on two days none. On those days the female's absences from the box included periods of 47, 68, 149 and 161 minutes, and on each of the longer absences she returned with a full crop. While it is possible the female was being fed by the male away from the barn, it is also possible that she

The female Kestrel is turning the eggs.   The male is settling down on the clutch.

was hunting for herself. Interestingly, on one occasion of prolonged female absence the male arrived at the box and, finding it empty, did not cover the eggs but rested, preened and slept on the adjacent perch for 35 minutes, then left: the female returned sometime later. On the morning of 15 May the male arrived on three occasions without prey. Finally, at 13.18 the female left, either to exchange prey outside the barn or to hunt for herself. It was the first time she had left the box since arriving back at 19.47 the previous evening: she had been with the eggs non-stop for 17hrs 31mins.

Following hatching the female's presence in the box declined steadily: from 3 June onwards, when the chicks were 9 days old, she was present in the box only during periods of feeding. In prey exchanges the male arrived on the platform, the female taking the prey from him and heading out of the barn. While on these exchanges the female was robust in claiming her meal, the male was left on the platform. However, there were several occasions when the female knocked the male off the platform during the exchange. These incidents coincided with days when the male had delivered fewer prey items, and while it may represent only the anxiety of a hungry female to obtain food, watching the exchange sequences (and lapsing into anthropomorphism) it is difficult to avoid the feeling that the female was administering a swift cuff with her wing to punish an erring partner.

What is intriguing in Fig. 96a is the time the male Kestrel spent in the box during the female's absences in 2017, in contrast to the male's behavior in 2018 and 2019. The male was always diligent in replacing the female in the box, on one occasion spending 192 minutes with the eggs. The times the male replaced the female in the box increased as the hatching approached and reaching a maximum during the 4-day period (19-22 May) immediately before hatching. During those four days the male spent 14hrs 38mins covering the eggs (15% of the total time). While it is generally considered that male Common Kestrels lack a brood patch (though as noted in the previous Chapter Packham (1985b) did observe a male Kestrel that had a brood patch and was incubating) it is difficult not to wonder if the male was incubating given such a time. However, once the first chick had hatched the male made no further attempt to spend time in the box apart from one occasion on 26 May when the chicks were 2-3 days old. Then, he arrived with prey to find the female absent. He went into the box and left 19 minutes later when the female returned. It is unfortunately not possible to say if he fed, or attempted to feed, the young chicks. Chick feeding by the male is known in the other British breeding falcons and so it would not be surprising if the male was feeding: the surprise, perhaps, would be that he was, why he did so on only one occasion. On all other occasions when he entered the box after hatching it was to deposit prey.

Chick-rearing
While the chicks were very young (1-3 days old, 25-27 May) only the male provisioned the family, the female feeding and brooding. By the 31 May, with the chicks 7 days old, the female was providing the bulk of prey deliveries, her absences suggesting she was probably hunting herself as well as receiving prey from the male. What is interesting is that as early as 26 May, with most of the brood only one day old, the female was taking significant time away from the box. On the following day she was present for longer periods, but by the time the video system was back in service after the unfortunate glitch, her absences were significant and becoming increasingly so. The perceived wisdom is that the female needs to brood the chicks for long periods until they are 7-10 days old and able to thermoregulate. In 2017, while the female was brooding during the night, she had largely abandoned daytime brooding when the chicks were 6 days old, and entirely abandoned it by the time they were 8 days old. It is, of course, the case that the weather was benign throughout the chick-rearing phase: during the first 7 days of chick life the daytime temperature rarely fell below 20°C. As an aside, personal observation of Kestrel chicks being raised in a wire basket set in a tree at the edge of a copse on high, Scottish moorland, and therefore exposed to both low ambient temperatures, wind and rain, noted the female brooding chicks of about 15 days of age.

While the female's long absences from the box did not harm or hinder the development of the chicks, it did contribute to the death of one chick, neither the female nor the male adequately policing the area around the nest site. Prey deliveries to the box often exceeded what the chicks could consume, photographs showing whole prey and scraps littering the box, mixed with pellets from the chicks once they could manage large prey sections. From early May Jackdaws had taken an interest in the box and its contents, visits becoming more frequent after hatching: the Jackdaws were stealing prey morsels and, on one occasion, a pellet. On 2 June, with both adult Kestrels

Above Jackdaw on the perch in 2019. Jackdaws visited the box in each year, but only entered it in 2017.

Below The injured chick in 2017.

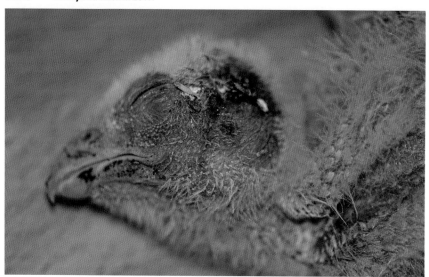

again absent, a Jackdaw entered the box, staying for almost 5 minutes. It left, but returned a minute later and spent a further 5 minutes in the box. It soon returned for a third time. Coincidentally, 2 June was a box photography day (photographs of the growing chicks being taken every 4-5 days) and four hours after the Jackdaw visit the box was briefly visited by a photographer. One chick had an obvious head wound and appeared dead. It was removed and taken to the farmhouse where it was found to be alive. Over the next few hours it recovered somewhat, but then sadly died. It seems probable that while exploring the feeding opportunities in the box the Jackdaw had pecked at one chick. The peck was not followed up and so presumably was not an attempt at predation. Neither does it appear to have been aggression as the other four chicks were not harmed. It seems, rather, to have been a random act: perhaps

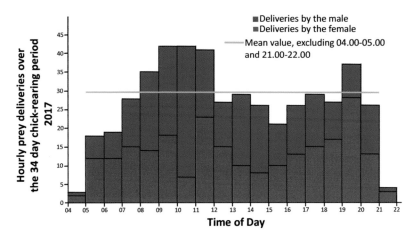

Figure 97. Histogram of prey delivery times by the female (red rectangles) and male (blue rectangles) in 2017. The trough in deliveries between 05.00-07.00 is statistically significant at the 90% level. The peak between 09.00-10.00 is statistically significant at 85%). The green line is the mean hourly delivery rate (29.6 items) for the period 05.00–21.00.

the chick moved suddenly, causing a reflex peck from the Jackdaw. The dead chick weighed 92g, consistent with it being 9 days old.

As well as unfortunate visitors, the Kestrels also had untidy, unwanted neighbours. Wood Pigeons were the most persistent would-be lodgers and, despite being sometimes chased off by the female Kestrel began carrying nesting material to the box top. As the box lid was semi-transparent the incubating female could watch the comings and goings as well as hear the scratching and cooing of her neighbours. Eventually, the male Kestrel attacked the box top, but was unable to dislodge the resident pigeon. Next the female tried, producing a flurry of feathers but being no more successful. Later, the female came to the box edge to watch. The pigeons were using the platform's perching pole as a stopping off point when bringing in sticks: on one delivery a pigeon dropped a stick on the female's head. This, it seems, was too much, and the female flew to the top where a brief scuffle produced another cloud of feathers and dislodged much of the accumulated nesting material. Satisfied, the female sat at the box edge, but was attacked and forced into the box by a pigeon which followed her in, the brief fight producing another cloud of feathers before the pigeon flew out. This contest proved decisive as the pigeons abandoned what little remained of the nest on the box roof and started another at the opposite end of the barn.

Hunting and prey
Fig. 97 shows the prey deliveries by the adult birds to the chicks until the day of first fledging. The colours indicating male and female deliveries are for interest only: near-simultaneous deliveries by male and female meant that both adults were hunting, so it is not possible to say with certainty which

Prey exchange during incubation.

In 2019 the female Kestrel brought in all the reptiles and amphibians, apart from the lizard (*left*) which the male caught. However, almost all the birds were brought in by the male (*right*), at one point he brought several fledglings in a short period.

adult made which kill. The deliveries between 04.00 and 05.00, and 21.00 and 22.00 were made during civil twilight (*i.e.* pre-sunrise and post-sunset respectively). If these data are excluded the delivery distribution has a mean of 29.6 deliveries/day and a standard error of 7.5. By comparison with data in Chapter 3, the capture pattern suggests less cyclical vole activity and Kestrel hunting than was observed in The Netherlands.

Fig. 95 shows that on 5th and 8th June (and to a lesser extent on 6 June) deliveries were lower than would have been expected. For most of the breeding season the weather was benign, but meteorological data from the nearby Middle Wallop station notes a prolonged period of heavy rain on 5 June and in the early hours of 6 June, and again early on 8 June. The wind speed was also 10-12m/s during much of 5-6 June, with gusts to much higher speeds, consistently reaching speeds greater than 13.9m/s and reaching 19.5m/s mid-afternoon on 6 June. Rain effects Kestrel hunting, for several reasons: it may limit rodent activity and may also inhibit the falcon's ability to pick up visual clues. Heavy rain may affect both visibility and make flight-hunting more difficult, particularly if the wind is strong. Sub-optimal wind speeds also inhibit flight-hunting, though perch-hunting is, of course, still an option, though as noted in Chapter 4, the hunting yield is reduced.

The meteorological data certainly suggests a correlation between high wind speed and rainfall, and reduced hunting success. However, it is worth noting that on 13 June when the wind barely reached 2m/s at any time and 18 June when the wind speed was mainly below 2m/s and always below 3m/s – had markedly differing delivery rates: on 13 June the prey delivery total was higher than average, while on 18 June it was lower than average. On 19 June there were no prey deliveries for 4 hours in the middle of the day, yet that day had neither significantly low wind speed nor any rainfall. Clearly, while weather is

Photographing fledgling Kestrels is not made easier when, after returning from a short comfort break, the photographer finds one fledgling has decided to use the hide as a perch.

an important factor in Kestrel hunting, other factors are also involved – chick hunger, adult bird condition *etc*.

Of the delivered prey, all items prior to the first hatch were mammals. Another 56 mammals were then delivered to the chicks before the first bird was seen (on 31 May when the chicks were 7-8 days old). From then on 8.1% of delivered prey was avian. Of the birds, it is perhaps ironic in view of the Jackdaw attack, that on one occasion the female brought a Jackdaw to feed her chicks. The Jackdaw could have been a fledgling or an adult, though the former is clearly more likely: the bird was so large, both physically and in weight, that the female struggled to get it over the small upstand into the box.

In 2017 one live trap was found to hold two half-eaten voles and a very irate weasel. No attempt was made to weigh the weasel.

The mammalian prey was predominantly Field Voles, though there were also a small number of Bank Voles. Surprisingly there were few mice (all of which were Wood Mice) and shrews. The male Kestrel brought in one Weasel. There have been many stories over the years of Kestrels attacking, and perhaps taking, Weasels, but the clear photographic evidence in this study appears to be the first incontrovertible evidence that Kestrels do indeed take them (see photo on p58). The live trapped mammals reflected the delivered prey, 90% being voles, again chiefly Field Voles, with the remainder being Wood Mice. One trap contained the half-eaten remains of two voles and a live, well-fed, but extremely displeased Weasel. The enthusiasm the Weasel showed for attacking the hand that was freeing it suggested that the male Kestrel must have been very sure of success before attacking the Weasel it dispatched. The male Kestrel also brought one young rabbit to the box. The carry clearly exhausted him, and he eventually abandoned attempts to lift the prey over the upstand into the box, taking it out of the barn and returning with it, in pieces, over the rest of the day.

Because of the two lost days in the prey delivery totals during the chick-rearing phase it is not possible to be precise regarding the number of deliveries between hatching of the full brood and the last day on which all four chicks raised to fledge were continuously present at the box. However, from the data it is reasonable to assume that on those two days about 20 deliveries would have been made. That being so, over the period 25 May-24 June 472 prey deliveries were made.

The trapped voles weighed 13-26g, with an average of 19.8g. The weight spread is less than that normally seen in adult voles (14-45g) perhaps reflecting the number of young animals in the population in late spring. From the videos it was only possible to define the delivered voles as being 'small', 'medium' or 'large' based on body size relative to the delivering falcon. Assuming a similar weight spread of delivered prey, the average weight of a vole delivered to the chicks would also be about 20g. One delivered vole was significantly bigger than any trapped animal, probably with a weight closer to 45g. The trapped Wood Mice weighed 25-37g with an average of 27.7g.

Prey delivery by the male Kestrel. The photograph was taken by the author, not off the video. It was very early on an overcast morning so the ISO had to be pushed.

Prey delivery by the male Kestrel to near-fledglings. When the male arrived (*above*), there were two chicks on the platform, the one closest to the camera thinking he had the best chance of getting the prey. But one of the two in the box had other ideas and using a wing to shut out its sibling (*below*) it quickly grabbed the vole (*opposite above*) and retreated to the box (*opposite below*).

The four surviving chicks in 2017, the day before they fledged.

Assuming the avian prey comprised the fledglings of smaller species, apart from the Jackdaw, and allowing for the young rabbit comprising a total of 5 prey deliveries, the total wet weight of prey delivered to the chicks was estimated at 9.5kg. However, from the data recorded in 2019 it is known that both adults retrieved prey from the box during chick-rearing when deliveries outpaced chick hunger. In addition, it is known that some of the prey deliveries by the male soon after hatching were taken away by the female and not brought back to the box – the male was obviously feeding the female as well as the chicks. It is therefore not possible to accurately define the percentage of the 9.5kg of prey used to rear five chicks from a hatch weight of about 16g to 92g and, subsequently four chicks from 92g to a fully-fledged weight of about 250g. All four fledglings were very strong fliers within a few days of leaving the box, compatible with having been very well-fed. Assuming all prey was involved in chick-rearing, delivered prey produced about 1kg of Kestrel. That is about 15% higher than noted by Steen *et al.* (2012) for Kestrels in Norway. Given the approximate nature of establishing individual prey item weights, and the unknown number of prey items diverted to feed the female, the difference is not considered significant.

In 2018 Barn Owls visited the nestbox on more than 10 occasions. Mostly these were early in the breeding season, but even after the female Kestrel chased one owl away the visits continued. This photograph was taken a few days after the fourth, and final, egg was laid on 16 May. The photograph was taken a little after midnight. The owl and the front of the nestbox are illuminated by a lightning flash which picks out some of the owl's plumage.

2018

2017 had made clear the need for extra observations and an improved external camera. An event recording camera was therefore added to the box and the external camera was updated to provide better images. Both cameras were infra-red (IR) sensitive, so when daylight fell below a set level, IR lights were turned on to allow night viewing.

Activity began in mid-March, a Barn Owl spending time in the box the day before a male Kestrel carried out his own inspection. A female Kestrel inspected the box the following day. As in 2017 the falcon pair were not ringed. A comparison of photographs of the birds from the two years was inconclusive, but the behaviour of the 2018 pair suggests that at least one of the Kestrel pair was new. This is not unexpected. It is known that the turnover of, for instance, female Kestrels is relatively quick. Riddle (2011) records one female breeding in three successive years (and one which bred in three years out of four, having been replaced in the second year), but the turnover is often more rapid.

The first prey exchange between male and female was seen on 31 March. On 9 further occasions between 31 March and 11 May the male was recorded as arriving with prey, but the female only took it at the box four times: on other occasions the male took the prey away, presumably exchanging outside the barn. Most of the prey the male brought to the box was rodent (almost exclusively voles), but on 10 May he brought an unidentifiable bird: he brought another the following day and a third on 16 May.

Throughout April the female regularly visited the box, often making scrapes in the gravel base and settling into them. She was also roosting close to the

The male Kestrel in the nestbox in 2018, apparently looking for eggs. He started doing this in late March and continued to do so right up to the time of laying of the first egg on 11 May.

box and on one occasion chased a Barn Owl out of the barn. The male also visited the box, often appearing to scan the base for eggs. On 24 February a huge Arctic air mass had arrived in Britain from the east. Dubbed the *Beast from the East* by the national press, the intense cold lasted almost three weeks and, after appearing to dissipate, reappeared (the *Mini Beast*) in mid-March. By this time we were concerned about the lack of eggs, particularly as reports from south-west England talked of breeding failures in Peregrines and other birds. But on 11 May (21 days later than the first egg of 2017), sometime between 01.28 and 04.21 an egg was laid. The female laid three more eggs over the next 6 days. The nest camera can define the laying time only when the female moves to reveal the egg, allowing the time interval to be established within limits – see Table 9.

Egg	Last Time at which egg was not *in situ* (hours from midnight 10-11 May)	Time at which egg was *in situ* (hours from midnight 10-11 May)	Time interval between eggs (hours)
First	1.47	4.35	
Second	47.48	51.90	46.78 ± 3.65
Third	95.35	96.33	46.15 ± 2.70
Fourth	140.82	141.95	45.55 ± 1.06

Table 9. Time of egg laying and interval between eggs in 2018.

Following the laying of the first egg the male arrived at the box with prey only on four more occasions, the last being on 16 May, the day the final egg was laid. Of those last four prey deliveries, three were birds. On 16 May the female took the prey and left the barn. She returned 2 minutes later presumably having cached it. No further prey exchanges were recorded. In 2017 during incubation the male would bring prey to the platform and the female would take it there and fly off, leaving the male to cover the eggs. In 2018 all but two prey exchanges after 16 May were outside the barn. Consequently, the number of prey items brought to the incubating female can only be inferred by noting the times the male arrived to replace the female in the box and the time the female was away. Assuming she required minimum of 10 minutes to consume prey, clean herself and, perhaps, defecate (see below), then on average the male was bringing 6 prey times daily in the period prior to the 7th day of incubation.

There was a further marked difference in behaviour from 2017, the male covering the eggs for short periods only after incubation had begun around mid-day on 16 May: on the morning of 15 May the male sat for two periods of 25 and 17 minutes, and the following day for four extended periods (14, 16, 20 and 38 minutes). Thereafter he did not sit for longer than about 10 minutes. From the Dutch data on the time to consume prey (Chapter 5 *Fig. 43*), it takes about 9 minutes for a Kestrel to eat a rodent weighing about 25g. The male was therefore covering the eggs only for as long as it took the female to take prey away and consume it, with minimal time for cleaning *etc*. She probably may also have preened briefly, though one characteristic of the 2018 female not shared by the 2017 female was that she often interrupted incubation to sit at the upstand entrance to the box and preen for several minutes. As an example, on 19 May the female was incubating from 12.27 to 15.33, but preened at the box edge for 4 minutes at 12.43, 3 minutes at 12.59, 3 minutes at 13.49, 6 minutes at 14.32 and 5 minutes at 14.48. The 21-minute total represents 11% of the 3hr 6min 'incubation' period.

Hatching was expected on 14-16 June, but by 20 June it was apparent that the eggs had failed. They were therefore lamped. This revealed no signs of embryo development and it was assumed that one of the Kestrel pair had failed to come into condition, resulting in sterile eggs. In favour of this was the effect the *Beast from the East* had on other south-western species.

However, when the camera data were analysed, the incubation behaviour being of value as the birds would not have known they were tending sterile eggs, something extraordinary was revealed. On 23 May, 7 days into incubation, the female left the clutch at 17.13. Thirty-three minutes later the male covered the eggs for 57 minutes, the longest period he sat in 2018. He then left the box, returning at 20.05 and sitting for another 26 minutes. But the female did not return. The male abandoned the eggs overnight but covered them again the following morning (24 May) at 06.14 for 25 minutes. The female then returned at 06.51 and sat for 2hr 13min. She then departed. During the rest

After the female Kestrel disappeared, the male would often arrive with prey at the platform and call for long periods. Occasionally he would also arrive without prey and sit on the box edge staring at the eggs. While anthropomorphism should always be avoided, it was very difficult to watch the videos without feeling very sorry for the male whose behaviour, often hanging his head, gave every indication of sadness.

of the day the male sat several times for periods of up to 33 minutes. He also sat on the perch beside the box and called, often and loudly. He did not cover the eggs overnight or roost on the perch. On 25 May the male returned to the box many times, once bringing prey (a bird) and frequently calling. On that day he went into the box several times, but never for more than a minute or so. He seemed to be looking for the female rather than making any realistic effort to cover the eggs. Then, at 21.04, 36 hours (all but 2 minutes) after she had last been to the box the female returned and restarted incubation. By then, assuming the male's minimal efforts had been in vain, the eggs had been uncovered for 50 of 52 hours, including all of two nights, and had undoubtedly been chilled even if they had originally been fertile.

Once the female had returned she incubated in much the same fashion as before, with occasional breaks for preening at the box edge. However, the male's behaviour was very different. For the first three days after the female's return (26-28 May), he did not cover the eggs during prolonged absences by the female, rarely spending more than a minute in the box if he entered at all: mostly during his time at the box he would sit on the perch and preen. On 29 May having not visited the box until 15.00, the male covered the eggs for 9 minutes. On 30 May he began to cover the eggs more frequently and for periods up to 30 minutes.

After the female's prolonged absences and the restart of incubation, for the second time in 2018 a third bird appeared in the barn. This time the bird perched close to the platform of the nestbox and several times made movements suggesting it intended to go into the box. It then flew to the top of the box. Its tail suggests a second female or a juvenile. While the incubating female had not noticed, or did not react to, the bird while it sat on the platform edge, she headed for the entrance as soon as the 'third bird' reached the box top. The third bird then flew out of the barn and did not re-appear.

The question, of course, is what happened to the female that caused her to disappear for more than two days (in total)? On 21 May, two days before the female's first night away from the eggs, a third Kestrel was caught on the external camera. The female of the breeding pair is perched at the platform, the male is on a perch which extends towards the camera, when a third Kestrel passes between them. Travelling too fast for a clear view, the sex of the third bird is indeterminate, though the tail suggests a female or juvenile. Then, on 28 May, three days after the female had restarted incubation, a third Kestrel lands on the platform and attempts to enter the box. The incubating female chases it away. Our immediate thought was that this might have been the first female, the one that had laid the eggs, returning after an extended trauma had kept her away and being chased off by a usurping female who had been in the area for most of the time. But is that credible? As neither female was ringed the only way of deciding was analysis of photographs of the first female and the 'second'. But the shots are inconclusive. Sadly, the mystery remains.

In the photograph (*top*) the male has just exchanged prey with the female and gone to inspect the box, looking for eggs. The female is soliciting copulation and the male flies to her (*above left*). To avoid the suggestion that the female only solicited copulation when presented with food, the photograph (*above right*) proves otherwise. In fact, the copulation with the female holding recently-presented prey was the only one such event filmed. As already noted in Chapter 6, the filmed copulations averaged 11.5s with a range of 4.3-15.3s at times varying from 06.50 to 18.34. All photographs from the 2019 video stream.

2019

Hoping for a better year than 2018 a second nest box camera was added in the hope of capturing a male brood patch.

The male was first recorded at the nest box on 14 April, the first female visit being the following day. As in previous years there was also interest from Wood Pigeons and Barn Owls, but by 18 April the female Kestrel was roosting on the perch to establish ownership. On the following day the first prey exchange between the male and female was recorded. Interestingly a Tawny Owl investigated the box on at midnight on 20-21 April, the only time a Tawny was recorded in the three years of observation: it did not return. On 22 April copulation was recorded at the box: it was recorded several more times on the perch. The first egg was laid on 25 April, the clutch of 5 eggs being completed on 3 May. The second within-box camera, set low in the box, allowed a more exact timing for egg laying – Table 10.

Egg	Time egg laid	Time interval between eggs (hours)
First	14.05-14.08 on 25/04	
Second	12.13-12.20 on 27/04	46.17 ± 0.17
Third	10.10-10.15 on 29/04	45.93 ± 0.43
Fourth	07.34-07.42 on 01/05	45.43 ± 0.43
Fifth	08.28-08.39 on 03/05	48.93 ± 0.16

Table 10. Time of egg laying and interval between eggs in 2019.

In comparing the data of Tables 9 and 10 it is notable that while in 2018 the female laid her last (fourth) egg faster than the previous three, in 2019 the female laid her last (fifth) egg after a much longer interval. The camera installations allow the interval between eggs to be measured with unprecedented accuracy, 3, 7, 5, 8 and 11 minutes across the five eggs. The evidence of behavioural differences strongly suggests the females of 2017 and 2019 are different birds. From Tables 9 and 10. the egg laying interval for female Kestrels is 46 hours, though later eggs may take up to 49 hours. While the second within-box camera allowed an accurate timing of egg laying, it did not detect any evidence of a male brood patch.

Incubation

Fig. 96c (p273) shows the incubation times, as a percentage of 24 hours each day, by the female, and egg covering periods by the male, from the laying of the first egg. The data is ambiguous in terms of the start of incubation. The first egg hatched sometime between 10.53 and 11.25 on 29 May. Assuming the incubation period of Village (1980), *i.e.* 30.8 days (consistent with the data of the 2017 study) then incubation began early on the afternoon of 29 April.

The Common Kestrel

Fig. 96c is consistent with that assumption, but shows a dip in incubation time on 30 April before a steady increase in incubation time in early May. It seems the 2019 female used a form of 'pulsed' incubation (see Chapter 7, *Fig. 81* and Wiebe *et al.*, 1988b).

A comparison of Figs. 96a and 96c suggests that the 2019 Kestrel pair was marginally less diligent than the 2017 pair in keeping the eggs covered. The 2019 male spent significantly less time on the eggs than had been the case in 2017. However, in both cases the male spent longer covering the eggs in the last days before hatching: why this occurs is open to debate.

Prey deliveries by the male during incubation are shown in Fig. 98. The first two days of incubation (29 and 30 April) have been ignored. On each of these days only one exchange was seen at the box. However, for long periods on those two days the female was absent and may well have been feeding herself or exchanging prey outside the box. The data for the remaining 28 days of incubation are again ambiguous as the male brought prey to the box, and then often covered the eggs in the female's absence, but also (our observations) occasionally called the female to a nearby telegraph pole where prey was exchanged. It has therefore only been possible to identify 'probable' prey exchanges by noting the female leaving the box and the male arriving in it soon after. On most occasions the female returned to the box 10-20 minutes later. On some occasions the female would leave the box and the male would not replace her. Such absences were both short and long. During the latter the female could have been hunting for herself. These incidents have been excluded from Fig. 98. Assuming 'probable' exchanges were real, the male brought a daily average of 5.4 prey items, chiefly rodents, to the female. Of the 112 definite prey items brought in by the male 83% were rodents, 17% birds. The male also brought in a single Vivaporous Lizard.

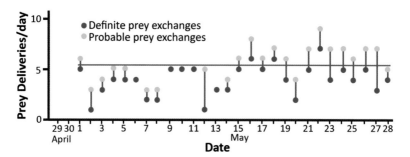

Figure 98. Prey deliveries by the male Kestrel during incubation in 2019. The blue circles are definite exchanges, male to female. The turquoise circles are probable exchanges, defined as the female leaving the nest box, being replaced by the male within a minute or two, and the female returning within 20 minutes. The red line indicates the mean deliver rate (5.4 prey items/day) over the 28-day period assuming all turquoise exchanges were definite.

Unusual bill-to-bill exchange of a mouse, from the 2019 video stream.

Fig. 99 is a histogram of prey deliveries by the male during incubation against time of day. The morning peak in deliveries is statistically significant at the 95% level. This reflects the finding in 2017 with respect to prey deliveries by the adult birds during chick-rearing, but again differs from the cyclic nature of Kestrel hunting noted in the work of Dutch researchers from the University of Groningen. The second peak, in the early afternoon, of Fig. 99 is also statistically significant, at the 99% level, but only if it is assumed that all prey was captured by the male, *i.e.* that probable prey exchanges are definite.

While egg laying and prey deliveries went well during incubation the female was troubled by the persistence of Wood Pigeons intent on occupying the box. Stand-offs occurred several times, but on one day the situation escalated, one pigeon pursuing the female into the box and pinning her to the ground, close to the five-egg clutch, in a battle lasting almost 4s. The pigeon then left the box, landed on the perch, shook itself down and after a few seconds went back into the box and continued the fight. After a further 90 minutes the pigeon returned, went straight into the box and there was a further short fight. Three days later there was a final short fight after which the pigeons decided to make a nest on a roof beam in the centre of the barn and there were no further problems. Images from the conflict are show on the following pages.

Figure 99. Histogram of 2019 prey delivery times by the male Kestrel during incubation. The trough in deliveries between 08.00-09.00 is statistically significant at the 95% level. The peak between 14.00-15.00 is statistically significant at 99%. In each case the significance level applies only if it is assumed that all 'probable' exchanges were real. The red line indicates the mean hourly delivery rate (9.5 prey items) for the period 05.00-21.00.

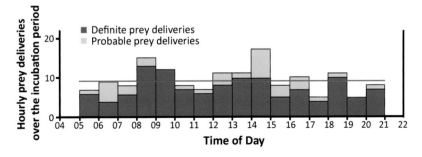

Images from the fights between the female Kestrel and a Wood Pigeon on 25 May 2019. The fight started with a stand-off (*top left*), continued with the female on her back each time, the pigeon on top of her. The eggs are visible in the photo (*above*). After the battle the female went back to incubating (*below*) amongst a large number of pigeon feathers.

Chick-rearing

Hatching times are given in Table 11.

Hatch Number	Time
1	10.53-11.25 29 May
2	14.12-16.41 29 May
3	06.44-07.30 30 May
4	07.30-09.21 30 May
5	13.15-14.10 30 May

Table 11. Hatch times of eggs in 2019.

Following hatching the female spent much of her time brooding/feeding the chicks during the subsequent three days, but by the time the chicks were 4 days old, while she brooded through the night (at least from 21.00 to 06.00) she rarely brooded them for other than short periods during the day. Night brooding continued until 9 June (chicks then 10 days old), though from 6 June the female was sat with the chicks, who huddled up to her, rather than being beneath her. From 10 June the female roosted outside the box. Once the eggs had hatched the male went into the box only to drop off prey. He covered the chicks once only, for just over one minute on the evening of 29 May when two chicks had hatched, *i.e.* there were still three eggs. The male attempted to feed the chicks on only two occasions. On 7 June he brought in a shrew which one chick took, but immediately dropped. On the evening of 11 June his attempt amounted to no more than offering several chicks, one at a time, the chance to take a whole vole from him: none of the chicks could. The male turned to leave with the vole intact and met the female coming into the box: she took the vole from him and did the job properly.

Chick-rearing went much smoother in 2019 than in 2017, with no intrusions from Jackdaws and all five chicks fledging successfully. The first chick fledged on 28 June, 30 days after the first hatch and 29 days after hatching of the complete clutch. Two chicks fledged the following day. The last two chicks also left the box that day. One flew up to a construction beam of the barn, the other flew down to a lower beam. During the afternoon, the higher chick, which had trekked along the beam and eventually managed to make it to a blind triangular construction joint, was seen scrambling up a beam apparently

The male Kestrel offers a shrew to one of the chicks. The chick did not take it, and the male left, leaving the uneaten shrew on the box floor.

Above   Having decided the chicks were not hungry the female Kestrel has placed a headless rat in the corner of the box for later feeding.

Below   The female Kestrel was feeding the chicks with a vole when the male arrived with another. The hunting success of the adult pair meant there were often uneaten prey items in the box and on the platform.

Bottom   On 14 June, with the chicks about 14 days old, the male left a bird for them, but they still seem puzzled about what to do with it.

Arriving at the nestbox with a Field Vole the male Kestrel found it empty, but then flew to a fledgling on a nearby beam. He advanced towards the youngster several times offering the prey, but each time pulled away as the fledgling grabbed for it. The motive seemed clear – if you want this prey you will need to follow me. The male then left and did not return.

First flights were often towards a sibling, but landings could be awkward for both youngsters.

The Common Kestrel

in order to gain height. The chick fell, colliding with its sibling some distance below: both chicks fell to the ground. The two were rescued and returned to the box as it was clear neither would likely fledge before nightfall. Both fledged the following morning. See photographs later in the Chapter.

Hunting and Prey

Fig. 100 shows the delivery rate of prey to the nest box during chick-rearing in 2019. During the period from hatching to fledging the male brought 69% of the prey fed to the chicks, either exchanging it with the female at the box or dropping it into the box. It is also possible, of course, that he delivered some of the remaining 31% to the female for her to take to the chicks. Fig. 101 is a comparison of prey deliveries in 2017 and 2019.

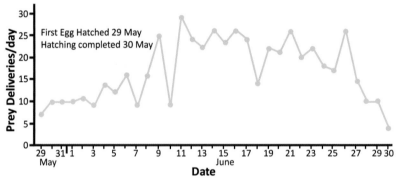

Figure 100. Prey deliveries by the adult Kestrels during chick-rearing in 2019. The first egg hatched on 29 May, all eggs had hatched by 30 May.

Fig. 102 is a histogram of prey deliveries by the adults during chick-rearing. Deliveries between 06.00 and 07.00 are significantly low at the 99% level. If grouped, the peak of deliveries between 10.00 and 13.00 is statistically significant at the 90% level, but the peak appears at a different position to that revealed by the male's hunting pattern during incubation. It is therefore not clear that any true cyclicity in terms of rodent availability is being shown by the data.

The interesting aspect of Fig. 100 is the decline in expected prey deliveries on the 7th, 10th and 18th of June. Meteorological data for 2019 and 2020 was collected from two sources during the breeding season. The Chilbolton Observatory near Stockbridge, about 6km NNE of the nestbox (as the Kestrel flies), provided hourly data on rain, wind speed and direction, while an out-station of Chilbolton at Sparsholt, 4.5km E of the nestbox, also supplied hourly rainfall data. These data were combined to compile Figs. 103-105 (overleaf). In each case the time of prey deliveries has been marked.

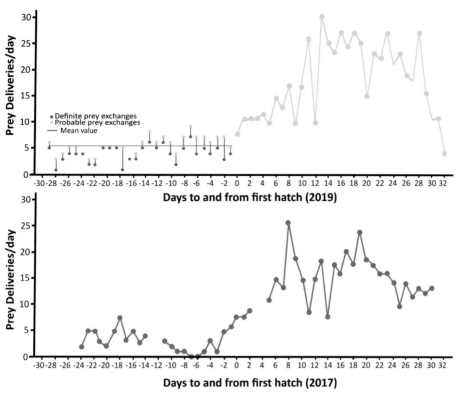

Above **Figure 101.** Comparison of prey deliveries during incubation and chick-rearing in 2017 and 2019.

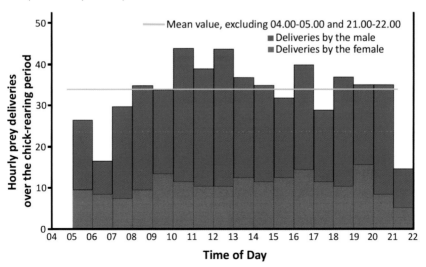

Below **Figure 102.** Histogram of 2019 prey delivery times by the adult Kestrels during chick-rearing. The trough in deliveries between 06.00 and 07.00 is statistically significantly at >99%. The peaks in deliveries between 10.00 and 11.00 and 12.00 and 13.00 are statistically significant at 90%. The green line is the mean hourly delivery rate (33.6 items) for the period 05.00–21.00.

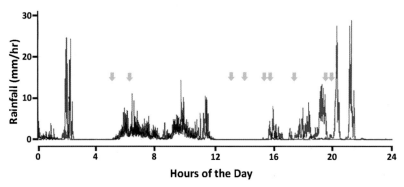

Figure 103. Rainfall data from Sparsholt (in red) and Chilbolton (in blue) for 7 June 2019. The orange arrows indicate prey deliveries by the adult Kestrels.

Fig. 103 shows that 7 June was a day of very heavy rainfall, though with wind speeds throughout daylight hours always in the range favoured by Kestrels. On that day there 9 prey deliveries to the chicks when 16-18 would have been expected (see Fig. 100). Eight of the prey deliveries (7 rodents and 1 bird) are assumed to have been taken by the male as in each case exchanges were outside the barn with the female leaving the chicks, but returning with prey within 1-2 minutes. The ninth delivery was a Slow Worm brought in

Figure 104. Rainfall data from Sparsholt (in red) and Chilbolton (in blue), together with wind speed data for 10 June 2019. The orange arrows indicate prey deliveries by the adult Kestrels.

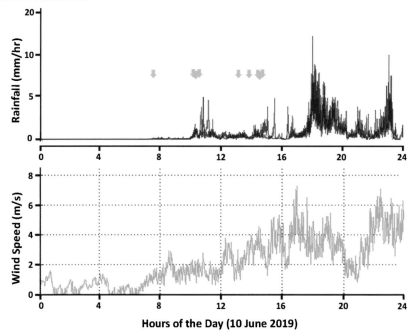

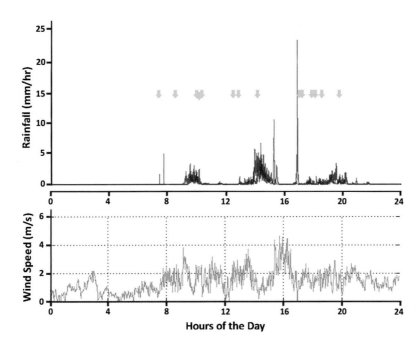

Figure 105. Rainfall data from Sparsholt (in red) and Chilbolton (in blue), together with wind speed data for 18 June 2019. The orange arrows indicate prey deliveries by the adult Kestrels.

by the female after an absence of 16mins. As the female delivered all the observed Slow Worms it is assumed she took this one herself. Allowing for a possible slight time differences in rain occurrence between the barn and the measuring sites, the data again clearly shows that heavy rainfall severely restricts Kestrel hunting. Sunrise was at 04.52 on 7 June, and the male Kestrel took obvious advantage of the lack of rain to hunt early, then chose a lull in rain to hunt again around 06.00, after which he was restricted to clear spells in the afternoon.

Fig. 104 shows that 10 June was another day of very heavy rainfall, with wind speeds below optimum for flight-hunting during the morning. The day's prey comprised 7 rodents, a bird and a Slow Worm, nine items when 25-30 would have been expected. From timings, it is likely the female took both the bird and Slow Worm, the male capturing the rodents. The rain-free morning period produced fewer successful hunts than would have been expected, with wind speed being continuously sub-optimal. A period of higher wind was then utilised while rainfall was minimal before heavy rain brought an end to the day's hunting.

Fig. 105 shows that 18 June was a day of occasional heavy rainfall, with wind speeds again below optimum during most daylight hours. About 25 prey deliveries were expected, but only 14 were seen. The prey comprised 9

Two photos of the female on the perch having delivered prey to the chicks at 15.25 on 10 June 2019. The imperative to feed the chick occasionally meant the adults had to hunt in poor weather.

rodents, 4 birds and a frog. The male delivered all the birds and at least 6 of the rodents. The female probably took the frog. Fig. 105 shows that the adults did their best in poor wind conditions, choosing times when the rain was light to work hard in sub-optimal wind. As noted earlier in the Chapter, perch-hunting is still possible, though the hunting yield would probably be reduced. A period of higher wind was then utilised while rainfall was minimal before heavy rain again brought an end to the day's hunting. Interestingly, during the one period of good wind and zero rain around 16.00 there were no prey captures.

As in previous years, live traps were set where it was known the adult falcons might hunt. While in earlier years voles were the main catch, in 2019 more mice than voles were trapped. The Kestrels reversed this trend, voles being the main prey. The weights of the trapped prey were as in previous years, voles being lighter than mice, but the mean weights being comparable (at 18g and 25g respectively). Again, from the video it was possible only to define the delivered voles as being 'small', 'medium' or 'large' based on body size relative to the delivering falcon.

The bulk of the prey taken comprised mammals, chiefly voles, though some mice and several juvenile rats were delivered. The Kestrels also caught one juvenile Rabbit and one shrew. A total of 548 prey items were delivered prior to fledging, 479 mammals, 68 birds, four Slow Worms and one Common Frog. Mammals therefore comprised 87.4% of the total, birds 12.4%. Of the birds, the male took 75%. Two of the avian prey tally were juvenile Jackdaws

The last acts of the breeding season for both adult Kestrels were to remove prey from the box. Here the female takes away a rodent which is well past its best-before date.

which were likely to have been taken by the female. However, 50% of the birds delivered to the chicks by the female were captured within 9 minutes of her leaving the barn (with several taken within 4 minutes) and were probably taken by the male and then exchanged. With that assumption, the male took almost 90% of captured birds. On one day he took three birds within one hour having apparently located a nest of near-fledglings. As previously noted, 17% of the prey taken by the male during incubation was avian. While the male was clearly not a bird-specialist, he does seem to have been more adept at taking them. The female took all the reptilian/amphibian prey during chick-rearing, but it will be recalled that the male took a lizard during the incubation phase.

The wet weight of prey that raised five fledglings was about 11kg. Again this weight has to be reduced because in the early stages (the first 4-5 days) of chick-rearing the female was consuming a significant fraction of the delivered prey. During the latter stages of fledging more prey was being delivered than the chicks could eat. There were several instances of a chick taking prey from an adult, and then leaving it on the platform or in the box without attempting to eat it. In the last days before fledging the female retrieved four rodents from the box/platform, several having been in position long enough for *rigor mortis* to have set in: the female's last act in the box was to take away one last vole. In 2019 the prey produced 1.25kg of Kestrels. With an assumption about the wet weight taken by the female, this is again comparable to the figure noted by Steen *et al.* (2012) for Kestrels in Norway.

This fledgling flew into a vertical beam and somehow managed to cling on for a short time (*above*) before attempting to jump down to a nearby diagonal beam (*below*). But the young falcon missed and was soon falling (*opposite*). It struck a sibling on a beam below and both landed on the ground. They were retrieved, replaced in the nestbox and fledged the following day.

It is a little after 05.00 on a day which promises sun, but little wind. In the photograph (*above*) the male Kestrel is silhouetted against the dawn light. It is the third time he has landed on the pole, but each time he has arrived without prey. The incubating female can watch his arrival from the nestbox and presumably each time she is hoping he is bringing prey which will allow her to eat breakfast and take a break from covering the eggs. On this third occasion she flew out to the pole (*below*). Unlike the photograph above which is as shot, the photograph below has been heavily processed to reveal the Kestrel pair. It is difficult not to imagine that the female is explaining the facts of life to a hunched, chastened partner. In view of the tragedy that will play out in a few weeks time, the photograph has a terrible poignancy.

2020

The Kestrel pair that arrived to inspect the box this year discovered that a pair of Stock Doves had taken up residence, the female dove having already laid her two-egg clutch. On 5 April, having presumably decided that the box offered good opportunities for breeding the female Kestrel flew into the box: the female dove was incubating her eggs. The fight that followed was the first of several over the next three days as the doves made attempts to reclaim the box (and their eggs). The eggs survived until the Kestrel chicks were mobile when they were crushed during one the regular excursions the youngsters made around the box.

Having taken possession from the doves, the female Kestrel did have to defend the box once more against an enthusiastic Barn Owl. The owl visited several times in the falcon's absence, then made the mistake of trying again when she was present: the female Kestrel won the short fight that followed, the owl not being seen again. As in previous years, the adult Kestrels were not ringed. From tail colour the 2020 female looked very similar to that of 2019, but differences in behaviour suggested that it was not the same bird. Copulation was first filmed on 5 April, and was filmed regularly prior to the laying of the first egg and between the laying of the third and fourth eggs. Surprisingly, it was filmed again after the clutch was completed on 3 May – twice on 5 May and once on 15 May (12 days after clutch completion). This appears to be the first time that copulation post-clutch completion has been recorded, the latter occurrence being particularly noteworthy. Copulations

The male Kestrel covering the clutch in 2020. At this stage the two, by now very cold, Stock Dove eggs were still intact.

were timed between 5.2 and 8.8 seconds, with a mean time of 6.7s, and occurred at times between 06.02 and 17.20, though most were recorded between 06.00-07.00. The first egg was laid at on 25 April: Table 12 gives the timings for the four eggs.

Egg	Time egg laid	Time interval between eggs (hours)
First	15.33-15.36 on 25/04	
Second	21.32-21.35 on 27/04	53.98±0.11
Third	12.50-12.55 on 01/05	87.19±0.17
Fourth	13.51-13.57 on 03/05	49.02±0.18

Table 12. Time of egg laying and interval between eggs in 2020.

The intervals of Table 12 are a sharp contrast to the more consistent laying periods noted in previous years. It is not unusual for Kestrels (indeed, all birds) to delay the laying of eggs if local conditions – weather, food availability – deteriorate. As resorbing of ovulated ova has not been proven (and may not be possible), so it is likely that the female delays ovulation, leading to short delay. In this case it was probably bad weather which caused the delay as 28 April, the expected date for the third egg, was a day of persistent, though not torrential, rain (Fig. 106).

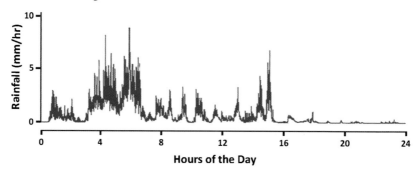

Figure 106. Rainfall data from Chilbolton for 28 April 2020. Due to equipment problems resulting from Covid-19 travel restrictions, no rain data was available from Sparsholt in 2020.

As in 2019, while the second within-box camera allowed an accurate timing of egg laying, it did not detect any evidence of a male brood patch.

Incubation

Fig. 106 shows the incubation times, as a percentage of 24 hours each day, by the female, and egg covering periods by the male, from the laying of the first egg. As in previous years the data are ambiguous in terms of the start of

Left Copulation continued until 12 days after clutch completion. On this occasion the Kestrels were facing away from the camera, allowing the position of both tails and the female's wings to be seen.

Right The evicted Stock Doves returned to the nestbox often during Kestrel egg laying, spending more time looking at the growing falcon clutch than their own, now cold, eggs.

incubation. The female began incubation in earnest only after laying the final (fourth) egg but spent 40% of day with the eggs from the laying of the third. The male spent longer covering the eggs as hatching approached than during the early days of incubation, behaviour which had also been seen in previous years.

With the exception of the first hatch, hatching times could not be established with the accuracy of those in 2019 as the 2020 female stayed over the eggs/brood as the chick freed itself so that the first glimpse of a new infant was usually after its down has lost the initial dampness. The exception was the first egg where the cameras caught the point when a shell section broke free and a wing emerged, at 16.02 on 31 May. The second hatch had happened by 04.50 on 1 June, the third by 12.58, the fourth by 17.00 on the same day. Hatching of the complete clutch therefore took less than 23hrs. If it is assumed that incubation began on 1 May (Fig. 107 overleaf) then incubation took 30.75 days until the first hatch and 31.75 days for the complete clutch, consistent with the timing of Village (1980), *i.e.* 30.8 days. The female's incubation was 'steady' as defined by Wiebe *et al.* (1988b) – see Chapter 7 *Fig. 81*.

16.02, 31 May. The egg shell cracks and a wing of the first hatchling of the year emerges, watched intently by the female.

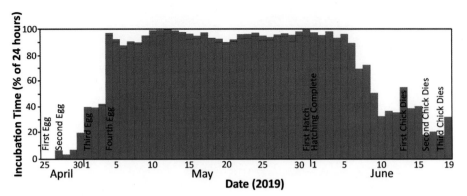

Figure 107. Percentage of time the female (red rectangles) and male (blue rectangles) covered the eggs during incubation and chick rearing in 2020. It seems probable that the female began incubation in earnest only after laying the final (fourth) egg, with the 40% of time spent with the eggs prior to laying of the fourth egg accounting for the 24hr time spread of hatching. As in earlier years the male spent longer covering the eggs as hatching approached than during the early days of incubation.

Fig. 108 shows the prey deliveries by the male to the female during the courtship and incubation. Fig. 109 has histograms of prey delivery times during the three phases of breeding.

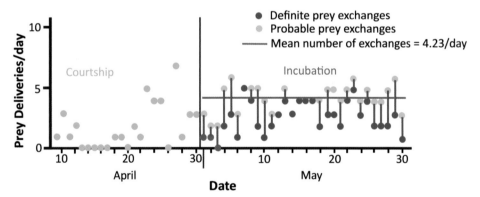

Figure 108. Prey deliveries by the male to the female during 'courtship' and incubation. The 'courtship' period includes only filmed exchanges, *i.e.* only those which occurred at the nestbox. In practice the filming noted significant times during which neither male or female was at the box: on 26 April the male was not recorded at the box at all, and after the female left the box perch at 05.52 she did not return until 20.28 for roosting. A significant number of prey exchanges were almost certainly occurring outside the barn. Five exchanges were recorded on two days, with 6 on one day. These numbers are consistent with other studies of Kestrels and with the known 'standard' intake of adult falcons. It is likely, therefore, that the male was delivering 4-6 prey items to the female on all 'courtship' days. During incubation definite and probable exchanges were recorded, 'probable exchanges' being as defined in Fig. 98.

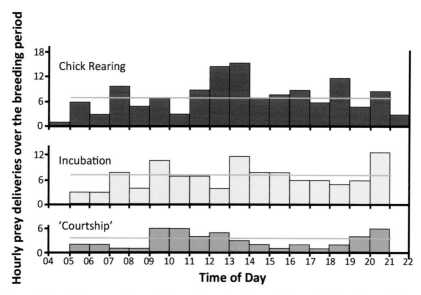

Figure 109. Histogram of prey delivery times by the male Kestrel during 'courtship' and incubation in 2020, and by the adult birds during chick rearing (to 19 June). The troughs in deliveries between 04.00-07.00 are statistically significant at the 90% level. The peak between 12.00-14.00 during chick rearing is statistically significant at 90%. The 20.00-21.00 peak during incubation is statistically significant at 90%. The histograms are markedly different from those of earlier years in not showing a clear statistically significant peak during the early morning. The green lines on all three histograms are the mean deliveries/hour for the period 05.00–21.00 hrs. (3.00 during 'courtship', 7.06 during incubation and 6.89 during chick rearing to 19 June).

Chick-rearing
Brood times of the female after hatching are shown in Fig. 107. The male made no effort to cover the chicks. Chick-rearing went smoothly until 14.32 on 9 June when the male arrived at the box with prey. The female was not there, but arrived 3s later. The male passed the prey to her at the box edge and departed. It was the last prey delivery he made, though it is possible that he passed a further prey item to the female outside the box which she delivered to the box at 15.38. Exactly why he disappeared is not known. On the next two days the female supplied a large number of prey items to the chicks. The male had been an excellent hunter, delivering far more prey than necessary to feed the brood immediately after hatching (see photo on following page). In addition to this surplus he may also have been caching prey. On 10 June the female brought in 3 prey items in 67mins, then paused for 2hrs before bringing in two more in 28mins. On 11 June the female brought in 6 prey items in 49mins. As well as her delivery rate, the time intervals between prey deliveries was also very short, as little as 4mins, implying retrieval from the platform and from a cache rather than hunting. What was also notable was that while the male brought in large prey frequently, when the female was

Above The male Kestrel was extremely good at catching rats, brining in several each day. The male's hunting skills are clear from the fact that 10 prey items have already accumulated on the platform.

Below

(*Left*) The female feeding the three-day old chicks with morsels of the freshly-delivered prey in the photo above.

(*Right*) The female now tries to feed one chick with a complete rat tail. Thankfully the chick dropped it rather than attempting to swallow it, though the female did try again before returning to morsel-feeding.

hunting she brought in small prey infrequently, and the majority of prey delivered on 10-11 June was large.

By the 12 June the prey deliveries had fallen well below the requirements of four growing chicks, suggesting male prey caches were exhausted. Not only was delivered prey smaller than before the male disappeared, but the spectrum changed: the male had delivered a high fraction of birds, the female brought only a couple; the male brought no reptiles, the female brought a Slow Worm. Late on the evening of 13 June a chick died. It was probably the smallest, least developed of the brood: it probably died of starvation. Just after midnight on 15/16 June a second chick died. Although the female delivered 9 prey items on 14 June after the death of the first chick, the rate dropped after the second death. A third chick died in the early evening of 18 June and by the evening of the 19 June the fourth chick was barely able to stand on its own. There is some, but no conclusive, evidence that the female tried briefly to feed dead chicks to the survivors (coverage of the box floor is not 100% event though there are two cameras). No remains of dead Kestrels were found at the end of the breeding phase but, of course, Jackdaws could have removed carcasses.

Following hatching the female spent 14 nights brooding, but, of course, the male had disappeared 8 days after hatching was completed. Having seen the first chick die, the female continued night brooding for two nights, but roosted outside the box on the third night: the second chick died an hour or so after midnight on that night. Nevertheless, the female spent the next two nights roosting away from the box. She began night brooding again only after the third chick had died. It is interesting that the female started to night brood the surviving chick, though any attempt to explain this would be speculative. The barn did not have a wi-fi connection so monitoring the position with the adults and chicks was by visit only. The Covid-19 crisis limited the author's ability to travel often and the appalling 2020 spring weather meant that Simon, the manager of the arable farm on which the barn stands (and a falconer), was behind with work and so was not checking the recorder as often as normal. His work meant he was occasionally close to the barn and was seeing Kestrels often enough to believe that the breeding season was continuing as usual. The tragedy unfolding did not become apparent until 19 June, the recorder not having been viewed for six days. In discussion between myself and Simon we agreed that we would supplement the food of the surviving chick in an effort to raise it to fledge. This was a difficult decision. Science says nature should be allowed to take its course: we should have continued to film the outcome even though we knew that would almost certainly mean the death of the fourth chick. Against the ethics of science are set the morality of watching something unfold knowing that it can be prevented. Given that the Kestrels were breeding in southern English farmland and not a remote, uninhabited area, the probable reason for the male Kestrel's disappearance was human activity – collision with a vehicle or perhaps (though much less likely) direct action – and on that basis we felt justified in taking action. Day-old chicks were therefore placed in the box. At first the chick ignored them, but the female utilised them as soon as she found them. Food supplements were then supplied through to the point at which the chick (probably a female) fledged in late June. The female and fledgling then became inseparable for several weeks, the adult protecting the youngster if anything, *e.g.* a corvid or Little Owl, threatened it. Since it is known that loss of the male will cause some females to abandon a brood, the female's decision to continue to fight (albeit a losing battle) for her chicks was heartening even as analysing the video feeds was heart-breaking.

Fig. 110 shows the delivery rate of prey during the chick rearing phase up to 19 June when supplementary feeding began. The low number of prey deliveries on 8 June may have been due to poor weather for hunting though sickness in the male (resulting in death) cannot be discounted. Prey deliveries in 2020 are indicated by orange circles. The red circles are the reduction in the required delivery rate as chicks died (again normalised from 2019 data and assuming all three chicks died at noon on the day of death). As can be seen the female achieved the 'red' rate on 19 June when, in response to the sight of the fourth chick being unable to stand, supplementary feeding began. While it is possible the female could have reared the fourth chick successfully the delivery rate on the previous two days was below that required and it is more likely the chick would have died had supplementary feeding not begun.

In a final, and welcome, chapter to the season, in late August Stock Doves returned to the box, laying two eggs and raising two chicks to fledge.

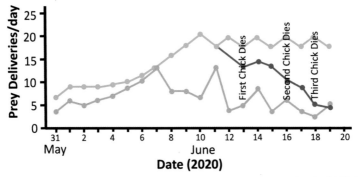

Figure 110. Prey deliveries by the adult Kestrels during chick-rearing in 2020 (orange circles). The first egg hatched on 31 May, hatching was complete by 1 June. The male Kestrel was not recorded after 14.32 on 9 June. There was only one further delivery that day when the female brought in a rodent (possibly exchanged between male and female). The green circles are prey deliveries derived from the previous year normalised for a four-chick clutch with the effect of poor hunting weather smoothed out. The number of prey deliveries by the female on the 10 and 11 June were almost certainly from cache. The green circles show that the inevitable consequence of the falling delivery rate in 2020 was chick starvation. The red circles are the reduction in the required delivery rate as chicks died.

Above   On 1 June, the day hatching was completed, the male brought his partner a Yellow Underwing Moth. Human sentimentality would see this is as a tender gesture, but the female was likely less impressed than she would have been had the male brought a plump rodent.

Below   The last video stream of the male. He arrived with a vole, but the female was away from the box. She arrived shortly after. The male gave her the vole and departed.

9   Movements and Winter Density

There is no hard evidence to suggest that parent birds engage in active, aggressive attempts to expel their brood, though Shrubb (1993a) does suggest that adults occasionally steal from their young and move them away from preferred hunting areas, either of which would encourage the juveniles to seek their own space. But whatever the reason, once adult Kestrels have stopped feeding their offspring, the young Kestrels disperse, usually taking up residence close to their parents' breeding territory. Riddle (2011) made several observations of Kestrels being seen in congregations, with groups of 15-25, of 30, and even 40 on one occasion. These aggregations seem to have been in highly-prized prey areas and are probably unusual.

    As time from cessation of feeding increases, the fraction of birds travelling greater distances also increases: as winter approaches only 20-25% of young are still within 25km of their natal site. In sedentary populations these birds, together with those that have travelled greater distances, will find territories which they will maintain for the winter. In migratory or partially migratory populations, the distances travelled by young birds as winter approaches may

When winter arrived, this female spent nights in a barn, hunting each day if the weather allowed.

indicate migration rather dispersal, particularly as early dispersal directions tend to be random, whereas later routes tend to be south-westward – but see the results of a Belgian study mentioned in the next few pages.

In a tabulation of data on recovered, ringed young accumulated over a 70-year period to 1984, Village (1990) found that in the first month after dispersal 74% of birds had moved less than 25km from their natal site. However, some had moved much greater distances, 4% exceeding 150km. Village's own observations included one Scottish bird which had flown over 300km from the parental breeding territory in only 24 days. Another British bird (Riddle, 2011) ringed on 19 June in Ayrshire died after flying into a pylon on Tenerife, 3076km away, on 11 October. That represents a daily flight of 27km if it assumed the bird died on the day of arrival. But such daily flights are trivial in comparison to the data accumulated by Dejonghe (1989), who quotes one bird flying at 243km/day for five days and another which averaged 200km/day for nine days, and in a study of Kestrels ringed in Belgium. There, Adriaensen *et al.* (1997) noted one bird found in Senegal (4501km from the point of ringing), one in Morocco (1968km) and one in Algeria (1602km). Clearly Kestrels are strong fliers. While all these long-distance recoveries are of

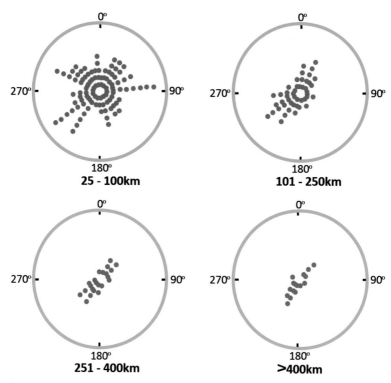

Figure 111. Direction of movement of Belgian juvenile Kestrels recovered by the end of March of the following year. Each dot is a single bird. The data are presented in 10° groups. Redrawn from Adriaensen *et al.* (1997).

birds heading south/south-west, Adriaensen and co-workers also noted one falcon flying west to Ireland, a distance of 1107km. This direction of travel is interesting as most recoveries suggest migrating Kestrels are heading south where, it would be anticipated, winter conditions would be more favourable in terms of both temperature and prey availability. But the data of Adriaensen *et al.* (1997) shows that this is not the case (Fig. 111)

Fig. 111 notes that for distances of travel less than 100km Kestrels are evenly distributed in terms of direction of travel, but as distances increase a distinct south-west or north-east direction emerges. North-east implies that some Belgian Kestrels are choosing to fly long distances towards Scandinavia for the winter. Interestingly, Cavé (1968) found something similar in a study of Dutch Kestrels. While most recoveries of first winter birds were found to have travelled south-west (at distances up to 1650km) a small number had travelled north-east including one which had flown 610km. For birds recovered during their first spring/summer, again the majority had moved south-west, but some had gone in the opposite direction over distances of 600-700km. For birds in their second winter, most had headed south-west, but again some had gone

Movements and Winter Density

This female Kestrel was ringed, and also given a colour ring, on 20 June 2013 near Stirling, Scotland. It was photographed on 30 August 2019 at Kaluga Airport near Moscow, 2556kms and 2262 days later. *Elvira Nikolenko*, with thanks to Dave Anderson.

north-east (up to 840km), while others had flown east (up to 510km). Neither Adriaensen *et al.* (1997) nor Cavé (1968) offer any explanation for these surprising movements, and they remains a curiosity. In Britain, Village (1990) notes that Kestrels migrating from Scotland chiefly headed south or south-east, though some birds moved much shorter distances in other directions. In his southern England study area, Village found a much more even spread of movement directions, though east and south-east were favoured.

Young Kestrels fledged from early nests disperse sooner and tend to stay closer to their natal territories than fledglings from later nests. This is not a great surprise as later birds searching for a territory are moving through areas which are more likely to be occupied. But the requirement to move further usually means that establishment of a territory is delayed. That, in turn, is reflected in both winter survival rates and in the ability of the young birds to breed in the following spring. By contrast to the significant movement of young birds, adults are more sedentary in areas where winter survival is possible. In a study in The Netherlands, Cavé (1968) found that over four successive winters female Kestrels which bred the following spring had moved less than 8.5km during the winter: females which were found not to be breeding in the following spring moved greater distances.

Further work on Belgian Kestrels, Adriaensen *et al.* (1998), also indicates that young Kestrels are driven to move further from the natal territory if prey is in short supply when they become independent. Using the energy content of the seeds (acorns and beech nuts) from the previous winter as an indicator of probable rodent population, Adriaensen and co-workers showed a linear relationship between this measure of prey availability and the average distance travelled from the natal territory by young birds. Again, extra distance travelled will be reflected in winter mortality and spring breeding rates. In both the work in the UK by Village (1990) and on Dutch Kestrels (Daan and Dijkstra 1988) it was found that early fledged birds were more likely to breed in their first year than were those that fledged later, the Dutch study suggesting by as much as a factor of 10. In his studies, Village (1990) found no correlation between the distances travelled by dispersing siblings, but a significant correlation in the direction of travel. He also found some evidence of siblings travelling together, though this does not appear to be a general rule.

For most Kestrel populations in Europe true migration, as opposed to the dispersal of juveniles at the end of the breeding season, begins in September and October, and is usually complete by November[1]. With the provisos noted earlier in the Chapter (see Fig. 111 and explanatory text), the general direction of travel is south-westward, at least initially, the falcons heading towards southern France and Spain. Birds reaching Spain may cross the Straits of Gibraltar to north Africa, then make their way south to west Africa. In Britain, the sedentary population is joined by birds arriving from Scandinavia[2]. All populations of Kestrels in north-eastern and eastern Europe that breed in areas with permanent winter snow-cover are migratory, some taking routes which cross southern Europe to reach Spain, then crossing to north Africa and heading south to west Africa or the coastal belt from Liberia to Nigeria. Others cross the Adriatic, Italy and the Mediterranean to north Africa and then head south across the Sahara to west Africa. Although data is scant, it seems Kestrels from south-central Asia take a route though Arabia to reach east Africa (see Village in Wernham *et al.*, 2002). Birot-Colomb *et al.* (2019) detail the counts recorded at the Défilé de l'Écluse in the French Jura. The totals seen reflected the cyclicity of rodents and the winter snow cover in Scandinavia during the years 1993-2010, but there was then a sudden increase in Kestrel numbers during 2013-2017, mirroring increases also seen at Falsterbo (in Sweden) and Organbidexka in Spain's northern Pyrenees. This increase is mentioned again in Chapter 11 when the European population of Kestrels is considered.

Migrant Kestrels are a small fraction of the British population, chiefly birds from Scotland and northern England. However, not all falcons from

---

[1] Global warming is altering the time of migration. In a study of British migratory species, Usui *et al.* (2017) noted that the start of spring migration was advancing by 2.1days/decade (1.2 days/°C). It is likely that a similar change, in this case a delay, is occurring in the timing of the autumn migration.

[2] In 2018 a female Kestrel ringed on 1 July 2017 at Trysil, Norway (north-east of Oslo, near the Swedish border) was photographed on 30 June on the Pentland Hills, south of Edinburgh, the photograph allowing the ring to be read. The female had flown 1099km in 354 days. She mated with an adult male and laid 6 eggs. Four chicks were fledged.

Female Kestrel hunting in the snow, central Europe. *Torsten Prohl.*

these areas migrate, raising the question of why some birds chose to travel while others from the same area do not. Those that remain in an area risk bad weather with its inherent peril of starvation: snow cover reduces the availability of voles and Britain is unpredictable in terms of snow duration, explaining why Kestrels breeding in areas of northern and eastern Europe where snow cover is both more certain and lasts predictably longer are entirely migratory. But remaining in an area means being able to maintain a territory and even, perhaps, a mate, and so enhances the chances of early breeding when spring finally arrives. Migration reduces the rigours of winter, but there is a trade-off, the journey is potentially dangerous and must be accomplished twice, on arrival the bird must acquire a winter territory and, possibly, there may be problems with finding a suitable spring territory after the return journey with, consequently, a later breeding date. Village (1990) found no evidence in his studies of migrants becoming residents in subsequent winters, or of residents choosing to become migrants. It seems that once a pattern of behaviour is chosen by an individual bird that pattern is maintained throughout life.

Wallin *et al.* (1987), studying Kestrels in Sweden, found that falcons from the north of the country migrated twice the distance of those from the south, but took the same time to complete their migration. As the migration route of the southern Kestrels lay entirely within that of the northern birds, this meant that those from the north were moving at twice the speed of their southern cousins. However, Wallin and co-workers found that during the return journey, the northern birds travelled at the same speed, and so reached their breeding grounds later, the more leisurely journey reflecting the fact that snow cover in the north took longer to clear. There is no evidence of a similar leap-frogging in migrating British Kestrels.

In many species of raptors, males migrate further than females (Newton, 2008), but this is not the case in Kestrels where not only do females travel further than males, but the fraction of females migrating is greater than in males (Wallin *et al.*, 1987; Village, 1990; Kjellén, 1992, 1994). During long migrations Kestrels use flapping flight rather than the soaring flight associated with many other raptors and rarely, therefore, form part of the spectacular concentrations of migrating raptors seen at various famous points adjacent to sea crossings (*e.g.* Gibraltar and Falsterbo in Sweden). However, in his study of migrating birds at Falsterbo, Kjellén (1992) did observe Kestrels, and noted that both females and juveniles migrated earlier than males (Fig. 112). The reasons for differential migration are still debated, Kjellén considering the most likely explanation being the timing of the moult. Female Kestrels complete their moult during incubation and so are in a better position to migrate earlier than males who must wait to moult as they are the primary food supplier to the brood and so cannot afford the reduction in hunting performance which might result from feather loss. This 'moult hypothesis' does not contradict the migration pattern seen in other falcon species in which males and females depart at the same time (*e.g.* Hobbies), as these are long distance migrators

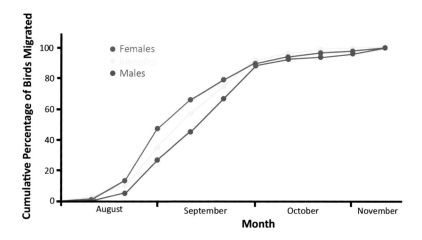

Figure 112. Differential autumnal migration of adult and juvenile Kestrels observed during 1986-1990 at Falsterbo Peninsula, southern Sweden. The curves show the cumulative percentage of falcons that had migrated at a certain date. Redrawn from Kjellén, 1992.

and can arrest the moult in autumn prior to migration, completing it at their winter quarters.

As already noted, the likely reason for males to shun migration and stay on the wintering grounds (assuming, of course, that staying is an option), is to occupy or protect a territory so as to be ready to mate and breed early when females return. For similar reasons, males that do migrate tend to return earlier in the spring. For British migrants Village (1985) notes not only that males arrive earlier than females at the breeding grounds, but that adults arrive earlier than first-year birds. Village (1990) notes a degree of winter site fidelity in migrants, pointing out that such behaviour is to be expected, since a bird finding available resources to sustain it over winter would likely head for the same area rather than seek out a new area unless a different, but good, area was identified by chance.

Spring migration may start as early as February, though is mainly seen in March and April, with some birds, usually those returning to northern Scandinavia, not leaving until early May. The routes followed by spring migrants may differ from the line taken on their outward journeys, wind direction altering the points at which coastlines are crossed. Though the tendency is for birds to return to, or close to, former breeding grounds or natal sites, birds ringed in other European countries have been recovered in Britain, and *vice versa*. As Village (1990) notes, while rare, this genetic exchange seems enough to have prevented the development of sub-species anywhere in Europe, as has happened in the extremely sedentary, and therefore isolated, populations on the Canary and Cape Verde islands.

The density of wintering Kestrels varies from zero birds in areas where the entire population migrates, to a density significantly larger than that of summer in areas where migratory arrivals outnumber the local population. In Britain the density at the start of winter may be larger than the spring density, juveniles adding to the breeding and non-breeding population. In southern areas, the number of northern birds arriving usually outnumbers local birds which migrate away adding a further population increase. But as winter deepens the population and density decreases as birds, particularly juveniles, die. These changes were monitored by Village (1990) in his study areas in southern Scotland and eastern England. In Scotland, the density followed the expected pattern, being much the same in autumn as it was during the breeding season as some juveniles stayed and were joined by juveniles from elsewhere, this influx outnumbering the breeding birds which migrated. Then as winter began, immigration ceased, and the density fell as birds died or moved in search of better resources. The population and density rose in the spring as migrants returned. In mid-winter, the density had fallen to 20-25% of the breeding density. By contrast, in his English study areas Village found that the winter density exceeded the breeding density, significantly so in his arable farmland area where the peak population was occasionally double that

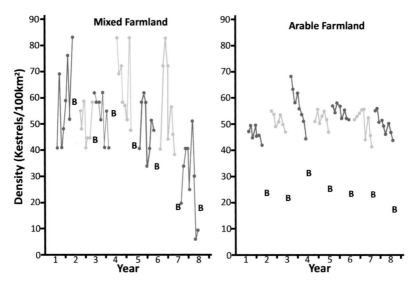

Figure 113. Monthly estimates of Kestrel density in winter (coloured dots) and breeding density in the following spring (B) in seven successive winters/following springs in the late 1980s in mixed and arable farmland in east-central England. The different colours only differentiate between successive years, the lines between the dots aiding differentiation. The falcon density usually declined in both habitats between autumn and spring, but while the difference between late winter and breeding densities was minimal on mixed farmland, it was very significantly lower on arable farmland. Redrawn from Village (1990).

recorded in spring (Fig. 113). Village saw the Kestrel population decrease more in mixed farmland than in his arable farmland area, confirming that the major factor influencing Kestrel winter density is food supply. However, while Village was able to show a positive correlation between Kestrel and vole numbers (as might be expected) in his mixed farmland area, there was no such correlation with vole numbers in the arable study area, and only a marginally significant correlation with mice numbers. This is counter-intuitive – though as Village notes, his study did not note the relative importance of non-rodent prey (birds and invertebrates) in the Kestrel diet – so the general conclusion must be that the winter density of Kestrels is primarily influenced by food supply, together with favourable weather so that the birds can hunt.

Winter roosting Kestrels choose spots which offer shelter against the elements – barns and ruins, rock clefts in natural or quarried cliffs, tree holes *etc.* Such roosts may also be used during the day if rain, snow or fog make hunting impossible. Shrubb (1993a) observed Kestrels sharing barn roosts with Barn and Little owls, which, since both predate the falcons, is an indication of the greater necessity to escape exposure. Shrubb noted that winter roosts may be traditional, telling the story of Kestrels which used a barn for 20 years until it was pulled down, then took up winter residence in the one erected as a replacement. Shrubb also notes that Kestrels are stealthy in their approach to roosts, pausing occasionally on a devious route to watch for danger, and that their preferred roosts in buildings were on or close to doors and windows which offered ease of escape. More interestingly he noted that on doors or beams the birds would usually roost, not across the beam as would be expected, but along it, in the manner of Nightjars (*Caprimulgus europaeus*), this allowing the bird to have its body against a wall which aided it in keeping warm. In the case of the Kestrels studied in Chapter 8, the males did not spend winter nights in the open barn which contained the breeding box, or in any of the other barns at the site: instead they chose an outside roost, high up close to the eaves of a separate barn, presumably accepting less shelter for a location which offered an easy escape if danger threatened.

Before leaving the topic of migration, it is worth noting an interesting analysis of changes in breeding and wintering ranges due to global warming. In the study of Potvin *et al.* (2016), the team used ring recovery data from 1960-2015 for 29 Finnish breeding birds including the Kestrel. The researchers found that only four of the 29 species showed a southward breeding range shift, but many more had shown a decrease in migration distance. The Kestrel, together with several other species, showed a northern shift in breeding range (by about 3.5km/year) and an increase in migration distance, by about the same amount.

10 FRIENDS AND FOES

Friends

Norrdahl *et al.* (1995), in a study area in western Finland, noted that the nests of Curlews (*Numenius arquata*) were closer to Kestrel nests than random scattering would indicate: on average, Curlew nests were 490m from a Kestrel nest, but 580m from a point chosen at random within the nesting area. While the scatter of separations overlapped, the difference between the two means was statistically significant, *i.e.* the position of a Kestrel nest was influencing the breeding decisions of the Curlews. The benefit the Curlews were deriving was found to be in the predation of their chicks – although Kestrels took 5.5% of the chicks in nests close to their own, Curlews in areas away from the falcons lost 9% of chicks to corvids. The Kestrels enthusiasm for chasing corvids from their territory was benefiting the Curlews, predation losses being significantly reduced.

While such commensal behaviour has been noted before, most famously between species of geese and Snowy Owls, this seems the only confirmed example of the behaviour involving Kestrels, though other observations of inter-species 'friendships' have been noted. Among the most curious was that observed by Kupko and Kübler (2007) in Berlin. The German team were watching Kestrel and Peregrine pairs nesting within 15m of each other on a tower. The female Peregrine had a single nestling and abandoned it in favour of chasing away the Kestrels and adopting their five nestlings. The male Peregrine continued to feed the single chick, raising it to fledge, while the female Peregrine successfully raised three of the five Kestrel nestlings. The authors note that after the fledglings had left the breeding site one young Kestrel was found close by, minus its head, though there was no indication of how or by what it had been decapitated.

Adult Curlew, southern Scotland. The local area had many nestboxes erected for use by Kestrels.

Peregrines rearing Kestrel chicks, Berlin. *Sonja Kübler and Stefan Kupko*.

While it might be assumed that such behaviour is unique, other instances have been reported. Ratcliffe (1962) records seeing a female Peregrine incubating Kestrel eggs. He believed the female had likely lost her clutch but, feeling the urge to incubate, had dispossessed a female Kestrel. Ratcliffe was not able to return to the nest to see if the Peregrine had successfully raised the Kestrel brood. He did, however, see another Peregrine female which had taken over a Kestrel clutch, and in this case the Peregrine pair had successfully raised five Kestrels to fledge (Ratcliffe 1963). It is worth noting that Ratcliffe's observations were carried out at a time when organochlorine pesticides were causing breakages of Peregrine eggs and so it may well be that in his cases female Peregrines had lost clutches and so were inclined to seek alternatives which they otherwise might not have sought. However, semi-adoption of Kestrel youngsters by adult Peregrines was also witnessed by Johnson (2008) at a coastal site in Devon many years after the problems of egg thinning had ceased. In this case the falcon pairs were nesting on a cliff ledge only 6m apart, though not visible to each other. When the fledglings were able to move about the ledge the two broods (five Kestrels who were older than the two Peregrines) met. They tolerated each other, and when the young Kestrels begged for food from the returning adult Peregrines they were fed apparently without distinction. Other close nesting of Kestrels and Peregrines has also been reported, *e.g.* Smith (1992) on the chimney of a working mill.

Close association of Kestrels with other predatory species has also been noted. In Austria Waba and Grüll (2009) observed Kestrel and Barn Owl pairs that had taken up residence in the same nestbox, both females laying full clutches (six Owl eggs and four Kestrel eggs). The Kestrels then abandoned their clutch which the Austrian observers removed. The owl chicks hatched, but the parent birds deserted the brood, four of the six hatchlings rapidly succumbing to starvation. At that point the female Kestrel returned and began to feed the remaining two owl chicks, Waba and Grüll considering that the owlets, which were already mobile, played an active role in the adoption. Assisted by the male, the female Kestrel raised the two owlets to fledge, both apparently leaving the box independently. The sharing of nestboxes by Kestrels and Barn Owls has also been noted in Britain (Tempest, 2009, Riddle, 2011), though the outcome for each species is unclear. Riddle (1992) records a nest site on a dam in which a Kestrel laid four eggs and a Barn Owl laid five. The Kestrel was the first user, but apparently abandoned the site after the owl arrived. The owl incubated all nine eggs, but the Kestrel eggs did not hatch having presumably chilled after the female Kestrel left. The five owl eggs hatched, and the chicks were successfully raised to fledge.

Barn Owls and Kestrels sharing a nestbox. *Gordon Riddle.*

Barn Owl chicks and unhatched Kestrel eggs in a box at a dam site. *Gordon Riddle.*

Foes

Such 'friendly' relations are, however, a rarity, Kestrels defending their territories/hunting ranges vigorously against other species (as well as other Kestrels in spring and summer, and against juvenile birds when the latter are attempting to consolidate winter territories). Kestrels hassle any raptors which appear within their territory, as well as other trespassers if the female is incubating or if there are chicks in the nest. Shrubb (1993a) records male attacks on Great Black-backed Gulls (*Larus marinus*), while Coath (1992) records a Kestrel and a first summer Red-footed Falcon locking talons and tumbling out of the sky over Essex. Corvids are attacked on sight, Kestrels showing a particular dislike for Magpies: as Hasenclever *et al.* (1989) record that in their study in Westphalia, Germany, Magpies were the primary predators of Kestrel chicks, this is perhaps not surprising. Magpies also pirate Kestrel prey: Sage (1957) noted a Kestrel dropping onto prey from telegraph wires and a Magpie immediately landing on its back, beating its wings so as to maintain balance as it pecked at the falcon's head. The Kestrel dropped its prey and was able to fly off, leaving the Magpie with its booty.

Shrubb (1993a) catalogued attacks by Kestrels on other diurnal raptors and noted that the Common Buzzard was the victim in most of these. He considered this was explained by the similar hunting techniques of both (flight-hunting and perch-hunting) and the overlap of their prey, Buzzards taking voles as well as larger prey. Buzzards also take earthworms on farmland. Shrubb noted that often he would see Buzzards hovering below Kestrels, a classic manoeuvre to move the higher bird away from an area: this technique is often used by Kestrels for the same purpose. As noted in Chapter 11, the Kestrel is a declining species in Britain, and it has been suggested that the increase in Buzzard numbers may be one reason for the reduction in numbers.

The behaviour of fledgling Kestrels at a colonial nest site of both Common and Lesser Kestrels in Spain observed by Bustamante (1994) has already been mentioned (Chapter 7). While Bustamante saw fledglings of both species apparently cohabiting in peace, there were occasional signs of aggression.

Kestrel chick killed by a Golden Eagle. *Gordon Riddle.*

There was also notable aggression between the adults of both species and of adults to the fledglings of the other species.

As well as attacking raptors which stray too close to their nests, Kestrels also suffer reciprocal attacks. Both Gamble (1992) and Ristow (2006) report attacks by Hobbies on Kestrels. As Kestrels have been observed attempting to predate Hobby chicks (Nicholson 2010), this suggests, perhaps, that predation by both species of the chicks of the other may occur. Adult Kestrels may also suffer direct predation from raptors. Predation by Buzzards seems limited, but Kestrel predation by Golden Eagles (*Aquila chrysaetos*), Goshawks and Peregrines have been recorded, as are kills by Sparrowhawks and Tawny Owls. Shrubb (1993a) notes one very interesting piece of information regarding flight-hunting Kestrels in his Welsh study area. He noted that the flight-hunting times were reduced from those at his study area in Sussex, and that the birds could be seen extending their necks to look behind them. A hovering Kestrel, intent on observing the ground below, is certainly vulnerable to attack from above, and Shrubb considers that although the altered behaviour may reduce the risk of attack, predation by Goshawks and Peregrines may still be a contributory factor to the decline of Kestrels in Wales. In Europe, Mikkola (1983) notes that Eagle Owls are significant predators of Kestrels, accounting for 91% of all the Kestrels taken by owls: the remaining 9% were taken by Tawny (7%), Barn (1%) and Little (1%) owls. Although Mikkola notes owls being taken by Eleonora's Falcons, Gyrfalcons, Lanner Falcons and Peregrines he reports no observation of owl predation by a Kestrel. Foxes are also known to have eaten Kestrels, though such predation presumably involves sick birds unable to take-off from the ground or chicks which have fallen from the nest.

The piracy of a Kestrel by a pair of Red Kites apparently acting in unison, was noted by Prŷs-Jones (2018) in Chapter 3. As interesting as the cooperative attack was, the subsequent continued pursuit of the falcon by, initially, both kites and then by a single kite was more compelling. The single kite chased the Kestrel, apparently attempting attacks several times, before the falcon disappeared into a cloud. Even then the kite circled below the cloud. As

Prŷs-Jones notes, the relentless pursuit seems to make sense only if the kite was attempting to predate the falcon. Prŷs-Jones notes that there have been instances where Kestrels have been seen in the diet of Red Kites, but it has always been assumed that the falcons had been picked up as carrion.

Kestrels react to humans intruding into their nest areas in a range of ways. Females will occasionally sit tight on eggs, moving only when forced to do so. Other females will vacate the nest and fly away. Each of the two options represents a risk with any predator – stay too long and the adult could become prey to the predator; go to early and development of the chicks could be compromised (or, of course, they could become prey). The time to flush seems to depend on brood size, female temperament and, perhaps, environmental issues. In most cases, females will leave at a time between the two extremes and then sit nearby and sound the alarm call, alerting the male of the pair and warning the intruding human to move away or risk attack. Attacks rarely occur, but have been known, with close passes and even occasional soft touches.

To conclude this section mention must be made of one remarkable interaction between a juvenile Kestrel and a potential prey in which the prey became foe rather than victim. Marco Lodder, a voluntary helper, was cleaning out owl boxes in The Netherlands when he saw the falcon attack a Starling some 20m out from the shore of a lake (Lodder, 2018). The capture resulted in the attached birds landing in the water. The Starling managed to

Red Kite. Possible predator of Kestrels?

The Common Kestrel

escape the talons and climbed onto the Kestrel's back from where it attempted to drown the falcon by pushing its head under water. The Kestrel paddled furiously towards the shore, eventually shaking itself free of its torturer. Marco then rescued the shocked and waterlogged bird, placing it on a post in the sun where over two hours it dried and recovered sufficiently to fly off. The following pages are photographs of this extraordinary encounter.

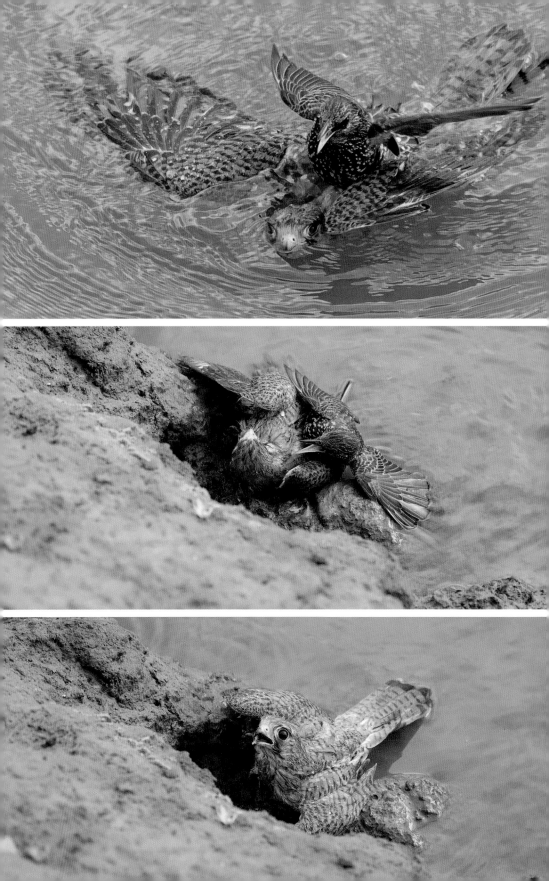

All images of the battle and of the recovering Kestrel above by *Marco Lodder*.

11 Survival and Population

In this Chapter I look firstly at the survival of Kestrels, both adults and chicks, the age profile of the species clearly being a dominant input to the population. Causes of Kestrel deaths are then considered before population estimates are set down.

Survival

The age of Kestrels at the time of their death can only be ascertained with any certainty for ringed birds, but use of that data has both advantages and disadvantages. The advantage is that the age is precise for birds ringed as nestlings. The major disadvantage is that 90% of ringed birds are never recovered, so that information for the total population must be inferred from a small sample. The problem of this inference is that it is mathematically complex and still debated. In recent years the use of wing tags and colour rings on birds, each of which allows identification of living birds, has aided data accumulation, but has also introduced further problems – do tags fall off and if so in what numbers and on what timescales? Does not seeing a colour ringed bird mean it is dead or merely that it has not returned to a particular area? A second disadvantage of data from ring recoveries is that recovered birds are likely to have died close to, or due to, human activity, which means

Survival and Population

It is sobering when photographing a delightful huddle of young Kestrels – there are five in this brood in a nestbox in southern Scotland – to remember that only two or three will survive their first winter.

that birds dying as a result of 'natural' causes are less likely to be recovered. Birds dying of starvation and/or disease may be recovered if they drop near human habitation, but otherwise are unlikely to be spotted, while the victims of predation are very unlikely to be recovered (though the finding of leg bones with rings attached in the nests of raptors is not unheard of). Nonetheless, ring recovery data does allow an estimate to be made of survival rates, and, with the caveat that it is only an approximation, is presented here.

Using data from 665 ring recoveries between 1912 and 1972 Village (1990) estimated the survival rate of juvenile Kestrels at 40% (*i.e.* 60% did not reach their first birthday), with 66% of second year birds and 69% of third year birds surviving. Village reports the longest-lived (recovered) bird being 15 years old, which is still a record for a UK Kestrel. Considering the mortality rate of nestlings, Village found that it was independent of brood size. Village also investigated the survival of the Kestrels in his English farmland study areas and found similar results (Table 13).

	Percentage surviving to age:			
	1	2	3	>3
Males	36 (50)	76 (45)	86 (28)	77 (22)
Females	17 (24)	63 (38)	68 (25)	67 (9)

Table 13. Winter survival of Kestrels on English farmland, as a percentage of the number recovered (as given in parenthesis).

Table 13 suggests that mortality is higher for females than for males at all ages, though the difference is not statistically significant because of the small sample sizes, particularly of females.

Daan and Dijkstra (1988) studied the age at death of Kestrels in the period 1967-1973 in The Netherlands, and prepared an age-histogram which is the basis of Fig. 114.

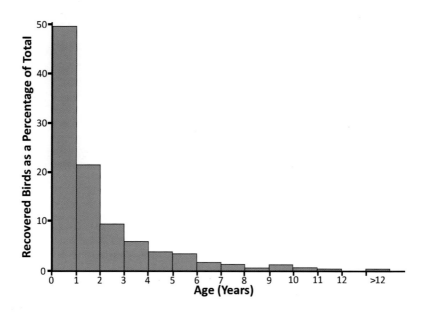

Figure 114. Percentage of Kestrels recovered dead which were of a known age during the period 1967-1973 in The Netherlands. Within each histogram the recovered birds were no older than the age on the horizontal axis, *i.e.* 49.8% of the birds were less than one year old *etc*. During the study period only one bird was recovered which was more than 12 years old – that bird was 14.6 years old. Drawn from data in Daan and Dijkstra (1988).

While it is not easy to directly compare the two data sets (UK and The Netherlands) as they assume different birth dates (Village assumed his birds were one-year old on 1 August, whereas the Dutch study assumed 1 June) it is clear from each that juvenile Kestrels have a low survival probability, presumably associated with lack of hunting experience. It would therefore be assumed that youngsters that fledge early should have a better chance of survival. And such is the case, this having been first identified by Cavé (1967) who noted that while 47% of young Kestrels fledged before 30 June survived their first winter, only 35% of those born later did so. Village (1990) noted the same effect of early fledging. Village divided his year into four periods (prior to 15 June, 15-21 June, 22-30 June and after 30 June), and then considered the recoveries made prior to 1 October, from 1 October to 31 December, and 1 January to 31 July. While the mortality increased in all his four fledge periods as winter progressed, it was most noticeable in the period 1 October to 31 December, with the percentage of recovered birds increasing from 18% to 28%, meaning that youngsters fledging after 30 June were 55% more likely to die than those fledged before 15 June. Clearly, while such factors as weather and prey availability are important in fledgling survival, the time a first-year Kestrel has to learn to fend for itself before the onset of winter is critical to its survival.

In their study, Daan and Dijkstra (1988) note that one bird was recovered at an age of 14.6 years, which compares well with the 15 years quoted by Village (1990) giving confidence that the histogram is a reasonable approximation to the Kestrel population age structure. Calculating an average age at death for a Kestrel from the data of Fig. 114 is difficult as both the mean and the median of the data suggest Kestrels do not live beyond two years. However, excluding one- and two-year old birds, each of which have a very low survival rate, the data imply that any bird reaching its third birthday has an approximately 50% chance of reaching its fifth birthday, *i.e.* the average life span of a mature Kestrel is 4-5 years, with birds rarely surviving to achieve ages in double figures. This age is consistent with the work of Newton *et al.* (2016) who carried out a meta-analysis of data collected on the survival of raptors and owls. Newton *et al.* concluded that there was a relationship between body weight and survival probability with small raptors (such as the Kestrel) having an annual survival of 60-70% once they survived the critical first year of life – the probability rising to over 90% for the heaviest raptors – and that there was no significant difference between the survival probabilities of the sexes.

Daan and Dijkstra (1988) also carried out a more detailed study of the effect of laying date on survival, calculating the 'fitness' of an egg laid in a Kestrel nest based on a number of factors including the probability of a chick surviving from hatching to fledging, and winter survival. From this the researchers were able to calculate an egg's reproductive value. Egg 'fitness' is obviously related not only to the hatching chick's likelihood to survive its first winter, but the probability that it will be in condition to attempt to breed the

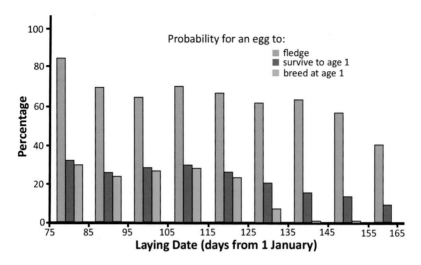

Figure 115. Probability (as a percentage) of Kestrel eggs with differing laying dates to hatch and survive to fledge; to survive to 1 June of the following year; and to breed in the following year. Redrawn from Dijkstra (1988).

following spring, and Daan and Dijkstra were able to compile data to explore this (Fig. 115).

What is clear from Fig. 115 is that the attrition rate of juvenile Kestrels is very high. But it is not uniquely so, being similar to that of the American Kestrel (Henny, 1972), the Sparrowhawk (Newton *et al.,* 1982), and to other British falcon species.

Intuitively it would be assumed that a young Kestrel would find its first winter difficult because limited hunting experience might mean finding enough food would be difficult, and limited life experience would make surviving winter's cold perilous. Village (1990) investigated the association between Kestrel survival and both vole numbers and winter temperature (Fig. 116).

Fig. 116 indicates that adult survival was only minimally dependent on either vole numbers or temperature, but for juveniles survival was positively correlated with both. Village noted that the vole population was itself correlated to winter temperature so that the exact relationship between juvenile survival, vole numbers and temperature is more complex than first sight suggests. Low temperatures indicate the possibility of snow cover which makes catching voles more difficult and may force the juvenile Kestrels to catch birds, which are a more difficult prey for an inexperienced bird, increasing the risk of starvation. In a study in Germany, in the Rhine valley close to Cologne, Kostrzewa and Kostrzewa (1991) also found that the mortality of Kestrels (of all ages) was correlated with winter temperature, with deaths from cold and starvation accounting for approximately 50%[1] of deaths in all winter months, but a

---

[1] The figure of 50% of Kestrel deaths being from starvation is consistent with the study by Newton *et al.* (1982) on Sparrowhawk and Kestrel deaths in the UK.

Survival and Population

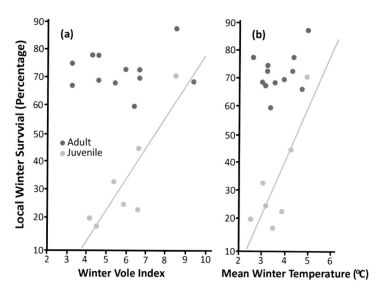

Figure 116. Survival rates of adult and juvenile Kestrels in relation to (a) winter vole index (a measure of the vole population density) and (b) mean winter temperature. The regression lines in each case are for juvenile birds only and are significant in each case. Adult survival was not correlated to either vole index or temperature. Redrawn from Village (1990).

higher percentage during the breeding season (Fig. 117 overleaf). Cold is less likely to be a factor during the breeding season, though both it and starvation may occur in the absence of parental care. However, as noted in Chapter 6 *Fig. 61*, the number of territorial Kestrel pairs was not closely correlated with vole numbers, suggesting that snow cover may have been more of a problem for adult Kestrels overwintering near Cologne than it was for the falcons in Village's study areas.

As noted in Table 13 Kestrels also survive their second winter less well than older birds. But having survived two winters, adult birds are better able to survive a third. This was confirmed in the study of Dutch Kestrels by Daan and Dijkstra (1988) who plotted annual adult survival rates as a function of vole population (Fig. 118 overleaf).

Fig. 118 shows that survival is better for Kestrels of all ages in good vole years, though the effect is most noticeable for juvenile Kestrels – but whatever the vole population, the survival rate of juveniles is very low, the Dutch data being consistent with the survival percentages noted by Village (1990) who found only 21% survival in his English arable farmland study area in 1984/5 and 38% in his English mixed farmland area in 1980/84.

Starvation is the most likely cause of death, a problem which, again, is more problematic for juveniles many lacking not only life skills, but having problems finding and defending territories. The territories they do acquire are also likely to be less productive than those of adults.

The Common Kestrel

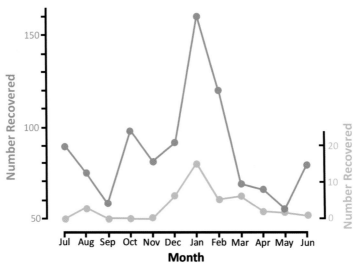

Figure 117. Kestrel monthly mortalities in northern Germany during the period 1909-1987 (a total of 1041 falcons). Data in orange relates to all deaths. Data in turquoise relates to deaths due to cold and starvation only (a total of 40 falcons). Redrawn from Kostrzewa and Kostrzewa (1991).

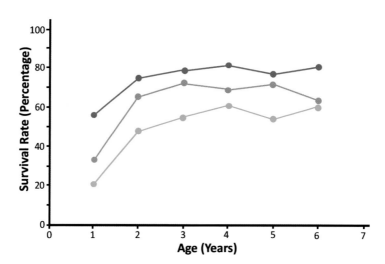

Figure 118. Age-specific annual survival rates of Kestrels in The Netherlands in years of high (purple), medium (orange) and low (green) vole population densities. Redrawn from Daan and Dijkstra (1988).

This Kestrel chick fledged on the day the photograph was taken.

Starvation may also lead to death by other causes. Weak birds may be less able to avoid traffic accidents, either because they are distracted by the possibility of food or because their reactions or flight capabilities are impaired. Newton (1986) also noted that in Sparrowhawks the use of fat reserves to counteract starvation releases accumulated pesticides into the bird's blood stream, adding another potential causal effect of death even if lethal doses are not created. While the use of the organochlorine pesticides which caused so much damage to British raptors ended in the late 1970s, the use of insecticides and rodenticides has continued, and in some instances the long-term effects of these are not, as yet, known. As an example, the recent controversy over the possible effect of neonicotinoids on bee populations shows that each time a new generation of pesticides is introduced it takes time before unwelcome side-effects can be identified, and then even more time before governments act to control their use. And each time the cycle begins again.

Kestrels are much less likely to die from poisoning than other British raptors as they rarely consume carrion. However, a hungry bird will take what it can and so may become 'collateral damage'. Kestrels are, though, as vulnerable as other raptors to indiscriminate killing by landowners and their agents. Prior to the Protection of Birds Act 1954, shooting was a significant cause of death in recovered Kestrels, amounting to 50% of the total. Following the passing of the Act shooting contributed a diminishing fraction in subsequent decades, some 27% in the 1950s; about 10% in the 1960s; but less than 5% by the 1980s. In a study of Kestrel deaths from 1962-1997 (1482 birds) Newton *et al.* (1999) found that only 1.8% had died from shooting[2]. But as deaths by shooting decreased, those by collision with vehicles increased: having been non-existent as a cause in the years before 1950, the fraction leapt up after the 1960s, and was 20% by the 1980s: in their study, Newton *et al.* found 34.9% of Kestrels were road casualties, implying that the number dying in collisions had risen again in the 1990s. Fig. 119 indicates the cause of death of the 1482 recovered Kestrels. Road traffic collisions were the major cause of identified accidental death in almost all years from the 1960s, and as many accidental deaths could only be classified as 'trauma' these might also include unidentified impacts. Newton *et al.* also separated the deaths of juvenile Kestrels from those of adults. Fig. 120 illustrates juvenile deaths as a fraction of all Kestrel deaths across 35 years: the fraction is essentially constant, and extremely high.

Fig. 119 indicates that accidents (usually impacts) are a significant cause of Kestrel deaths. To the standard list of traffic, windows and other obstacles, recent years have seen a further addition. Studying bird deaths at an on-shore windfarm at Campo de Gibraltar in southern Spain, Barrios and Rodríguez (2004) found the death rate of Kestrels was higher than for any other species at 0.19 deaths/turbine/year. Most deaths occurred in summer when the local

---

[2] Although reduced, shooting deaths have not fallen to zero. On 30/01/20 the BirdGuides website reported the illegal shooting of three Kestrels (www.birdguides.com/news/three-kestrels-killed-since-christmas/).

Survival and Population

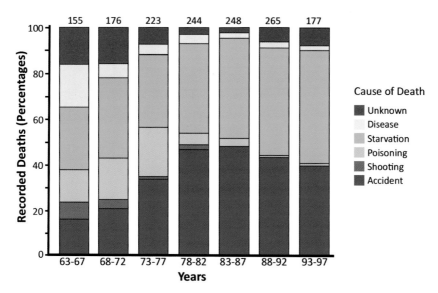

Figure 119. Cause of death of 1482 Kestrels recovered during 1963-1997 grouped in five-year periods. The number of recovered birds in each period is given at the head of the relevant histogram. Redrawn from Newton *et al.* (1999).

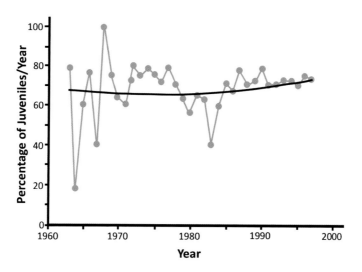

Figure 120. Percentage of recorded Kestrel deaths that were juvenile (first-year) birds during the period 1963-1997. Redrawn, data and regression line, from Newton *et al.* (1999).

Two casualties of windfarms. The male on the left was found beneath a turbine and appeared to have been struck a glancing blow by a blade. The bird on the right had been cut in half. The tip of a 36m turbine blade rotating at 20rpm travels at 270kph (170mph). *Gordon Riddle*.

population was maximal, but all the deaths occurring during the 12-month study period were juveniles. The next highest death rate, 0.12/turbine/year, was for resident Griffon Vultures (*Gyps fulvus*), these mainly occurring during the autumn and winter when the absence of thermals meant the vultures used slopes on which the turbines sat for updraft. The high Kestrel death rate was attributed to the local area around the turbines being a favoured hunting habitat. In a study of the causes of windfarm collisions and potential mitigating strategies, during which the authors examined more than two hundred reports, Marques *et al.* (2014) speculate that hovering birds might lose sight of the turbines, or might be pushed towards the blades by unexpected gusts, and that this might be the reason for the higher death rate observed by Barrios and Rodríguez (2004). It is also the case that the windfarms at Campo de Gibraltar are set on mountain ridges which encourage flight-hunting in rising updrafts: the private nature of such sites would also provide good rodent habitat. Marques *et al.* (2014) note work which suggests that American Kestrels are unable to distinguish moving turbine blades in certain lighting conditions (McIsaac, 2001). Marques *et al.* also note the high death rate of Kestrels at another site, Candeerios, Portugal, and the attempt to reduce mortality by the planting of scrub species and a reduction of shrub areas by goat grazing to move hunting Kestrels away from the turbines (Cordeiro *et al.*, 2013).

The above studies were in Spain, but there is evidence from across the globe that windfarms represent a hazard to birds – not just raptors, but birds of all forms. It is estimated that in the USA alone 140,000-328,000 birds (Loss *et al.*, 2013) and as many as 1.6 million bats (Arnett and Baerwald 2013) are killed annually in collisions. As a consequence, there have been attempts to understand how fatalities might be minimised. In one study, Watson *et al.* (2018), an international team, considering only raptor deaths, suggested that better siting criterion and the possibility of turning turbines off at critical times could mitigate fatalities. The latter suggestion was also addressed by Spanish researchers (Sebastián-González *et al.*, 2018) who used the mathematical technique of network analysis to study bird fatalities in two regions of Spain to highlight turbines which were causing the bulk of deaths for an indicator species. In the study the researchers used Griffon Vultures as indicators for large-sized, Kestrels for medium-sized, and Swifts for small, birds, and Pipistrellus *spp.* for bats. In each case the Spanish team were able to identify problem turbines, a finding which would allow these to be removed or switched off at critical times with, they claim, minimal economic cost for maximum mitigation of fatalities. One other mitigating factor also needs to be noted – in a study in Mexico, Cabrera-Cruz and Villegas-Patraca (2016) found that raptors migrating along a flyway which had been intercepted by a large windfarm had altered their routes to avoid the site, *i.e.* birds might be learning to avoid the turbines.

While there is currently no evidence to support the contention that windfarms are resulting in high death rates in UK Kestrels, the decline in the British population has coincided with an increase in on-shore windfarms. However, it must be pointed out that other factors may also be involved and may be much more significant. The decline in the population of British Kestrels is considered below.

Disease is also likely to be a significant killer of Kestrels, though it may more often be a proximate factor as the likelihood of a bird succumbing to disease increases if it is starving and/or suffering the exposure effects of prolonged inclement weather. Newton *et al.* (1999) found that 46.2% of their 1483 recoveries had died from 'natural causes', but attributed most of these to starvation (86%, *i.e.* 39.8% of total deaths), with only 11.2% (5.2% of total deaths) to disease – see Fig. 119. In a study in The Netherlands, Smit *et al.* (1987) examined the cause of death of 481 dead Kestrels – 67% had died of 'non-infectious diseases', a heading which included flying into structures, being shot and 'exhaustion'. The latter, which probably includes many cases of starvation as emaciated birds are often considered to have died of exhaustion when the underlying cause is starvation, contributed 39% of this category (26% of the total). However, 33% of the examined birds died of other causes. Of this total (159 birds) 39% died of bacterial infections (including 19 cases of avian tuberculosis), 32 succumbed to 'local inflammation', while 11 deaths were labelled 'diverse'. In their study of the microbial burden of 663 wild raptors admitted to a wildlife rehabilitation centre in Catalonia, Spain between 1996 and 2014, Vidal *et al.* (2017) noted that 69% of the birds carried microorganisms, these including *Eschericha coli* (the most common, usually seen in birds which had died from septicaemia or respiratory problems; Staphylococcus *spp.* which are associated with Bumblefoot; *Mycobacterium avium*, which is connected with avian tuberculosis – one such case was a Kestrel; and *Pseudomonas aeruginosa*, which is associated with oral infections.

In the study of Smit *et al.* (1987) mentioned above, those deaths not related to starvation or infection, were due to parasites. Of these the major cause was coccidiosis (67%), with other worm infestations contributing 27%, the remainder being labelled 'diverse'. Of the worms (collectively known as helminths) the most common are various nematodes. Smith (1996) mentions that one nematode, an ascarid, had been known to create an obstruction in the intestinal tract of a Eurasian Kestrel, though he unfortunately does not give a reference for this. Caryospora *spp.* protozoans have also been found in free-living Kestrels (Martínez-Padilla and Millán 2007) and caryospora infection was certainly implicated in the death of a juvenile Kestrel captured when sick in Berlin – Krone (2002): the bird had a heavy infection which resulted in severe haemorrhagic enteritis and eventual death. Krone notes that Kestrels may be infected with both *Caryspora neofalconis* and *C. kutzeri* each of which may use earthworms as a paratenic host: as noted in Chapter 3, Kestrels often hunt earthworms. Another protozoan, *Trichomonas gallinae*, may also occur in Kestrels. Well-known in pigeon/dove species, where it is responsible for canker, it can also occur in raptors, causing a disease known as 'frounce' among falconers. In each case, the protozoan causes swellings and lesions in the mouth/nasal areas, making swallowing difficult and, ultimately, causing death.

Kestrels may also carry blood parasites, Korpimäki *et al.* (1995) finding the haematozoan *Haemoproteus tinnunculi* in 40% of tested females and 25% of males, and *H. brachiatus* in 13% of females and 10% of males. The haematozoans are carried by Culicoides biting midges. Korpimäki *et al.* noted that females mated with males carrying higher parasite loads laid eggs later and had smaller clutches than those mated with 'healthier' males. Kestrels are also host to other parasites – Trematoda flukes, Cestoda tapeworms and nematodes have been found in Slovenian birds (Šumrada and Hanžel 2012, and references therein: see, also, Krone (2007)) – which, though not necessarily fatal, may contribute to a weakening of the bird and accelerate death from other causes. While the potentially harmful effects of internal parasites are obviously significant, it is worth noting that Barton and Houston (2001) in a post-mortem study of 379 birds of six raptor species, including 76 Kestrels, found that only 20% of the specimens of all six species had one or more internal parasites, a low percentage which surprised the researchers.

Kestrels may also carry ectoparasites such as louse flies (Hippoboscidae), many forms of these being associated with falcons (Maa, 1969). Several species were found in a study of birds close to Moscow (Matyukhin *et al.*, 2012). Further body parasites are also probable, other falcons being known to carry

Carnus haemapterus infesting the underwing of a ringing-age Kestrel. It is usually assumed that near-fledge chicks move away from siblings and the nest site to be first in the queue when prey is delivered, but an alternative suggestion is that they move away in the hope of reducing the parasite burden.

infestations of screw-worm flies and chewing bird lice. As noted in Chapter 2, Philips (2000) found a total of 21 families of parasitic mites infesting falcons, including at least ten species which infested the feathers and nasal cavities of Kestrels, or which were found subcutaneously. In a study of raptor nests as a habitat for invertebrates, Philips and Dindal (1977) found the nests were a potential home to parasites which may infect adult birds and nestlings, as well as animal saprovores (invertebrates which feed on carrion, excreta, pellets *etc.*) and humus fauna (invertebrates which feed on nest material). The study was carried out in North America and so did not specifically look at Eurasian Kestrels, but did include the American Kestrel, the nests of which were found to include eight species of arthropods from six families. There is no reason to believe that British Kestrel nests will not also carry a burden of potential parasites. Indeed, Fargallo *et al.* (2001) who noted an increase in chick infestation with *Carnus haemapterus* (a dipteran blood-sucking parasite) in nestboxes in Spain, found the average fly count was five-times higher than for chicks in natural nests.

In later work in Finland, Sumasgutner *et al.* (2014b) found that Kestrel pairs using nestboxes which had been left uncleaned from the previous year nested earlier than those in nestboxes which had been cleaned, and conjectured that prey remains and pellets from previous occupation might indicate 'public information' used in the following year to indicate advantageous sites. The researchers found no difference in the breeding success of Kestrels in cleaned or uncleaned boxes. There was, however, a difference in the ectoparasite burden of the chicks (particularly *Carnus haemapterus*) which can successfully overwinter among box detritus. In work in eastern China, Mingju E *et al.* (2019) also

Kestrel chick defecating out of the nestbox. While older chicks are able to do this, younger chicks cannot. Combined with pellets and prey remains, nestboxes quickly become unwholesome.

noted that Kestrels preferred to occupy nestboxes used in previous years and so were picking up information on nest-site quality (though the Chinese work did not include assessing parasite burden). See Chapter 6 for further details.

Contaminant burden is another reason for poor health and reduced hunting ability. However, in a series of studies into environmental pollutants in the eggs of birds of prey in Norway – a work programme which became part of the *Program for Terrestrisk naturovervåking (TOV)* (Programme for Terrestrial Monitoring) in 1992 – one recent report (Nygård and Polder, 2012) noted that the organochlorine contamination of Kestrels was much lower than that seen in raptors favouring avian prey, although mercury had risen from 'not discernible' prior to 1989 to levels of order 200ppb by 2005-2010. But while the results are, to an extent, encouraging, data produced in The Netherlands were more worrying. Movalli *et al.* (2017) took breast feathers from male Kestrel specimens collected over the years 1901-2001, and held in the Dutch Naturalis Biodiversity Centre, and searched for a total of 50 elements including heavy metals. The data represents a benchmark for studying concentrations taken from future samples, but indicated that at times the levels of cadmium, chromium, selenium and arsenic had been above those known to have adverse effects on birds. Mercury is a particular problem as it arises from numerous industrial processes (including the burning of fossil fuels), is transported by atmospheric processes and so may be deposited at distance from any source, and readily converts to a more toxic form (methylmercury) in the presence of dissolved organic matter which is found naturally throughout the environment. Given present concerns over rising levels of cadmium and mercury in the environment, the data suggests that the insidious nature of these and other elements mean that Kestrels may be susceptible to contaminants that were thought to be of little danger to the species.

Population

The Kestrel is Amber listed as a *Bird of Conservation Concern* in both the United Kingdom and the Republic of Ireland due to its population decline over the last several decades. It is a *SPEC 3* species, defining it as one with unfavourable conservation status in Europe, but with a global population concentrated outside Europe.

In the first *Atlas of British Breeding Birds* (Sharrock 1976) the population of Kestrels was estimated at 100,000 pairs. Newton (1984) lowered the estimate to 84,000 pairs. The *New Atlas of British Breeding Birds* (Village in Gibbons *et al.*, 1993) noted that Kestrels could be found in most parts of the British Isles, but that the species' range had diminished. Saxby (1874) included Kestrels in the birds of the Shetlands, noting that it was migratory, arriving in small numbers to nest on sea cliffs, especially those of Unst and Fetlar. But as Shetland's human population declined, with a presumed reduction in the population of mice, Kestrels had to compete with Merlins for small birds. Failing to do so, Kestrels eventually disappeared from the islands. They are still seen regularly as spring and autumn migrants, but are no longer considered a breeding species, though pairs did breed in 1905 and 1992. The species is also absent as a breeding species from the northern Outer Hebrides, and from much of western Scotland. It is found throughout most of England and Wales, though is scarce across much of the Principality and in England's South-West Peninsula. The scarcity in western Scotland, Wales and western England could, perhaps, be due, in part, to high rainfall in those areas which inhibits hunting more than it does in eastern Britain, though the *New Atlas* notes abundant Kestrels in the Lake District which is also an area of high rainfall. Kestrels are also found on the Isle of Man and the Channel Islands. It is scarce in Ireland, where its primary prey, the Field Vole, is absent. The *New Atlas* revised the estimate of population downwards again, suggesting 50,000 pairs bred across the British Isles, though accepting that the likely error on the calculation meant that the range could be 25,000-89,000 pairs.

Shrubb (1993a) reduced the best estimate further, considering the range to be 35,000-40,000 pairs (though this was for Great Britain rather than the British Isles). Baker *et al.* (2006) then suggested the population had stabilised, their estimate being 36,800 pairs for the United Kingdom, with a further 8,400 pairs in the Republic of Ireland. Baker and co-workers consider their estimate to be '2' on a scale of '1' (good) to '3' (poor) deriving it by the extrapolation of a combination of sample surveys and the data of the Common Birds Census (CBC) and/or Breeding Bird Survey (BBS). The latter are surveys of bird population carried out by the British Trust for Ornithology (BTO). The CBC ran from 1962 to 2000 and involved around 250 sample areas annually which were monitored by volunteers. The sample areas were mainly in England, with few sites in the other UK countries. The CBC was replaced by the BBS which had initially run in parallel since its inception in

1994. The BBS increased the number of sample areas, though the coverage was still greater in England than elsewhere.

Clements (2008) extrapolated studies made in several counties across England to define breeding densities in differing farmlands, urban and moorland areas, and used this to derive a total population of 53,000-57,500 pairs for Britain. This overall population estimate broke down as 43,000 pairs (best estimate) in England with ranges of 2,500-3,500 pairs in Wales and 7,500-11,000 pairs in Scotland. Even allowing for the possible limitations on data gathering – a small number of sample areas and some counties missing – which makes such a low figure very unlikely to be realistic, it is considered that Clements overestimates the Welsh breeding population and that the Welsh Ornithological Society's view that the species is a 'rather scarce breeding resident' together with the Kestrel's red-listing in the Principality are a better view of the situation. Musgrove *et al.* (2013) lowered Clements' estimate to 45,000 for Great Britain, with a further 1000 pairs in Northern Ireland.

The latest BBS report (Harris *et al.,* 2018) notes that the Kestrel had declined consistently over the period 1995-2016, with a 35% drop in population. The drop over the decade to 2016 had been 31%, with 27% over following year (*i.e.* to 2017). However, for Scotland the situation had improved (Fig. 121). Riddle (2011) notes that the decline there has been greater than that seen in England quoting a 54% decline between 1995 and 2008 and a further 64% decline 2008-2009, figures which are consistent with Harris *et al.* But Harris and co-workers suggest a doubling of the Kestrel population in Scotland from 2016-2017. Gordon Riddle, one of the foremost authorities on the Kestrel in Scotland, considers this unlikely and is sceptical of the claim (G. Riddle pers. comm.). Though there are areas in which the population does appear to have increased, *e.g.* the Pentland Hills, Riddle considers these are isolated areas of success, to be set against a continuing story of decline. The position in Wales is probably worse than in England or in Scotland.

Fig. 121 notes an overall decline from 100% (2010 population) to about 75% by 2015. The '100%' level was replicated in 1997 (Fig. 121), with a slightly higher figure for 1994 (about 102%). The Welsh Bird Report for 2017 (Birds in Wales, 2018) suggested a population decline from 100% (1994) to about 15% in 2017. If Clements (2008) was correct that there were about 3,000 pairs in Wales in 2008 then there would have been about 3400 pairs in 1994. That would mean a decline to only about 500 pairs in 2017.

For the Republic of Ireland, Lynas *et al.* (2007) Amber listed the Kestrel because of concerns over the population decline, concerns amplified by Crowe *et al.* (2010) who noted that in their study of 52 species in the Republic of Ireland over the period 1998-2008, the Kestrel was one of four showing the greatest rate of decline (the other three being Swift, Skylark and Mistle Thrush (*Turdus viscivorus*)). The study divided the country into 10km x 10km squares which were investigated by 450 field workers. Kestrels were one of a small number of species that were not seen in 70% of the squares. Crowe and

Survival and Population

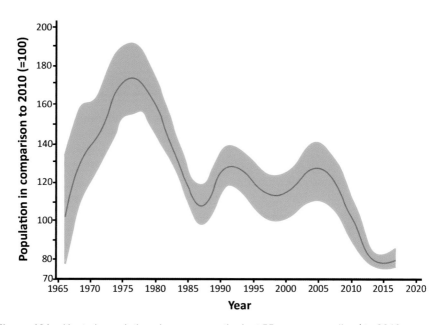

Figure 121. Kestrel population changes over the last 55 years normalised to 2010 when the population was of order 50,000 pairs. The purple line represents the best estimate, the shaded green area being the probable range between upper and lower limits. Redrawn from CBC/BBS data combined with earlier estimates and normalised by the author. The text addresses the contentious apparent upswing in population seen in the 2017 BBS data.

co-workers do not quote a population figure for the Republic, but suggest a 3% decline over the period 1998 to 2008 implying a population of about 7500 pairs if the Baker *et al.* (2006) estimate is correct. However, Mee (2012) in a review of data from both Eire and Northern Ireland which did not reference the work of Crowe *et al.* (2010) considered the total population of the island of Ireland (*i.e.* Eire and Northern Ireland combined) was 10,000[3]. If the estimate of 1000 pairs in Northern Ireland is assumed, this implies a Republic population of about 9000 pairs. In a report by Calhoun and Cummins (2013) looking ahead over a five-year period (2014-2019) the Amber status of the Kestrel was confirmed, but no population data was given.

Fig. 121 notes the increase in Kestrel numbers following the decline which principally resulted from the effects of organochlorine pesticides. What has caused the persistent decline since?

There is evidence to suggest that the mortality of mammal-eating predators is still being influenced by pesticides. Walker *et al.* (2013) measured the concentrations of second-generation anticoagulant rodenticides (SGARs) in the livers of Barn Owl, Red Kite and Kestrel corpses collected across

---

[3] Mee (2012) does not say whether the quoted population is pairs or individuals. Since pairs are usually stated, that has been assumed.

Britain. SGARs were developed to combat the pesticide resistance developed by Brown Rats, and include several poisonous compounds, most often difenacoum and bromadiolone. Walker *et al.* state SGARs were detected in 94% of Red Kites and 84% of Barn Owls. While the concentrations were not shown to have been the direct cause of death in any of these birds, some of the Red Kites showed signs of haemorrhaging which may have been due to high liver concentrations. The study also indicated that the incidence of SGARs in the livers of the two predators was rising. The position for Kestrels differs because while SGARs were found in the livers of 95% of dead birds, the sample size was small (n=20). There was, therefore, no evidence that either the incidence or concentration of SGARs had increased over previous studies. Nevertheless, the rise observed in the other two species implies that rodenticide concentrations in the livers of mammal-eating predators are on the increase and may be contributing to bird mortality.

There is also evidence that the population of Field Voles, the main prey of Kestrels, is becoming less cyclic, raising concerns over the potential loss of ecosystem function. In a cooperative study by researchers across Europe (Cornulier *et al.*, 2013) data on the annual variation of the population of Microtus species and other rodents were aggregated from sites which included northern England, central Europe (from western France to Poland), and northern Europe (Finland and northern Fennoscandia) over time since 1995. When analysed, the data indicated that in 83% of studied rodent populations there had been a two-fold decline in the spring population cyclic amplitude, while in 67% there had been a decline in both spring and autumn cyclic amplitudes, though in these cases the spring decline was the most marked. Only in eastern Germany and Poland was there no evidence of a general pattern that vole population cycles were being dampened. Superimposed on this cycle dampening there was a general decline in vole numbers. Cornulier *et al.* note that there are three basic theories of why such cycle dampening occurs, and that all of these invoke environmental change, one involving random changes, the other two requiring sustained change. While Cornulier *et al.* are careful to avoid confirming a correlation between climate change and vole cycle decline, they do note that the evidence is suggestive of a climatic driver. Whatever the proximate cause of the decline, it has coincided with a time of global environmental change and it is known that a degradation of wintering conditions affects winter reproduction in voles and vole survival, while a reduction in winter population leads to a reduction in spring breeding and a dampening of any spring cycle. Cornulier and co-workers note that the decline, with its frequent prolonged periods of low amplitude changes in vole (and other rodent) populations would not preclude occasional peaks, such as that seen in Fennoscandia in 2011, it would very likely result in a decline in the number of rodent predators, and that, inevitably, would mean a decline in Kestrel numbers. In Fennoscandia rodent populations live beneath the snow which creates a 'blanket' shielding them from much colder ambient air temperatures as well as offering some protection from predation while they forage. Warmer winters have caused the snow blanket to become less efficient, the thaw-freeze conditions which are now more normal creating a more difficult habitat with food occasionally encased in ice (see, for instance, Korslund and Steen, 2006). While such winter conditions are not common in Britain, warmer, wetter winters are also likely to create difficult conditions for voles and so reduce winter survival. However, in their analysis of the effect of climatic variables on 68 breeding bird species in England, Pearce-Higgins and Crick (2019) believed that climate change would overall influence the Kestrel population in a positive way.

The evidence for the reduction in vole population cyclicity might also help to explain why the Kestrel is a declining species across mainland Europe. The European Bird Census Council (PECBMS, 2012) notes a 38% reduction in the European Kestrel population since 1980, but a 42% reduction since 1990, the European population showing a similar trend to that of British Kestrels. At present the European decline is considered 'moderate', meaning less than 5% per annum. Interestingly, one country in which the trend has been reversed is Finland where the Kestrel population almost doubled from 1998 to 2009 (Honkala *et al.*, 2012). There was a slight fall in 2010/2011, but this seems to have been part of a three-year cycle which the population has shown consistently over the last 20 or so years, following a similar rodent population cycle. The reduction in pesticide levels is considered to account for the initial rise in Kestrel numbers, while the later continued increase resulted from the nationwide deployment of nestboxes to which the birds have rapidly become accustomed. A further thought on the rise of the population in Finland will be noted below when we consider Kestrel predation by Goshawks.

In relation to population, the observations of Birot-Colomb *et al.* (2019) at the Défilé de l'Écluse in the French Jura, already mentioned in Chapter 9 with regard to migration routes, are worth noting again. While counts in 1993-2010 reflected the cyclicity of rodents and the winter snow cover in Scandinavia, there was then a sudden increase in Kestrel numbers during 2013-2017, mirroring increases also seen in Sweden (Falsterbo) and Organbidexka (northern Spain). The increase suggests a healthy, probably increasing, population in Scandinavia, despite the population in Denmark having decreased by about 10%. The autumnal numbers also increased in the Jura, consistent with the Swiss population of the falcons having increased by 138% between 1999 and 2017 (Schmid *et al.*, 2018).

The recent decline in Kestrel numbers in Great Britain has coincided with an increase in Buzzard, Peregrine and Goshawk numbers. Shrubb (2003) noted the increase in Buzzard numbers in Wales and considered these to have increased pressure on Kestrels as the larger birds are resident and so tend to claim the better territories, and as they take larger prey are better able to make a living in the landscape, moving Kestrels away from better hunting areas. The

Above   Adult Buzzard, southern Scotland.
Opposite   Recently fledged Buzzard. Our house adjoins a small wood in which buzzards nested in 2020. My wife christened the adult(s) *Busby*, so the single chick they raised was, inevitably, called *Matt*. Our house is remote from others and *Matt* spent a great deal of time in the garden, grubbing for worms and, eventually, taking seeds and peanuts from the bird table, to the consternation of the usual clientele. *Matt* also tried, not too successfully, to use the bird bath.

increase in Buzzard numbers is not restricted to Wales. Stevens *et al.* (2019) suggest that the Buzzard population of central southern England increased by 50% between 2011 and 2016, while Walls and Kenward (2020) suggest a 46% increase for the UK as a whole since 2000. If Buzzards are outcompeting Kestrels in Wales as Shrubb (2003) suggests, they will presumably be doing the same in the rest of the UK.

The remains of a female Kestrel found beneath the tree-mounted nestbox she had taken over for breeding in south-west Scotland. Gordon Riddle, who found the remains and also took the photograph noted that the primary feathers, which are 'notoriously difficult to remove', had been plucked out, suggesting the culprit was a Goshawk.

Petty *et al.* (2003), in a study in the Kielder Forest, Northumberland, noted that Kestrel numbers declined over a 23-year period despite voles remaining abundant (though with cyclic fluctuations). Allowing for changes in vole population, the decline in Kestrel numbers was positively correlated with the increase in Goshawk numbers, the larger raptor having moved into the area in 1973 with numbers increasing until 1989 and then remaining stable (Fig. 122a). The positive correlation between Kestrel numbers and Goshawk numbers (Fig. 122b) suggests that the Goshawks took a progressively larger fraction of a declining Kestrel population.

The number of Short-eared Owls, diurnal vole hunters as Kestrels are, also declined (though to a lesser extent), while the numbers of Long-eared and Tawny owls (nocturnal vole hunters) stayed constant. Petty and co-workers therefore examined the diet of the local Goshawks. The hawks fed mainly on birds, including six raptors – Kestrel, Long-eared, Short-eared and Tawny Owls, Sparrowhawk and Merlin. Raptors comprised only 4.5% of the Goshawk diet, but Kestrels formed 58.2% of the raptor total, *i.e.* they contributed more to the diet than all other raptors combined. Kestrels were predated more in spring and autumn than during the summer (Fig. 123 overleaf). This means the Goshawks were killing unpaired Kestrels looking for territories prior to the commencement of breeding, predation which would have the greatest impact on the population, and during the autumn when the number of juveniles was high, predation which would have an impact on the population the following spring.

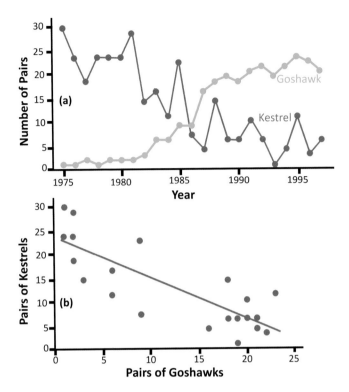

Figure 122.
(a)   Variation in numbers of pairs of Kestrels and Goshawks over a 23-year period in Northumberland's Kielder Forest.
(b)   Variation in the number of pairs of Kestrels and Goshawks over the same period. The correlation in the decline of Kestrels with the increase in Goshawks (as indicated by the regression line) is statistically significant.
Redrawn from Petty *et al.* (2003).

Kestrels seem particularly vulnerable to attack, which, as Petty and co-workers note, is likely to be as a result of flight-hunting, which allows a 'blind-side' attack. Though taken less often than Kestrels, Short-eared Owls are also vulnerable being diurnal, and having a hunting method – slow quartering – which is open to a similar 'blind-side' attack. Before deciding that Goshawks were the definite cause of the Kielder Kestrel population decline, Petty *et al.* considered three other possible factors. Lack of nest sites was dismissed as they believed the number of potentially usable stick nests had increased and that this outweighed the possible loss of cliff sites which had become overgrown. Next the researchers considered a regional decline which had impacted the Kielder population, but could find no evidence for one. Finally, they considered a reduction in prey population, but found no evidence for a long-term decline in vole numbers.

The Common Kestrel

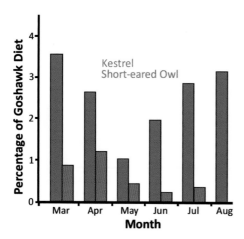

Figure 123. Variation in the fraction (as a percentage) of Kestrels and Short-eared Owls in the diet of Goshawks from March to August over a 22-year period in Northumberland's Kielder Forest. The difference from the expected diet fraction, given the populations of the species, was statistically significant, *i.e.* the Goshawks were preferentially hunting the diurnal vole eaters at particular times (spring and late summer). There was no such preference shown in the Goshawk take of nocturnal Long-eared and Tawny owls. The take (dietary fraction) of these owls was about five-times lower. Redrawn from Petty *et al.* (2003).

Petty *et al.* leave little doubt that Goshawk predation of Kestrels is significant in reducing numbers in areas where both are found, and it is no surprise to find that both Shrubb (2004) and Riddle (2011) noted the potential significance of Goshawk predation following population increases of the species in Wales and Scotland respectively. As noted earlier in this Chapter, Goshawks are found in Finland where the number of Kestrels is increasing, probably due to the installation of nestboxes. So why are Kestrels managing to overcome Goshawk predation and increase their population? The answer appears to be twofold – the two raptors have cohabited for centuries which means Kestrels have been able to adapt their behaviour in some way. And it is also the case that the increase in Kestrel numbers has occurred at a time when the Finnish Goshawk population is in decline (P. Byholm, pers. comm.).

The population of Peregrines in Britain has also increased in recent years, and the larger falcons may both outcompete Kestrels for nesting sites, particularly in urban areas (examples of the larger falcon ousting the smaller species from historic nesting sites are known), and reduce Kestrel numbers by direct predation. Red Kite numbers have also increased, creating pressure on Buzzards which may, in turn, have caused the latter to compete with Kestrels for territory as well as, perhaps, increasing direct predation.

However, as Kestrel numbers have declined, the population of Hobbies in Britain has increased, and while there is clearly no one-to-one comparison between the two smaller falcons in terms of habitat, diet *etc.* the implication is that predation by, or competition from, larger raptors cannot be the entire story. We must therefore look for other factors which may have influenced the Kestrel population decline.

Human use of the countryside for recreation is increasing, as is the human population of Britain, and there is evidence that disturbance of wildlife affects behaviour. In a simulation study of Golden Eagles in the USA, Pauli *et al.*

Above An adult male Goshawk which has mated with a juvenile female dispute a kill. *Torsten Prohl*.

Below Goshawk chicks, Scotland. These beautiful birds may grow up to be a danger to Kestrels.

Given the threat of potential Goshawks to flight-hunting Kestrels, these fledglings may spend their lives looking skyward.

(2017) noted that increases in human recreation caused significantly lower and more variable growth rates in the eagle population and in territory occupancy. The researchers concluded that even though the eagle was a long-lived bird, adaptation to human disturbance could not compensate for the rate of increase in recreational activities. There is no reason to believe that smaller raptors with shorter lifespans would be any better at adapting. However, for the Kestrel it is likely that changes in agriculture are a more fundamental concern.

The effect of vegetation on the hunting success of Kestrels has been known for many years. Kestrels usually flight-hunt at heights of 8-20m, Shrubb (1993a) noting that the height chosen varied with the season, seemingly as a result of vegetation length, the falcons flight-hunting above the usual height range in summer when vegetation was thicker or taller. Noting that in summer Kestrels were flight-hunting high over unmown fields, but much lower as soon as these were mown, Shrubb conjectured whether the height allowed a wider range of view. This would be correct if visual signals (either UV or visible light) are the primary/sole source of hunting cues as picking up unnatural (*i.e.* not wind-generated) vegetation movements might be easier over a longer distance, allowing a move to a closer vantage point for more decisive information. Cavé (1968), studying Kestrels in The Netherlands, also noted that the nature of flight-hunting – the search for prey on the ground – means that tall vegetation makes the technique difficult. He illustrated this by noting that Kestrels did not hunt above the reed banks on polders, choosing the shorter vegetation areas beside them. He also noted that one of the male Kestrels in his study area flight-hunted over fields of Barley (*Hordeum vulgare*), but only over fields of Lucerne (or Alfalfa (*Medicago sativa*)) after the crop had been mowed: Alfalfa forms much denser ground cover. More recently, Won et al. (2016), studying Kestrel hunting in various habitats in South Korea, also noted that Kestrels used perch-hunting rather than flight-hunting over taller vegetation, with the reverse being true over shorter vegetation. While in part this may have been due to the fact that insects formed a greater fraction of the prey than is the case in Britain (and insects were more abundant over all habitats in the South Korean study), the Kestrels flight-hunted for mammals as their chicks aged, and the preference for shorter vegetation was clear.

Riddle (2011) considered that the reduction in land quality in Scotland, due to increased numbers of sheep and deer, increased sheep numbers in Wales and afforestation in both countries limited Kestrel habitat, while farming practices (limiting the acreage of mixed farmland) in England had further reduced available habitat. From their work in the Republic of Ireland, Crowe *et al.* (2010) agreed, noting that farming practices were having an impact on the populations of all avian farmland specialists: while Kestrels inhabit a range of habitats, from towns to moorland, many of the falcons are indeed farmland specialists. Shrubb (1993a) concurred, noting that intensive sheep farming (the number of sheep in Wales, for instance, has almost doubled since the 1960s) reduces the availability of rough grazing, the preferred vole habitat and hence reduced the abundance of both voles and Kestrels. In later work (Shrubb 2003) noted the conclusions of Harris *et al.* (1995) who found that after myxomatosis had substantially reduced their numbers, the population of Rabbits had increased, with a subsequent decline in vole numbers.

The major changes in agriculture over recent decades that have been detrimental to Kestrels include the removal of hedgerows; the improvement of rough grassland; the increase in agricultural monocultures; the change from hay to silage production which means earlier and more frequent cuts; and the change from spring to autumn sowing. Each of these may have caused a decline in vole habitat. The general intensification of agriculture will also have reduced vole habitat and numbers – it is now common to see crops harvested and, within a day or two, fields ploughed and then reseeded, allowing little time for small mammals to benefit from harvesting spillage. Modern machinery also means that spillage had been significantly reduced. These changes have had an impact on all farmland birds. Krebs *et al.* (1999), in an article whose title reflects that of Rachel Carson, the American marine biologist and conservationist whose book (*Silent Spring*) on the effect of pesticides on species populations in 1962 is often considered to have stimulated environmentalism, estimates that 10 million birds from ten species disappeared from the British countryside in the period 1979-1999, a decline which was reflected across Europe. Krebs *et al.* note that in Europe 116 farmland bird species are now listed as being of conservation concern, and consider the decline is due in large part to the intensification and industrialisation of agriculture, adding land drainage and a reduction in traditional rotation to the list of detrimental effects noted above.

The idea that agriculture was changing the face of the British countryside has been taken up in a series of studies. While noting the intensification of agriculture Vickery *et al.* (2001) drew attention to the decline in grassland and its effect on bird species. The researchers noted that grassland was now intensively managed, with greater numbers of sheep, a two-fold increase in the use of inorganic nitrogen and, as noted above, a switch from hay to silage and an increase in monocultures. The effect of these changes was to decrease the availability of the land for both avian feeding and breeding. Close mowing reduced the invertebrate population, the timing of mowing reduced seed availability, short grass meant reduced cover and so raised the risk of predation, and increased sheep numbers meant ground nests were prone to trampling.

Recognising that field margin management had been proposed as a way forward in the protection of bird populations, Vickery *et al.* (2002) studied various management ideas and their effect on invertebrate populations in summer, and seed and green material availability in summer and winter. The research team concluded that the best solution was naturally regenerated set-aside margins, though uncropped wildlife strips and grass/wildflower strips were almost as effective. Grass-only strips were identified as the least effective.

Attempts to arrest and reverse the decline in farmland species come under the general heading of Agri-environment Schemes (AESs) which were supported by monies from the Common Agricultural Policy of the European Union and the British government (before the UK's departure from the EU at the end of January 2020). Such schemes began some 20 years ago with the intention of halting and then reversing the decline in 20 species of farmland birds, of which the Kestrel was one. The idea behind the first AESs in England were studied by Vickery *et al.* (2004). The team noted that the requirements of each of the birds on the list differed, making it difficult for a single scheme to aid the population of all 20 species. Consequently, while the quantity of AESs seemed adequate, more work was needed to define the required balance between 'broad and shallow' and 'narrow and deep' options so that rarer species could benefit along with the more common ones. In a later study (Vickery *et al.*, 2009) it was noted that despite their limitations most arable field margins were still sown-grass strips which had already been identified as being least effective.

The assessment of Vickery *et al.* (2009) tallies well with the study of Aschwanden *et al.* (2005) in Switzerland. There, the researchers examined the results of conservation measures put in place after a reduction in vole population had led to a decline in both Kestrel and Long-eared Owl populations. A Swiss AES required the creation of ecological compensation areas which constituted 7% of the country's farmland. Aschwanden *et al.* studied five habitat types, three defined by the AES – wildflower strips (which constituted 0.4% of studied acreage), herbaceous strips (0.3%) and low intensity (mowed only after 15 June) meadow (3.6%) – and two conventional field types – 'artificial' grassland (part of a crop rotation system, mowed often from April to October – 49.0%) and autumn-sown wheat (8.5%). All other field types – maize, potatoes *etc.* – were classified as 'other' and constituted the remaining 38.2%. The Swiss team found that the vole population density in wildflower and herbaceous strips was eight-times that of low-intensity meadows and artificial grassland. However, despite this increase the Kestrels (and the owls) did not preferentially hunt these prey-rich areas (Fig. 124).

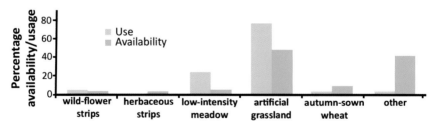

Figure 124. Comparison of habitat type availability and its usage by hunting Kestrels (as percentages). Redrawn from Aschwanden *et al.* (2005).

That the birds were trading the benefit of accessibility against prey availability became clear when mowing was considered, the birds transferring their attention to areas with greater accessibility (Fig. 125 overleaf).

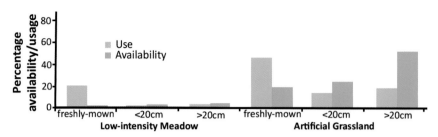

Figure 125. Comparison of habitat availability, in terms of grass height, and its usage by hunting Kestrels (as percentages). Redrawn from Aschwanden *et al.* (2005).

The existence of wildflower or herbaceous strips adjacent to low-intensity meadows or artificial grassland also made a difference, voles migrating into these areas and making them more attractive to Kestrels (Fig. 126).

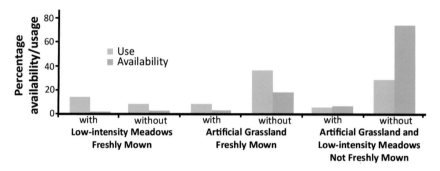

Figure 126. Comparison of habitat availability, in terms of whether it is with or without adjacent wildflower or herbaceous strips, and its usage by hunting Kestrels (as percentages). Redrawn from Aschwanden *et al.* (2005).

Aschwanden *et al.* also noted that while the Kestrels found the wildflower and herbaceous strips less useful during the summer, they were preferred hunting areas during winter and early spring when less dense vegetation allowed the falcons greater access to the higher vole densities. The Swiss researchers also found that aside from the actual details of the advantages of the AES to Kestrels, the benefit of the compensation areas lay in the overall ecology of the sites.

In western France, Meunier *et al.* (2000) surveyed almost 3000km of motorways and other roads assessing the importance of roadside verges (particularly the verges of motorways which were mowed less frequently) to diurnal raptors. The French team noted the importance of the habitats to Kestrels and Buzzards (and to Black Kites (*Milvus migrans*)), though they concluded verges were less important to harrier species. Kestrels used the verges throughout the year, but much more frequently in winter, the team considering this had less to do with the availability of prey than the abundance of posts which could be used for perch-hunting which is less energy-demanding than flight-hunting. Meunier *et al.* conclude that the management of verges was an important element in Kestrel conservation.

In a study of Kestrels in northern Italy Costantini *et al.* (2014) noted that falcons breeding in areas of intense cultivation (areas with limited temporary or permanent grassland) bred later and had offspring with poorer body condition (measured by body mass and size) – Fig. 127. Such young Kestrels are less able to cope with the rigours of their first winter.

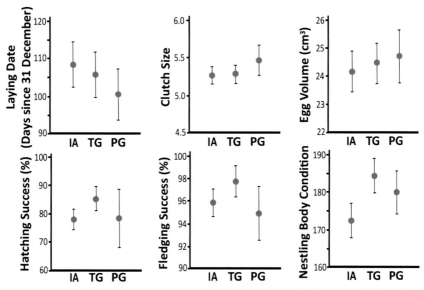

Figure 127. A comparison of the success of Kestrels breeding in areas of Intensive Agriculture (IA), Temporary Grassland (TG) and Permanent Grassland (PG) in terms of laying date, clutch size, egg volume, hatching success, fledging success and nestling body condition (based on a number of factors including body size). The black symbols indicate standard errors. Redrawn from Costantini *et al.* (2014).

In an equally interesting study, Newson *et al.* (2009) took data from the BBS 1994-2006 survey for 23 Red- and Amber-listed species, including the Kestrel, and looked at the change of population across 12 carefully selected habitats which covered the range of woodland, upland and lowland, various farmland types, wetlands and urban landscapes. For each habitat a 'preference index' was identified for each species which allowed the researchers to decide if the habitat was preferred or avoided. For habitats relevant to each species they then calculated the change in population over the 12-year period. In all cases there was, in general, a decline in population (of up to 10%) on individual habitats, though in some cases there was no change, or a small positive increase. For the Kestrel, there had been a 7% decline in urban and suburban habitats, a smaller decline in human rural habitats, a small increase

Female Kestrel with vole. The tail length and reddish tinge to the pelage would suggest a Bank Vole.

in mixed farmland and little or no change in other habitats. Newson and co-workers note that town living has an 'avoid preference' index for Kestrels, while all other habitats had indices implying degrees of preference. For the Kestrel, Newson and co-workers note, this could mean disappearance from a less-favoured habitat, coupled with the start of a more widespread decline as the range of habitats available to the falcons contracts.

Vegetation management for the conservation of the Kestrels has already been considered in Chapter 4 (*e.g.* the work of Garratt *et al.*, 2011, Garratt *et al.*, 2012 and Peggie *et al.*, 2011) and above. From the studies considered above, chiefly those relating to AESs, management is critically important. The three identified reports noted here are chapters of a PhD thesis devoted to the subject and its benefit to the Kestrel (Garratt, 2011). Garratt *et al.* (2011) note that the cutting of field margins as an aid to wildlife is often restricted to late July to avoid damage to ground-nesting species and that this is too late for Kestrels whose breeding season is over by then. In considering the options which might be possible within future AESs for the benefit of Kestrels, Garratt (2011) notes the success of roadside verge management in providing a preferred hunting habitat for the falcons. Garratt concludes by suggesting that the most profitable management for the benefit of the falcons, as well as other wildlife, including plants, might be to leave one half of field margins – the half adjacent to the field hedge or other boundary – uncut for ground-nesting species and as a 'reservoir' for small mammals, and cutting the remaining half at different times for different fields, so that there is close-cut grass available across a farm at various times from spring to autumn. Individual margins should be cut no more than twice as excessive cutting decreases plant diversity. Garratt notes that the decision on whether to leave grass cuttings *in situ* or to remove them is difficult. Leaving them benefits mammals and, probably, invertebrates, but increases soil fertility which can attract dominant plant species such as the Common (or Stinging) Nettle (*Urtica dioica*). Garratt's suggestion is to leave discrete areas of lying cut grass. While these suggestions are carefully considered and have great merit, the work required to bring them to fruition is considerable which is hardly likely to be welcomed by farmers who are already under time pressure.

Overall, while the importance of vegetation management cannot be overstated, it is probable that the decline of Kestrels in Britain is due to a combination of several factors, including those mentioned above (and others, see, for instance, the comment on wind turbine collisions in *Survival* earlier in this Chapter). The deep decline in Kestrel numbers since the last fluctuation peak of 2005 suggests that something radical is occurring and that the population will need more careful observation in the future to identify the need for further, perhaps more dramatic, conservation measures if the decline continues. It is very much to be hoped that solutions can be found so that future generations are not deprived the sight of a windhovering Kestrel.

REFERENCES

Titles in parentheses are translations of the originals.

Adair, P., (1891). The Short-eared Owl (*Asio accipitrinus* [sic]) and the Kestrel (*Falco tinnunculus*) in the vole plague districts, *Ann. Scot. Nat. History*, 219-231.

Adair, P., (1893). Notes on the disappearance of the Short-tailed Vole (*Arvicola agrestis* [sic]), and some on the effects of the visitation, *Ann. Scot. Nat. History*, 193-202.

Adriaensen, F., Verwimp, N. and Dhondt, A.A., (1997). Are Belgian Kestrels *Falco tinnunculus* migratory: an analysis of ringing recoveries, *Ring. Migr.* 18, 91-101.

Adriaensen, F., Verwimp, N. and Dhondt, A.A., (1998). Between cohort variation in dispersal distance in the European Kestrel *Falco tinnunculus* as shown by ringing recoveries, *Ardea*, 86, 147-52.

Albuquerque, J.L.B., (1982). Observations on the use of rangle by the Peregrine Falcon (*Falco peregrinus tundrius*) wintering in southern Brasil [sic], *Raptor Res.*, 16, 91-92.

Andersson, M. and Norberg, R.Å., (1981). Evolution of reversed sexual size dimorphism and role partitioning among predatory birds with a size scaling of flight performance, *Biol. J. Linne. Soc.*, 15, 105-130.

Aparicio, J.M., (1994a). The seasonal decline in clutch size: an experiment with supplementary food in the Kestrel *Falco tinnunculus*, *Oikos*, 71, 451-458.

Aparicio, J.M., (1994b). The effect of variation in the laying interval on proximate determination of clutch size in the European Kestrel, *J. Avian Biol.*, 25, 275-80.

Aparicio, J.M., (1998). Individual optimization may explain differences in breeding time in the European Kestrel *Falco tinnunculus*, *J. Avian Biol.* 29, 121-128.

Aparicio, J.M., (1999). Intraclutch egg-size variation in the Eurasian Kestrel: Advantages and disadvantages of hatching from large eggs, *Auk*, 116, 825-830.

Arnett, E.B. and Baerwald, E.F., (2013). Impacts of wind energy development on bats: implications for conservation, in Adams, R.A. and Peterson, S.C. (Eds.), *Bat Evolution, Ecology, and Conservation*, Springer, New York, 435–456.

Aschwanden, J., Birrer, S. and Jenni, L., (2005). Are ecological compensation areas attractive hunting sites for Common Kestrels (*Falco tinnunculus*) and Long-eared Owls (*Asio otus*)?, *J. Orn.* 146, 279-286.

Askew, G. N. and Ellerby, D. J., (2007). The mechanical power requirements of avian flight, *Biol. Lett.* 3, 445-448.

Avilés, J.M., Sánchez, J.M. and Parejo, D., (2001). Breeding rates of Eurasian Kestrels (*Falco tinnunculus*) in relation to surrounding habitat in south-west Spain, *J. Raptor Res.*, 35, 31-34.

Baker, H., Stroud, D.A., Aebischer, N.J., Cranswick, P.A., Gregory R.D., McSorley, C.A., Noble, D.G. and Rehfisch, M.M., (2006). Population estimates of birds in Great Britain and the United Kingdom, *Brit. Birds*, 99, 25-44.

Baker, R.R. and Bibby, C.J., (1987). Merlin *Falco columbarius* predation and theories of the evolution of bird coloration, *Ibis*, 129, 259-263.

Balfour, E., (1955). Kestrels nesting on the ground in Orkney, *Bird Notes*, 26, 245-253.

Barrios, L. and Rodríguez, A., (2004). Behavioural and environmental correlates of soaring-bird mortality at on-shore wind turbines, *J. Appl. Ecol.*, 41, 72-81.

Barton, N.W.H. and Houston, D.C., (2001). The incidence of intestinal parasites in British birds of prey, *J. Raptor Res.*, 35, 71-73.

Bennett, A.T.D. and Cuthill, I.C., (1994). Ultraviolet vision in birds: what is it for?, *Vis. Res.*, 34, 1471-1478.

Beukeboom, L., Dijkstra, C., Daan, S. and Meijer, T., (1988). Seasonality of clutch size determination in the Kestrel *Falco tinnunculus*: An experimental approach, *Ornis Scand.*, 19, 41-48.

Bird, D.M., Weil, P.G. and Lague, P.C., (1980). Photoperiodic induction of multiple breeding seasons in captive American Kestrels, *Can. J. Zool.*, 58, 1022-1026.

Birot-Colomb, X., Giacomo, C., Maire, M. and Matérac, J-P., (2019). (Annual numbers (1993-2017) of raptor autumnal migration at the Défilé de l'Écluse (Haute-Savoie and Ain, France)), *Nos Oiseaux*, 66, 101-126.

Boileau, N. and Bretagnolle, V., (2014). Post-fledging dependence period in the Eurasian Kestrel (*Falco tinnunculus*) in western France, *J. Raptor Res.*, 48, 248-256.

Boileau, N. and Hoede, C., (2009). (Variations in egg size in the Common Kestrel *Falco tinnunculus*), *Alauda*, 77, 21-30.

Bonin, B. and Strenna, L., (1986). (On the biology of the Kestrel *Falco tinnunculus* in Auxois), *Alauda*, 54, 241-262.

Boratyński, Z. and Kasprzyk, K., (2005). Does urban structure explain shifts in the food niche of the Eurasian Kestrel (*Falco tinnunculus*), *Buteo*, 14, 11-17.

Both, C., Bom, R. and Samplonius, J., (2013). Peregrine Falco *Falco peregrinus* steals vole from Kestrel *Falco tinnunculus*, *De Takkeling*, 21, 226-228.

Bright, J.A., Marugán-Lobón, J., Cobb, S.N. and Rayfield, E.J., (2016), The shapes of bird beaks are highly controlled by nondietary factors, *Proc. Nat. Acad. Sci. USA*, 113, 5352-5357.

Brockman, H.J, and Barnard, C.J., (1979). Kleptoparasitism in birds, *Anim. Behav.*, 27, 487-514.

Brua, R.B., (2002). Parent-embryo interactions, in *Avian Incubation: Behaviour, Environment, and Evolution* (D.C. Deeming, Ed), Oxford University Press, 88-99.

Bruderer, B. and Boldt, A., (2001). Flight characteristics of birds: 1. Radar measurements of speeds, *Ibis*, 143, 178-204.

Bruderer, B., Peter, D., Boldt, A. and Liechti, F., (2010). Wing-beat characteristics of birds recorded with tracking radar and cine camera, *Ibis*, 152, 272-291.

Bundle, M.W., Hansen, K.S. and Dial, K.P., (2007). Does the metabolic rate–flight speed relationship vary among geometrically similar birds of different mass?, *J. Exp. Biol.*, 210, 1075-1083.

Bustamante, J., (1994). Behavior of Common Kestrels (*Falco tinnunculus*) during the post-fledging dependence period in south-western Spain, *J. Raptor Res.* 28, 79-83.

Cabrera-Cruz, S.A and Villegas-Patraca, R., (2016). Response of migrating raptors to an increasing number of wind farms, *J. App. Ecol.*, 53, 1667-1675.

Cade, T.J., (1982). *The Falcons of the World*, Cornell University Press, Ithaca and Collins, London.

Calhoun, K. and Cummins, S., (2103). Birds of Conservation Concern in Ireland 2014-2109, *Irish Birds*, 9, 523-544.

Carrillo, J. and González-Dávila, E., (2010). Geo-environmental influences on breeding parameters of the Eurasian Kestrel (*Falco tinnunculus*) in the western Palearctic, *Ornis Fennica*, 87, 15-25.

Carrillo, J., González-Dávila, E. and Ruiz, X., (2017). Breeding diet of Eurasian Kestrels (Falco tinnunculus) on the oceanic island of Tenerife, *Ardea*, 105, 99-111.

Cavé, A.J., (1968). The breeding of the Kestrel *Falco tinnunculus* L., in the reclaimed area Oostelijk, Flevoland, *Netherlands J. Zool.*, 18, 313-407.

Charter, M., Izhaki, I., Bouskila, A. and Leshem, Y., (2007). Breeding success of the Eurasian Kestrel (*Falco tinnunculus*) nesting on buildings in Israel, *J. Raptor Res.*, 41, 139-143.

Chávez, A.E., Bozinovic, F., Peichl, L. and Palacios, A.G., (2003). Retinal spectral sensitivity, fur coloration, and urine reflectance in the Genus *Octodon* (Rodentia): implications for visual ecology, *Invest. Opthalmol. & Vis. Sci.*, 44, 2290-2296.

Chesser R. T. and 11 others, (2012). Fifty-third supplement to the American Ornithologists' Union check-list of North American birds, *Auk*, 129, 573-588.

Clements, R., (2008). The Common Kestrel population in Britain, *Brit. Birds*, 101, 228-234.

Coath, M., (1992). Talon locking between Kestrel and Red-footed Falcon, *Brit. Birds*, 85, 496.

Combridge, M.C. and Combridge, P., (1992). Red-footed falcon robbing Kestrels, *Brit. Birds*, 85, 496.

Cordeiro, A., Mascarenhas, M., and Costa, H., (2013). Long term survey of wind farms impacts on Common Kestrel's populations and definition of an appropriate mitigation plan, in *Conference in Wind Power and Environmental Impacts, Stockholm 5-7 February 2013,* Book of Abstracts, Report 6546.

Cornulier, T., Yoccoz, N.G., Bretagnolle, V., Brommer, J.E., Butet, A., Ecke, F., Elston, D.A., Framstad, E., Henttonen, H., Hömfeldt, B., Huitu, O., Imholt, C., Ims, R.A., Jacob, J., Jędrzejewska, B., Millon, A., Petty, S.J., Pietiäinen, H., Tkadlec, E., Zub, K. and Lambin, X., (2013). Europe-wide dampening of population cycles in keystone herbivores, *Science,* 340, 63-66.

Costantini, D., Dell'Omo, G., La Fata, I. and Casagrande, S., (2014). Reproductive performance of Eurasian Kestrel *Falco tinnunculus* in an agricultural landscape with a mosaic of land uses, *Ibis,* 156, 768-776.

Cresswell, W. (1993). Wintering raptors and their avian prey: a study of the behavioural and ecological effects of predator-prey interactions, PhD thesis, University of Edinburgh.

Crowe, O., Coombes, R.H., Lysaght, L., O'Brien, C., Choudhury, K.R., Walsh, A.J., Wilson, J.H. and O'Halloran, J., (2010). Population trends of widespread breeding birds in the Republic of Ireland 1998–2008, *Bird Study,* 57, 267-280.

Csermely, D., Bonati, B. and Romani, R., (2009). Predatory behaviour of Common Kestrels (*Falco tinnunculus*), *J. Ethol.,* 27, 461-465.

Daan, S. and Dijkstra, C., (1988). Date of birth and reproductive value of Kestrel eggs: on the significance of early breeding – Chapter 5 of Dijkstra, C., Reproductive tactics in the Kestrel (*Falco tinnunculus*): A study in evolutionary biology, PhD thesis, University of Groningen, 1988.

Daan S., Masman, D. and Groenewold, A., (1990a), Avian basal metabolic rates: their association with body composition and energy expenditure in nature, *Am. J. Physiol.,* 28, R333-R340.

Daan, S., Dijkstra, C. and Tinbergen, J.M., (1990b). Family planning in the Kestrel (*Falco tinnunculus*): the ultimate control of covariation of laying date and clutch size, *Behaviour* 14, 83-116.

Daan, S., Deerenberg, C. and Dijkstra C., (1996). Increased daily work precipitates natural death in the Kestrel, *J. Anim. Ecol.,* 65, 539-544.

Deane, C. D., (1962). Life of the wild, *Belfast Telegraph,* 19 September 1962.

Deerenberg, C., Pen, I., Dijkstra, C., Arkies, B-J., Henk Visser, G. and Daan, S., (1995). Parental energy expenditure in relation to manipulated brood size in the European Kestrel *Falco tinnuculus, Zoology,* 99, 39-48.

Dejonghe, J.-F., (1989). (Importance, structure, origins, biometrics and population dynamics of Kestrels during pre-nuptial migration in Cape Bon (Tunisia)), *Alauda,* 57, 17-45.

Desjardin, D., Maruniak, J.A. and Bronson, F.H., (1973). Social rank in House Mice: Differentiation revealed by ultraviolet visualisation of urinary marking patterns, *Science,* 182, 939-941.

Dial, K. P., Biewener A. A., Tobalske B. W. and Warrick D. R., (1997). Mechanical power output of bird flight, *Nature,* 390, 67-70.

Dickson, R.C., (1987). Kestrels copulating in winter, *Scottish Birds,* 14, 215.

Dickson, R.C. and Dickson, A.P., (1993). Kestrels feeding on road casualties, *Scottish Birds,* 17, 56.

Dietz, M.W., Daan, S. and Masman, D., (1992). Energy requirements for molt in the Kestrel *Falco tinnunculus, Physiological Zoology,* 65, 1217-1235.

Dijkstra, C., Vuursteen, L., Daan, S. and Masman, D., (1982). Clutch size and laying date in the Kestrel *Falco tinnunculus:* effect of supplementary food, *Ibis,* 124, 210-213.

Dijkstra, C., Bult, A., Bijlsma, S., Daan, S., Meijer, T. and Zijlstra, M., (1990a). Brood size manipulations in the Kestrel (*Falco tinnunculus*): Effects on offspring and parent survival, *J. Anim. Ecol.,* 59, 269-285.

Dijkstra, C., Daan, S. and Buker, J. B., (1990b). Adaptive seasonal variation in the sex ratio of Kestrel broods. *Functional Ecology* 4, 143-8.

Dijkstra, C., Daan, S., Meijer, T., Cavé, A.J. and Foppen, R.P.B., (1988). Daily and seasonal variations in body mass of the Kestrel in relation to food availability and reproduction, *Ardea*, 76, 127-140, 1988.

Dolnik, V.R., (1995). (*Energy and Time Resources in Birds in Nature*), Nauka, St. Petersburg, Russia. (In Russian).

Doody, J.S., Stewart, B., Camacho, C. and Christian, K., (2012). Good vibrations? Sibling embryos expedite hatching in a turtle, *Anim. Behav.* 83, 645-651.

Drent, R.H. and Daan, S., (1980). The prudent parent: energetic adjustments in avian breeding, *Ardea*, 68, 225-252.

Duncan, J.R. and Bird, D.M., (1989). The influence of relatedness and display effort on the mate choice of captive female American Kestrels, *Anim. Behav.*, 37, 112-117.

Duquet, M. and Nadal, R., (2012). (The capture of bats by raptors), *Ornithos*, 19, 184-195.

Ericson, P.G.P., Anderson, C.L., Britton, T., Elzanowski, A., Johansson, U.S., Källersjö, M., Ohlson, J.I., Parsons, T.J., Zuccon, D. and Mayr, G., (2006). Diversification of Neoaves: integration of molecular sequence data and fossils, *Biology Letters*, 2, 543-547.

Ermolaev, A.I., (2016). Common Kestrel (*Falco tinnunculus*, Falconiformes, Falconidae) in Colonial Settlements of Rook (*Corvus frugilegus*, Passeriformes,Corvidae) in Steppe Ecosystems of the Manych Valley, *Biology Bulletin*, 43, 870-875.

Espie, R.H.M., Oliphant, L.W., James, P.C., Warkentin, I.G. and Lieske, D.J., (2000). Age-dependent breeding performance in Merlins (*Falco columbarius*), *Ecology*, 81, 3404-3415.

Fairley, J.S., (1973). Kestrel pellets from a winter roost, *Irish Nat. J.*, 17, 407-409.

Fairley, J.S. and McLean, A., (1965). On the summer food of the Kestrel in Northern Ireland, *Brit. Birds*, 58, 145-148.

Fargallo, J.A., Blanco, G. and Soto-Largo, E., (1996). Possible second clutches in a Mediterranean montane population of the Eurasian Kestrel (*Falco tinnunculus*), *J. Raptor Res.*, 30, 70-73.

Fargallo, J.A., Blanco, G., Potti, J. and Viñuela, J., (2001). Nestbox provisioning in a rural population of Eurasian Kestrels; breeding performance, nest predation and parasitism, *Bird Study*, 48, 236-244.

Fox, R., Lehmkuhle, S.W. and Westendorf, D.H., (1976). Falcon visual acuity, *Science*, 192, 263-265.

Frere, H.T., (1886), Changes of plumage in the Kestrel, *The Zoologist*, Series 3, Vol. 10, 180.

Fritz, H., (1998). Wind speed as a determinant of kleptoparasitism by Eurasian Kestrel *Falco tinnunculus* on Short-eared Owl *Asio flammeus*, *J. Avian Biol.*, 29, 331-333.

Fuchs, J., Johnson J.A. and Mindell, D.P., (2015). Rapid diversification of falcons (Aves: Falconidae) due to expansion of open habitats in the Late Miocene, *Mol. Phylogenet. Evol.*, 82, 166-182.

Gamble, D., (1992). Aggression of Hobby towards other raptors, *Brit. Birds*, 85, 497.

Garratt, C.M., (2011). Managing vegetation for the conservation of the Common Kestrel *Falco tinnunculus* on farmland in England, PhD thesis, Newcastle University.

Garratt, C.M., Hughes, M., Eagle, G., Fowler, T., Grice, P.V. and Whittingham M.J., (2011). Foraging habitat selection by breeding Common Kestrels *Falco tinnunculus* on lowland farmland in England, *Bird Study*, 58, 90-98.

Garratt, C.M., Minderman, J. and Whittingham, M.J., (2012). Should we stay or should we go now? What happens to small mammals when grass is mown, and the implications for birds of prey, *Ann. Zoo. Fennici*, 113-122.

Gentle, L., Gooden, D. and Kettel, E., (2013). Attempted predation by Common Kestrel at a House Sparrow nestbox, *Brit. Birds*, 106, 412-413.

Gibbons, D.W., Reid, J.B. and Chapman, R.A., (1993). *The New Atlas of Breeding Birds in Britain and Ireland: 1988-1991*, T&AD Poyser, London.

Gil-Delgado, J.A., Verdejo, J. and Barba, E., (1995). Nestling diet and fledgling production of Eurasian kestrels (*Falco tinnunculus*) in eastern Spain, *J. Raptor Res.*,29, 240-244.

Glutz von Blotzheim, U.N., Bauer, K.M., and Bezzel, E., (1971). *Handbuch der Vögel Mitteleuropas Vol. 4*, Akademische Verlag, Frankfurt.

Gray, A.P., (1958). *Bird Hybrids*, Commonwealth Agricultural Bureaux, Farnham Royal, Buckinghamshire.

Greenewalt, C.H., (1962). Dimensional relationships for some flying animals, *Smithsonian Miscellaneous Collection*, 144, No. 2.

Greenwood, P.J., (1980). Mating systems, philopatry and dispersal in birds and mammals, *Anim. Behav.*, 28, 1140-1162.

Groombridge, J.J., Jones, C., Bayes, M.K., van Zyl, A.J., Carrillo, J., Nichols, R.A. and Bruford, M.W., (2002). A molecular phylogeny of African Kestrels with reference to divergence across the Indian Ocean, *Mol. Phylogenet. Evol.*, 25, 267-277.

Hackett, S. J. and 17 others, (2008). A phylogenomic study of birds reveals their evolutionary history, *Science*, 320, 1763-1768.

Hagen, Y., (1952). (*Birds of prey and other wildlife*), Gyldendal Norsk, Oslo.

Hakkarainen, H. and Korpimäki, E., (1996). Competitive and predatory interactions among raptors: an observational experimental study, *Ecology*, 77, 1134-1142.

Hakkarainen, H., Korpimäki, E., Mappes, T. and Palokangas, P., (1992). Kestrel hunting behaviour towards solitary and grouped *Microtus agrestis* and *M. epiroticus* – a laboratory experiment, *Ann. Zool. Fennici*, 29, 279-283.

Hakkarainen, H., Korpimäki, E., Huhta, E. and Palokangas, P., (1993). Delayed maturation in plumage colour: evidence for the female-mimicry hypothesis in the Kestrel, *Behav. Ecol. Sociobiol.*, 33, 247-251.

Hakkarainen H., Hunhta, E., Lahti, K., Lundvall, P., Mappes, T., Tolonen, P. and Wiehn, J., (1996). A test of male mating and hunting success in the Kestrel: the advantages of smallness?, *Behav. Ecol. Sociobiol.*, 39, 375-380.

Hambly, C., Harper, E.J. and Speakman, J.R., (2004). The energy cost of loaded flight is substantially lower than expected due to alterations in flight kinematics, *J. Exp. Biol.*, 207, 3969-3976.

Harris, S., Morris, P., Wray, S. and Yalden, D., (1995). A review of British mammals: population estimates and conservation status of British mammals other than cetaceans, *Joint Nature Conservation Committee*.

Harris, S.J., Massimino, D., Gillings, S., Eaton, M.A., Noble, D.G., Balmer, D.E., Procter, D., Pearce-

Higgins, J.W. & Woodcock, P., (2018). The Breeding Bird Survey 2017. *BTO Research Report 706*, The British Trust for Ornithology, Thetford.

Hart, N.S., Mountford, J.K., Davies, W.I.L., Collin, S.P. and Hunt, D.M., (2016). Visual pigments in a palaeognath bird, the emu *Dromaius novaehollandiae*: implications for spectral sensitivity and the origin of ultraviolet vision, *Proc. R. Soc.*, 283, 20161063.

Hasenclever, H., Kostrzewa, A. and Kostrzewa, R., (1989). (The breeding biology of the Kestrel *Falco tinnunculus* in eastern Westphalia 1972-1987), *J. Orn.*, 130, 229-237.

Henny, C.J., (1972). An analysis of the population dynamics of selected avian species, Bureau of Sport, Fisheries and Wildlife, Research Report 1, US Government, Washington DC.

Heukelen, C. van and Heukelen, E. van., (2011). (Kestrel *Falco tinnunculus* nest with 10 eggs), *De Takkeling*, 19, 125-128.

Hewitt, S., (2013). Avian drop catch play, *Brit. Birds*, 106, 206-216.

Hirsch, J., (1982). Falcon visual sensitivity to grating contrast, *Nature*, 300, 57-58.

Honkala, J., Saurola, P. and Valkama, J., (2012). *Linnut Vuosikirja 2011 (Bird Yearbook 2011)*, Finnish Museum of Natural History, Helsinki.

Honkavaara, J., Koivula, M., Korpimäki, E., Siitari, H. and Viitala, J., (2002). Ultraviolet vision and foraging in terrestrial vertebrates, *Oikos*, 98, 505-511.

Hudson, D.M. and Bernstein, M.H., (1983). Gas exchange and energy costs in the White-necked Raven, *Corvus cryptoleucus*, *J. Exp. Biol.*, 103, 121-130.

Huhta, E., Hakkarainen, H. and Lundvall, P., (1998). Bright colours and predation risk in passerines, *Ornis Fennica*, 75, 89-93.

Huitu, O., Helander, M., Lehtonen, P. and Saikkonen, K., (2008). Consumption of grass endophytes alters the ultraviolet spectrum of vole urine, *Oecologia*, 156, 333-340.

Huitzing, D., (2002). Probable kleptoparasitism between Kestrel *Falco tinnunculus* and Hobby *F. subbuteo*, *De Takkeling*, 10, 255.

Huth, H. H. and Burkhardt, D., (1972). (The visual range of the Violet Hummingbird), *Naturwissenschaften*, 59, 650.

Hyuga, I., (1955). (Breeding colonies of Japanese Kestrels), *Tori*, 14, 17-24.

Håstad, O., Victorsson, J. and Ödeen, A., (2005). Differences in colour vision make passerines less conspicuous in the eyes of their predators, *Proc. Nat. Acad. Sci. USA*, 102, 6391-6394.

Härmä, O., Kareksela, S., Siitari H. and Suhonen, J., (2011). Pygmy Owl (*Glaucidium passerinum*) and the usage of ultraviolet cues of prey, *J. Avian Biol.*, 42, 89-91.

Ingram, W.M., Goodrich, L.M., Robey, E.A. and Eisen, M.B., (2013). Mice infected with low-virulence strains of *Toxoplasma gondii* lose their innate aversion to cat urine, even after extensive parasite clearance, *PLoS ONE 8(9): e75246. Doi:10.1371/journal.pone.0075246*.

Itämies, J. and Korpimäki, E., (1987). Insect food of the Kestrel, *Falco tinnunculus*, during breeding in western Finland, *Aquilo Ser. Zool.*, 25, 21-31

James, P.C. and Oliphant, L.W., (1986). Extra birds and helpers at the nests of Richardson's Merlin, *Condor*, 88, 533-534.

Jarvis, E.D. and 104 others, (2014). Whole-genome analyses resolve early branches in the tree of life of modern birds, *Science*, 12 December, 1320-1331. The article was the first in a series of ten which filled that issue of *Science* and which were all related to the work of the Avian Phylogenomics Consortium.

Johansen, H., (1957). (The Birds of western Siberia Part III [*Accipiter-Aegypius*]), *J. Orn.* 98, 397-419. This was one of a series of articles on the birds of western Siberia published by Johansen. The papers were later collected into a 260pp book which was published in Leipzig.

Johnson, P., (2008). Peregrine Falcons feeding Common Kestrel chicks, *Brit. Birds*, 101, 327.

Jones, M.P., Pierce, K.E. and Ward, D., (2007). Avian vision: a review of form and function with special consideration to birds of prey, *J. Exot. Pet Med.*, 16, 69-87.

Kangas, V-M, Carrillo, J., Debray, P and Kvist, L., (2018), Bottlenecks, remoteness and admixture shape genetic variation in island populations of Atlantic and Mediterranean Common Kestrels *Falco tinnunculus*, *J. Avian Biol.*, e01768.

Karyakin, I.V., (2005). Anomalous late event of breeding the Kestrel in Volvograd District, *Raptors Conservation*, 2, 61.

Karyakin, I.V. and Nikolenko, E.G., (2010). Note of the Kestrel coming back on its birthplace and breeding in the nest where it was born, *Raptors Conservation*, 19, 201-204.

Kellie, A., Dain, S.J. and Banks, P.B., (2004). Ultraviolet properties of Australian mammal urine, *J. Comp. Physiol. A*, 190, 429-435.

Kettle, A., (1990). Red-footed Falcon attacking and robbing Kestrel, *Brit. Birds*, 83, 548.

Kirkwood, J.K., (1981). Bioenergetics and growth in the Kestrel (*Falco tinnunculus*), PhD thesis, University of Bristol.

Kirkwood, J.K., (1983). A limit to metabolizable energy intake in mammals and birds, *Comp. Bioch. And Phys.*, 75A, 1-3.

Kjellén, N., (1992). Differential timing of autumn migration between sex and age groups in raptors at Falsterbo, Sweden, *Ornis Scand.*, 23, 420-34.

Kjellén, N., (1994). Differences in age and sex ratio among migrating and wintering raptors in southern Sweden, *Auk* 111, 274-84.

Klemola, T., Korpimäki, E. and Norrdahl, K., (1998). Does avian predation risk depress reproduction of voles?, *Oecologia*, 115, 149-153.

Kochanek, H-M., (1984). The calls of the Kestrel *Falco tinnunculus*, *Charadrius* 20, 137-54.

Koivula, M. and Viitala, J., (1999). Rough-legged Buzzards use vole scent marks to assess hunting areas, *J. Avian Biol.*, 30, 329-332.

Koivula, M. and Korpimäki, E., (2001). Do scent marks increase predation risk of microtine rodents?, *Oikos*, 95, 275-281.

Koivula, M., Koskela, E., and Viitala, J., (1999a). Sex and age-specific differences in ultraviolet reflectance of scent marks of Bank Voles *Clethrionomys glareolus*), *J. Comp. Physiol. A*, 185, 561-564.

Koivula, M., Viitala, J. and Korpimäki, E., (1999b). Kestrels prefer scent marks according to species and reproductive status of voles, *Ecoscience*, 6, 415-420, 1999b.

Korpimäki, E., (1983). (Results from a nestbox experiment with Kestrels (*Falco tinnunculus*), *Lintumies*, 18, 132-137.

Korpimäki, E., (1984a). Food piracy between European Kestrel and Short-eared Owl, *Raptor Res*. 18, 113-115.

Korpimäki, E., (1984b). Population dynamics of birds of prey in relation to fluctuations of small mammal populations in western Finland, *Ann. Zool. Fennici*, 21, 287-293.

Korpimäki, E., (1985a). Prey choice strategies of the Kestrel *Falco tinnunculus* in relation to available small mammals and other Finnish birds of prey, *Ann. Zool. Fennici*, 22, 91-104.

Korpimäki, E., (1985b). Diet of the Kestrel *Falco tinnunculus* in the breeding season, *Ornis Fennica* 62, 130-137.

Korpimäki, E., (1985c). Rapid tracking of microtine populations by their avian predators: possible evidence for stabilizing predation, *Oikos* 45, 281-284.

Korpimäki, E., (1986a). Reversed size dimorphism in birds of prey, especially Tengmalm's Owl *Aegolius funereus*: a test of the 'starvation hypothesis', *Ornis Scand*. 17, 326-332.

Korpimäki, E., (1986b). Diet variation, hunting habitat and reproductive output of the Kestrel *Falco tinnunculus* in the light of the optimal diet theory, *Ornis Fennica*, 63, 84-90.

Korpimäki, E., (1987). Dietary shifts, niche relationships and reproductive output of coexisting Kestrels and Long-eared Owls, *Oecologia*, 74, 277-285.

Korpimäki, E., (1988). Factors promoting polygyny in European birds of prey – a hypothesis, *Oecologia*, 77, 278-285.

Korpimäki, E., ((1994). Rapid or delayed tracking of multi-annual vole cycles by avian predators? *J. Anim. Ecol.*, 63, 619-28.

Korpimäki, E. and Norrdahl, K., (1991). Numerical and functional responses of Kestrels, Short-eared Owls, and Long-eared Owls to vole densities, *Ecology*, 72, 814-26.

Korpimäki, E. and Rita, H., (1996). Effects of brood size manipulations on offspring and parental survival in the European Kestrel under fluctuating food conditions, *Ecoscience* 3, 264–273.

Korpimäki, E., Tolonen, P. and Bennett, G.F., (1995). Blood parasites, sexual selection and reproductive success of European Kestrels, *Ecoscience*, 2, 335-343.

Korpimäki, E., Lahti, K., May, C.A., Parkin, D.T., Powell, G.B., Tolonen, P. and Wetton, J.H., (1996). Copulatory behaviour and paternity determined by DNA fingerprinting in Kestrels: effects of cyclic food abundance, *Anim. Behav.*, 51, 4, 945-955.

Korpimäki, E., May, C.A., Parkin, D.T., Wetton, J.H. and Wiehn, J., (2000). Environmental- and parental condition-related variation of sex ratio of Kestrel broods, *J. Avian Biol.*, 31, 128-134.

Korslund, L. and Steen,H., (2006). Small rodent winter survival: snow conditions limit access to food resources, *J. Anim. Ecol.*, 75, 156-166.

Kostrzewa, R. and Kostrzewa, A., (1990). The relationship of spring and summer weather with the density and breeding success of the Buzzard *Buteo buteo*, Goshawk *Accipiter gentilis* and Kestrel *Falco tinnunculus*, *Ibis*, 132, 550-559.

Kostrzewa, R. and Kostrzewa, A., (1991). Winter weather, spring and summer density, and subsequent breeding success of Eurasian Kestrels, Common Buzzards, and Northern Goshawks, *Auk*, 108, 342-347.

Kostrzewa, R. and Kostrzewa, A., (1997). (Breeding success of the Kestrel *Falco tinnunculus* in Germany: results 1985-1994). *J. für Ornithologie* 138, 73-82.

Krebs, J.R., Wilson, J.D., Bradbury, R.B. and Siriwardena, G.M. (1999). The second Silent Spring? *Nature* 400: 611-612.

Kreiderits, A., Gamauf, A., Krenn, H.W. and Sunasgutner, P., (2016). Investigating the influence of local weather conditions and alternative prey composition on the breeding performance of urban Euraisn Kestrels *Falco tinnunculus*, *Bird Study*, 63, 369-379.

Krone, O., (2002). Fatal caryospora infection in a free-living juvenile Eurasian Kestrel (*Falco tinnunculus*), *J. Raptor Res.* 36, 84-86.

Krone, O., (2007). Pathology: Endoparasites, in *Raptor Research and Management Techniques*, Bird, D.L. and Bildstein, D.R. (Eds.), Hancock House, Blaine, WA, USA, 318-328.

Krüger, O., (2005). The evolution of reversed sexual size dimorphism in hawks, falcons and owls: a comparative study, *Evolutionary Ecology*, 19, 467-486.

Kübler, S., Kupko, S. and Zeller, U., (2005). The Kestrel (*Falco tinnunculus* L.) in Berlin: investigation of breeding biology and feeding ecology, *J.Orn.*, 146, 271-278.

Kupko, S. and Kübler, S., (2007). (Adoption of Common Kestrel nestlings *Falco tinnunculus* by a Peregrine Falcon *F. peregrinus*), *Vogelwelt*, 128, 33-37.

Kuusela, S., (1979). (The Kestrel still hovers), *Suomen Luonto* 38, 162-165.

Lehikoinen, A. plus 22 others. (2018). Declining population trends of European mountain birds, *Global Change Biology*, 25, 577-588.

Lind, O., Mitkus, M., Olsson, P. and Kelber, A., (2013). Ultraviolet and colour vison in raptor foraging, *J. Exp. Biol.*, 216, 1819-1826, and corrigendum (to Table S1).

Lockwood W.B., (1984). *The Oxford Book of British Bird Names*, Oxford University Press.

Lodder, M., (2018). Staring *Sturnus vulgaris* almost drowns Kestrel *Falco tinnunculus*, *De Takkeling*, 26, 154-160.

López-Idiáquez, D., Vergara, P., Fargallo, J.A. and Martínez-Padilla, J., (2016). Female plumage coloration signals status to conspecifics, *Anim. Behav.*, 121, 101-106.

Loss, S.R., Will, T. and Marra, P.P., (2013). Estimates of bird collision mortality at wind farms in the contiguous United States, *Biol. Conserv.* 168, 201-209.

Lynas P., Newton, S.F. & Robinson, J.A., (2007). The status of birds in Ireland: an analysis of conservation concern 2008–2013. *Irish Birds* 8: 149–166.

Maa, T.C., (1969). A revised checklist and concise host index of Hippoboscidae (Dipteria), *Pacific Insects Monographs*, 2, 261-299.

Marques, A.T., Batalha, H., Rodrigues, S., Costa, H., Pereira, M.J.R., Fonseca, C., Mascarenhas, M, and Bernardino, J., (2014). Understanding bird collisions at wind farms: an updated review on the causes and possible mitigation strategies, *Biological Conservation*, 179, 40-52, 214.

Martin, A.P., (1980). A study of a pair of breeding Peregrine Falcons (Falco peregrinus peregrinus) during part of the nesting period, BSc dissertation, University of Durham), quoted in Ratcliffe, D., The Peregrine Falcon, T&AD Poyser, London, 1993.

Martínez-Padilla, J. and Millán, J., (2007). The prevalence and intensity of intestinal parasitation in a wild population of nestling Eurasian Kestrel *Falco tinnunculus*, *Ardeola*, 54, 109-115.

Martínez-Padilla, J. and Viñuela, J., (2011). Hatching asynchrony and brood reduction influence immune response in Common Kestrel *Falco tinnunculus* nestlings, *Ibis*, 153, 601-610.

Martínez-Padilla, J., Vergara, P. and Fargallo, J.A., (2017). Increased lifetime reproductive success of first-hatched siblings in Common Kestrels *Falco tinnunculus*, *Ibis*, 159, 803-811.

Masman, D., Gordijn, M., Daan, S. and Dijkstra, C., (1986a). Ecological energetics of the Kestrel: field estimates of energy intake throughout the year, *Ardea*, 74, 24-39.

Masman, D., Daan S, and Beldhuis, H., (1986b). Energy allocation in the seasonal cycle of the Kestrel, *Falco tinnunculus*, Chapter 6 of Masman, D., The annual cycle of the Kestrel (*Falco tinnunculus*), PhD thesis, University of Groningen.

Masman, D. and Klaassen, M., (1987). Energy expenditure during free flight in trained and free-living Eurasian Kestrels (*Falco tinnunculus*), *Auk*, 104, 603-616.

Masman, D., Daan, S., and Dijkstra, C., (1988a). Time allocation in the Kestrel (*Falco tinnunculus*), and the principle of energy minimization, *J. Anim. Ecol.*, 57, 411-432.

Masman, D., Daan, S., and Beldhuis, J.A., (1988b). Ecological energetics of the Kestrel: daily energy expenditure throughout the year based on the time-energy budget, food intake and doubly labelled water methods, *Ardea*, 76, 64-81.

Masman, D., Dijkstra, C., Daan, S., and Bult, A., (1989). Energetic limitation of avian parental effort: Field experiments in the Kestrel (*Falco tinnunculus*), *J. Evol. Biol.*, 2, 435-455.

Massemin, S., Korpimäki, E., Pöyri, V. and Zorn, T., (2002). Influence of hatching order on growth rate and resting metabolism of Kestrel chicks, *J. Avian Biol.*, 33, 235-244.

Mather, J., (1986). *The Birds of Yorkshire*, Croom Helm, London.

Matyukhin, A.V., Zabashta, A.V. and Zabashta, M.V., (2012). Louse flies (Hippoboscidae) of Falconiformes and Strigiformes in Palearctic, in Birds of Prey in the Dynamic Environment of the Third Millenium: Status and Prospects, Gavriliuk, M.N. (ed), *Proc. 6th Int. Conf. on Birds of Prey and Owls of North Eurasia, Kryvyi Rib.*

McIsaac, H.P., (2001). Raptor acuity and wind turbine blade conspicuity, in *National Avian-Wind Power Planning Meeting IV*, Resolve Inc., Washington, DC, 59-87.

Mead, C.J. and Pepler, G.R.M., (1975). Birds and other animals at Sand Martin colonies, *Brit. Birds*, 68, 89-99.

Mee, A., (2012). An overview of monitoring for raptors in Ireland, *Acrocephalus*, 154/155, 239-245.

Meijer, T., (1989). Photoperiodic control of reproduction and molt in the Kestrel *Falco tinnunculus*, *J. Biol. Rhythms*, 4, 351-364.

Meijer, T. and Schwabl, H., (1988). Effects of food availability on hormonal development in breeding and non-breeding Kestrels: Field and Laboratory studies, Chapter 6 of Meijer, T.H., Reproductive decisions in the Kestrel *Falco tinnunculus*, PhD thesis, University of Groningen.

Meijer, T., Daan, S. and Dijkstra, C., (1988). Female condition and reproduction: effect of food manipulation in free-living and captive Kestrels, *Ardea*, 76, 141-154.

Meijer, T., Masman, D. and Daan, S., (1989). Energetics of reproduction in female Kestrels, *Auk*, 106, 549-559.

Meijer, T., Daan, S. and Hall, M. (1990). Family planning in the Kestrel (*Falco tinnunculus*): the proximate control of covariation of laying date and clutch size, *Behaviour*, 114, 1-4.

Messenger, D., Duckels, A.S., Pennington, M. and Taylor, J.,(1988). Kestrel taking Leach's Petrel, *Brit. Birds*, 81, 395.

Mester, H., (1980). (On the pairing and sexual behaviour of Kestrels), *Orn. Mitt.* 32, 150-152.

Meunier, F.D., Verheyden, C. and Jouventin, P., (2000). Use of roadsides by diurnal raptors in agricultural landscapes, *Biological Conservation*, 92, 291-298.

Mingju E, Tuo Wang, Shangyu Wang, Ye Gong, Jiangping Yu, Lin Wang, Wei Ou and Haitao Wang, (2019). Old nest material functions as an informative cue in making nest-site selection decisions in the European Kestrel (*Falco tinnunculus*), *Avian Research*, 10:43, https://doi.org/10.1186/s40657-019-0182-5

Mitchell, J., Placido, C. and Rose, R., (1975). Notes on a short-tailed vole plague at Eskdalemuir, Dumfriesshire, Transactions of the Dumfriesshire and Galloway Natural History and Antiquarian Society, Series III, Vol. 51, 11-14.

Mikkola, H., (1983). *Owls of Europe*, T&AD Poyser, London.

Mikula, P., Hromada, M. and Tryjanowski, P., (2013). Bats and Swifts as food for the European Kestrel (*Falco tinnunculus*) in a small town in Slovakia, *Ornis Fennica*, 90, 178-185.

Morales-Betancourt, J.A. and Castaño-Villa, G.J. (2018). Males in seemingly female-like plumage do not mimic females: UV reflectance reveals temporal cryptic dimorphism in a manakin species exhibiting delayed plumage maturation, *J. Avian Biol.*, 2018: e01467.

Movalli, P., Bode, P., Dekker, R., Fornasari, L., van der Mije, S, and Yosef, R., (2017). Retrospective biomonitoring of mercury and other elements in museum feathers of Common Kestrel *Falco tinnunculus* using instrumental neutron activation analysis (INAA). *Environ. Sci. Pollut. Res.*, 24: 25986–26005.

Musgrove, A., Aebischer, N., Eaton, M., Hearn, S., Newson, S., Noble, D., Parsons, M., Risely, K and Stroud, D., (2013). Population estimates in Great Britain and the United Kingdom, *Brit. Birds*, 106, 64-100.

Negro, J.J., Ibáñez, C., Pérezjordá, J.L. and Delariva, M.J., (1992). Winter predation by Common Kestrel *Falco tinnunculus* on Pipistrelle Bats *Pipistrellus pipistrellus* in southern Spain, *Bird Study*, 39, 195-199.

Nelson, R.W., (1970). Some aspects of the breeding behaviour of Peregrine falcons on Langara Island, British Columbia, MSc thesis, University of Calgary.

Newson, S.E., Ockendon, N., Joys, A., Noble, D.G. and Baillie, S.R., (2009). Comparison of habitat-specific trends in the abundance of breeding birds in the UK, *Bird Study*, 56, 233-243.

Newton, I., (1979). *Population Ecology of Raptors*, T&AD Poyser, London.

Newton, I., (1984). Raptors in Britain: a review of the last 150 years, *BTO News*, 131, 6-7.

Newton, I., (1986). *The Sparrowhawk*, T&AD Poyser, London (reprinted in 2010).

Newton, I., (2008). *The Migration Ecology of Birds*, Academic Press, London.

Newton, I., Bell, A.A. and Wyllie, I. (1982), Mortality of sparrowhawks and kestrels, *British Birds*, 75, 195-204.

Newton, I., Wyllie, I. and Dale, L., (1999). Trends in the numbers and mortality patterns of Sparrowhawks (*Accipiter nisus*) and Kestrels (*Falco tinnunculus*) in Britain, as revealed by carcass analyses, *J. Zool. Lond.*, 248, 139-147.

Newton, I., McGrady, M.J. and Oli, M.K., (2016). A review of survival estimates for raptors and owls, *Ibis*, 158, 227-248.

Nicholson, D., (2010). Common Kestrel attempt to predate Hobby chicks at the nest, *Brit. Birds*, 103, 244.

Noguera, J.C. and Velando, A., (2019). Bird embryos perceive vibratory cues of predation risk from clutch mates, Nature Ecol.&Evol., 3, 1225-1232, https://doi.org/10.1038/s41559-019-0929-8.

Norrdahl, K. and Korpimäki, E., (1996). Do nomadic avian predators synchronize population fluctuations of small mammals? A field experiment, *Oecologia* 107, 478-83.

Norrdahl, K., Suhonen, J., Hemminki, O. and Korpimäki, E., (1995). Predator presence may benefit: Kestrels protect Curlew nests against nest predators, *Oecologia*, 101, 105-109.

Nygård, T. and Polder, A., (2012). Environmental pollutants in eggs of birds of prey in Norway: current situation and time trends, *Norsk Institutt for Naturforskning (NINA) Report 834*.

Olsen, J., (2013). Reversed sexual dimorphism and prey size taken by male and female raptors: a comment on Pande and Dahanukar, *J. Raptor Res.*, 47, 79-81.

Oakley-Martin, D., (2008). Juvenile Common Kestrel diving at female, *Brit. Birds*, 101, 383.

Orihuela-Torres, A., Perales, P., Rosado, D. and Pérez-García, J. M., (2017). Feeding ecology of the Common Kestrel (*Falco tinnunculus*) in the south of Alicante (SE Spain), *Revista Catalana d'Ornitologia*, 33, 10-16.

Packham, C., (1985a). Bigamy by the Kestrel, *Brit. Birds* 78, 194.

Packham, C., (1985b). Role of male Kestrel during incubation, *Brit. Birds* 78, 144-145.

Palokangas, P., Korpimäki, E., Hakkarainen, H., Huhta, H., Tolonen, P. and Alatalo, R.V., (1994). Female Kestrels gain reproductive success by choosing brightly ornamented males, *Anim. Behav.*, 47, 443-448.

Pande, S. and Dahanukar, N., (2012). Reversed sexual dimorphism and differential prey delivery in Barn Owls (*Tyto alba*), *J. Raptor Res.*, 46, 184-189.

Pande, S. and Dahanukar, N., (2013). Reversed sexual dimorphism and prey delivery: response to Olsen, *J. Raptor Res.*, 47, 81-82.

Parker, A., (1979). Peregrines at a Welsh coastal eyrie, *Brit. Birds*, 72, 104-114.

Parr, D., (1967). A review of the status of the Kestrel, Tawny Owl and Barn Owl in Surrey, *Surrey Bird Report*, 15, 35-42.

Pauli, B.P., Spaul, R.J. and Heath, J.A., (2017). Forecasting disturbance effects on wildlife: tolerance does not mitigate effects of increased recreation on wildlands, *Anim. Conserv.*, 20, 251-260.

Pearce-Higgins, J.W. and Crick, H.Q.P., (2019). One-third of English breeding bird species show evidence of population responses to climatic variables over 50 years, *Bird Study*, 66, 159-172.

PECBMS (Pan-European Common Bird Monitoring Scheme), (2012). *Population trends of common European breeding birds 2012*, Czech Society for Ornithology, Prague.

Peggie, C.T., Garratt, C.M. and Whittingham, M.J., (2011). Creating ephemeral resources: how long do the beneficial effects of grass cutting last for birds? *Bird Study* 58, 390-398.

Pennycuick, C. J., (1988). Empirical estimates of body drag of large waterfowl and raptors, *J. Exp. Biol.*, 135, 253–264.

Pennycuick, C.J., (2008). *Modelling the Flying Bird*, Academic Press, London.

Pennycuick, C.J., Fast, P.L.F., Ballerstädt, N. and Rattenborg, N., (2012). The effect of an external transmitter on the drag coefficient of a bird's body, and hence on migration range, and energy reserves after migration, *J. Orn.*, 153, 633-644.

Pérez-Camacho, L., Martínez-Hesterkamp, S., Rebollo, S., García-Salgado G. and Morales-Castilla, I., (2018), Structural complexity of hunting habitat and territoriality increase the reversed sexual size dimorphism in diurnal raptors, *J. Avian Biol.*, 49 (10): e01745.

Peter, H-U., and Zaumseil, J., (1982). (Population ecology of the Kestrel (*Falco tinnunculus*) in a colony near Jena), *Ber. Vogelwarte Hiddensee*, 3, 5-7.

Pettifor, R.A., (1990). The effects of avian mobbing on a potential predator, the European Kestrel, *Falco tinnunculus*, *Anim. Behav.*, 39, 821-827.

Petty, S.J., Anderson, D.I.K., Davison, M., Little, B., Sherratt, C.J and Lambin, X., (2003). The decline of Common Kestrels (*Falco tinnunculus*) in a forested area of northern England: the role of predation by Northern Goshawks (*Accipiter gentilis*), *Ibis*, 145, 472-483.

Philips, J.R., (2000). A review and checklist of the parasitic mites (Acarina) of the Falconiformes and Strigiformes, *J. Raptor Res.*, 34, 210-231.

Philips, J.R. and Dindal, D.L., (1977). Raptor nests as a habitat for invertebrates: a review, *Raptor Res.*, 11, 86-96.

Piault, R., van den Brink, V. and Roulin, A., (2012). Condition-dependent expression of melanin-based coloration in the Eurasian Kestrel, *Naturwissenschaften*, 99, 391-396.

Piechocki, R., (1982). *Der Turmfalke*, Lutherstadt Ziemsen, Wittenburg.

Pike, G.V., (1981). Nesting Kestrels tolerating excessive disturbance, *Brit. Birds*, 74, 520-521, 1981.

Pikula, J., Beklová, M., and Kubík, V., (1984). The nidobiology of *Falco tinnunculus*, *Acta Scientiarum Naturalium Brno* 18, 1-55.

Plesník, J. and Dusík, M., (1994). Reproductive output of the Kestrel *Falco tinnunculus* in relation to small mammal dynamics in intensively cultivated farmland. In Meyburg, B-U. and Chancellor, R.D. (eds), *Raptor Conservation Today, Proceedings of the IV World Conference on Birds of Prey and Owls*,WWGBP/The Pica Press, Mountfield, East Sussex, pp. 61-5.

Ponting, E.D., (2002). Common Kestrel taking Canary from cage, *Brit. Birds*, 95, 23.

Potapov E., and Sale R.G., (2012). *The Snowy Owl*, T&AD Poyser, London.

Potier, S. (2020). Olfaction in raptors, *Zoo. J. Linne. Soc.*, 189, 713-721.

Potier, S., Lieuvin, M., Pfaff, M. and Kelber, A., (2020). How fast can raptors see?, *J. Exp. Biol.* 223, jeb.209031.

Potvin, D.A., Välimäki, K. and Lehikoinen, A., (2016). Differences in shifts of wintering and breeding ranges lead to changing migration distances in European birds, *J. Avian Biol.*, 47, 619-628.

Probst, R., Pavlicev, M. and Viitala, J., (2002). UV reflecting vole scent marks attract a passerine, the Great Grey Shrike *Lanius excubitor*, *J. Avian Biol.*, 33, 437-440.

Prŷs-Jones, R., (2018). Kleptoparasitism by Red Kites on Common Kestrel, *Brit. Birds*, 111, 479-480.

Pyle, P., (2013). Evolutionary implications of synapomorphic wing-molt sequences among falcons (Faclconiformes) and parrots (Psittaciformes), *Condor*, 115, 593-602.

Qninba A., Benhoussa A., Radi M., El Idrissi A., Bousadik H., Badaoui B. and El Agbani, M. A., (2015), (A very particular predation mode of Eleonora's Falcon (*Falco eleonorae*) on Morocco's Essaouira Archipelago), *Alauda* 83, 149-150.

Ratcliffe, D.A., (1962). Peregrine incubating Kestrel eggs, *Brit. Birds*, 55, 131-32.

Ratcliffe, D.A., (1963). Peregrine rearing young Kestrels, *Brit. Birds*, 56, 457-460.
Ravi, S., Noda, R., Gagliardi, S., Kolomenskiy, D., Combes, S., Liu, H., Biewener, A. A. and Konow, N., (2020). Modulation of flight muscle recruitment and wing rotation enables hummingbirds to mitigate aerial roll perturbations, *Curr. Biol.* 30, 187–195.
Rebecca, G.W., Cosnette, B.L. and Duncan, A., (1988). Two cases of a yearling and an adult Merlin attending the same nest, *Scottish Birds*, 15, 45-46.
Rejt, Ł., Rutkowski, R. and Gryczyńska-Siemiątkowska, A., (2004). Genetic variability of urban Kestrels in Warsaw – Preliminary data, *Zoologica Poloniae*, 49, 199-209.
Rejt, Ł., (2005). Utilisation of unhatched eggs by urban Kestrels (*Falco tinnunculus*), *Buteo*, 14, 31-35.
Reymond, L., (1985). Spatial visual acuity of the eagle *Aquila audax*: a behavioural, optical and anatomical investigation, *Vision Res.*, 25, 1477-1491.
Rich, D., (2016). Common Kestrel attempting to predate House Martin nest, *Brit. Birds*, 109, 63.
Ricklefs, R. E., (1968). Patterns of growth in birds, *Ibis*, 110, 419-451.
Ricklefs, R. E., (1973). Patterns of growth in birds II: Growth rate and mode of development, *Ibis*, 115, 177-201
Riddle, G.S., (1987). Variation in the breeding output of kestrel pairs in Ayrshire 1978-85, *Scottish Birds* 14, 138-45.
Riddle, G., (1992). *Seasons with the Kestrel*, Blandford, London.
Riddle, G., (2011). *Kestrels for Company*, Whittle Publishing, Dunbeath.
Riedstra, B. and Dijkstra, C., (2017). A runt clutch and presumed delayed incubation of Kestrels *Falco tinnunculus*, *De Takkeling*, 25, 251-257.
Riegert J. and Fuchs, R., (2011). Fidelity to roost sites and diet composition of wintering male urban Kestrels *Falco tinnunculus*, *Acta Ornithologica*, 46, 183-189.
Riegert, J., Fainová, D. and Bystřická, D., (2010). Genetic variability, body characteristics and reproductive parameters of neighbouring rural and urban Common Kestrel (*Falco tinnunculus*) populations, *Popul. Ecol.* 52, 73-79.
Rijnsdorp, A., Daan, S., and Dijkstra, C., (1981). Hunting in the Kestrel, *Falco tinnunculus*, and the adaptive significance of daily habits, *Oecologia*, 50, 391-406.
Ristow, D., (2006). A European Hobby *Falco subbuteo* attacks a European Kestrel *Falco tinnunculus*, *Ornithologischer Anzeiger*, 45, 175-176.
Romanowski, J., (1996). On the diet of urban Kestrels (*Falco tinnunculus*) in Warsaw, *Buteo*, 8, 123-130.
Roulin, A., Brinkhof, M.W.G., Bize, P., Richner, H., Jungi, T.W., Bavoux, C., Boileau, N. and Burneleau, G., (2003). Which chick is tasty to parasites? The importance of host immunology vs. parasitic life history, *J. Anim. Ecol.*, 72, 75-81.
Rozenfeld, F.M., Le Boulangé, E. and Rasmont, R., (1987). Urine marking by Bank Voles (*Clethrionomys glareolus* Screber, 1780: Microtidae, Rodentia) in relation to their social rank, *Can. J. Zool.*, 65, 2594-2601.
RUG/RIJP (The raptor group of the University of Groningen and the Rijkdienst voor de IJsselmeerpolders, Lelystad, Holland), (1982). Timing of vole hunting in aerial predators, *Mammal Review*, 12, 169-181.
Sage, B., (1957). Magpie robbing Kestrel, *Brit. Birds*, 50, 353.
Sale, R., Newberry, L., Newberry, S. and Sale, N., (2020). Breeding behaviour of Common Kestrels in southern England, *Brit. Birds*, 113, 217-226.
Salvati, L., (2001). Does high population density affect reproductive value? Evidence from semicolonial Kestrels *Falco tinnunculus*, *Vogelwelt*, 122, 41-45.
Salvati, L., (2002). Spring weather and the breeding success of the Eurasian Kestrel (*Falco tinnunculus*) in urban Rome, Italy, *J. Raptor Res.*, 36, 81-84.
Santing, J., (2010). (Second clutch in Kestrel *Falco tinnunculus*?), *De Takkeling*, 18, 150.
Saxby, H.L. (edited by S.H. Saxby), (1874). *The Birds of Shetland*, MacLachlan and Stewart, Edinburgh.

Schmid, H., (1990). (The breeding biology of Kestrels in Switzerland), *Ornithologische Beobachter*, 87, 327-349.

Schmid, H., Kestenholz, M., Knuas, P., Rey, L. and Sattler, T., (2018). (Status of the avifauna of Switzerland: special edition linked to the Atlas of nesting Birds), Station Ornithologique Suisse, Sempach.

Schmitz, A., Ondreka, N., Poleschinski, J., Fischer, D., Schmitz, H., Klein, A., Bleckmann, H. and Bruecker, C., (2018). The Peregrine Falcon's rapid dive: on the adaptedness of the arm skeleton and shoulder girdle, *J. Comp. Physiology A*, 204, 747-759.

Schoenjahn, J., Pavey, C.R. and Walter, G.H., (2020). Why female birds of prey are larger than males, *Biol. J. Linne. Soc.*, 129, 532-542.

Sebastián-González, E., Pérez-García, J.M., Carrete, M., Donázar, J.A. and Sánchez-Zapata, J.A., (2013). Using network analysis to identify indicator species and reduce collision fatalities at wind farms, *Biol. Conserv.*, 224, 209-212.

Sepp, T., McGraw, K.J., Kaasik, A. and Giraudeau, M., (2018). A review of urban impacts on avian life-history evolution: does city living lead to a slower pace of life?, *Global Change Biology*, 24, 1452-1469.

Sharrock, J.T.R., (1976). *The Atlas of Breeding Birds of in Britain and Ireland*, Poyser, Berkhamsted.

Shrubb, M., (1993a). *The Kestrel*, Hamlyn, London.

Shrubb, M. (1993b). Nest sites in the Kestrel, *Falco tinnunculus*, Bird Study, 40, 63-73.

Shrubb, M., (2003). The Kestrel (*Falco tinnunculus*) in Wales, *Welsh Birds*, 3, 330-339.

Shrubb, M., (2004). The decline of the Kestrel in Wales, *Welsh Birds*, 4, 65-66.

Slagsvold, T. and Sonerud, G.A., (2007). Prey size and ingestion rate in raptors: importance for sex roles and reversed sexual size dimorphism, *J. Avian Biol.*, 38, 650-661.

Smallwood, J.A., Dudajek, V., Gilchrist, S. and Smallwood M.A., (2003). Vocal development in American Kestrel (*Falco sparverius*) nestlings, *J. Raptor Res.*, 37, 37-43.

Smit, T., Bakhuizen, T. and Jonkers, D.A., (1987). (Causes of death in Kestrels *Falco tinnuculus* in the Netherlands), *Limosa*, 60, 175-8.

Smith, M.B., (1992). Peregrines nesting beside Kestrels on urban chimney, *Brit. Birds*, 85, 498.

Smith, S.A., (1996). Parasites of birds of prey: their diagnosis and treatment, *Seminars in Avian and Exotic Pet Medicine*, 5, 97-105.

Sodhi, N.S. and Olliphant, L.W., (1993). Prey selection by urban-breeding Merlins, *Auk*, 110, 727-735.

Sodhi, N.S., Warkentin, I.G., James, P.C. and Oliphant, L.W., (1991). Effects of radiotagging on breeding Merlins, J. Wildl. Manage., 55, 613-616.

Sonerud, G.A., Steen, R., Løw, L.M., Røed, L.T., Skar, K., Selås, V. and Slagsvold, T., (2013). Size-based allocation of prey from male to offspring via female: family conflicts, prey selection, and evolution of sexual size dimorphism in raptors, *Oecologia*, 172, 93-107.

Sonerud, G.A., Steen, R., Løw, L.M., Røed, L.T., Skar, K., Selås, V. and Slagsvold, T., (2014a). Evolution of parental roles in raptors: prey type determines role asymmetry in the Eurasian Kestrel, *Anim. Behav.* 96, 31-38.

Sonerud, G.A., Steen, R., Selås, V., Aanonsen, O.M., Aasen, G-H., Fagerland, K.L., Fosså, A., Kristiansen, L., Løw, L.M., Rønning, M.E., Skouen, S.K., Asakskogen, E., Johansen, H.M., Johnsen, J.T., Karlsen, L.I., Nyhus, G.C., Røed, L.T., Skar, K., Sveen, B-A., Tveiten, R. and Slagsvold, T., (2014b). Evolution of parental roles in provisioning birds: diet determines role asymmetry in raptors, *Behav. Ecol.*, 25, 762-772.

Spedding, G.R., (1987a). The wake of a Kestrel (*Falco tinnunculus*) in gliding flight, *J. Exp. Biol.*, 127, 45-57.

Spedding, G.R., (1987b). The wake of a Kestrel (*Falco tinnunculus*) in flapping flight, *J. Exp. Biol.*, 127, 59-78.

Steen R., Sonerud, G.A. and Slagsvold, T., (2012). Parents adjust feeding effort in relation to nestling age in the Eurasian Kestrel (*Falco tinnunculus*), *J. Orn.*, 153, 1087-1099.

Stevens, M., Murn, C. and Hennessey, R., Population change of Common Buzzards (*Buteo buteo*) in central southern England between 2011 and 2016, *Bird Study*, 66, 378-389.

Suh, A., Paus, M., Kiefmann, M., Churakov, G., Franke, F.A., Brosius, J., Kriegs, J.O. and Schmitz, J., (2011). Mesozoic retroposons reveal parrots as closest living relatives of passerine birds, *Nat. Comms.*, 2, 443 do1: 10.1038/ncomms1448

Sumasgutner, P., Nemeth, E., Tebb, G. Krenn, H.W. and Gamauf, A., (2014a). Hard times in the city - attractive nest sites but insufficient food supply lead to low reproduction rates in a bird of prey, *Frontiers in Zoology*, (http://www.frontiersinzoology.com/content/11/1/48 13pp).

Sumasgutner, P., Vasko, V. and Varjonen, R., (2014b). Public information revealed by pellets in nest sites is more important than ecto-parasite avoidance in the settlement decisions of Eurasian Kestrels, *Behav. Ecol. Sociobiol.*, 68, 2023-2034.

Šumrada, T. and Hanžel, J., (2012). The Kestrel *Falco tinnunculus* in Slovenia – a review of its distribution, population density, movements, breeding biology, diet and interactions with other species, *Acrocephalus*, 33, 5-24.

Tempest, J., (2009). Owls and Kestrel share tenancy of nestbox, *Brit. Birds*, 102, 645.

Thomas, A.R.L. and Balmford, A., (1995). How Natural Selection Shapes Birds' Tails, Am. Nat., 146, 848-868.

Tinbergen, L., (1940). (Observations on the division of labour in Kestrels (*Falco tinninculus*) during the breeding season), *Ardea*, 29, 63–98.

Tobalske B. W., (2007). Biomechanics of bird flight, *J. Exp. Biol.*, 210, 3135-3146.

Tobalske B. W., Hedrick, T. L., Dial, K. P. and Biewener A. A., (2003). Comparative power curves in bird flight, *Nature*, 421, 363-366, 2003.

Tolonen, P. and Korpimäki, E., (1994). Determinants of parental effort: a behavioural study in the Eurasian Kestrel *Falco tinnunculus*, *Behav. Ecol. Sociobiol.*, 35, 355-362.

Tomás, G., (2015). Hatching date vs laying date: what should we look at to study avian optimal timing of reproduction?, *J. Avian Biol.*, 46, 107-112.

Treleaven, R.B., (1977). *Peregrine: the private life of the Peregrine Falcon*, Headline Publications, Penzance.

Trollope, C.E., (2012). Hobbies take advantage of the gravy train, *Brit. Birds*, 105, 221.

Tucker, V.A., (1968). Respiratory exchange and evaporative water loss in the flying Budgerigar. *J. Exp. Biol.*, 48, 67-87.

Tucker V.A., (1972). Metabolism during flight in the Laughing Gull (*Larus atricilla*), *Am. J. Physiology*, 222, 237-245.

Tucker, V.A. and Heine, C., (1990). Aerodynamics of gliding flight in a Harris' Hawk, *Parabuteo unicinctus*, *J. Exp. Biol.*, 149, 469-489.

Tucker, V.A. and Parrott, G.C., (1970). Aerodynamics of gliding flight in a falcon and other birds, *J. Exp. Biol.*, 52, 345-367.

Usherwood, J.R., Cheney, J.A., Song, J., Windsor, S.P., Stevenson, J.P.J., Dierksheide, U., Nila, A. and Bomphrey, R.J., (2020). High aerodynamic lift from the tail reduces drag in gliding raptors, *J. Exp. Biol.* 223, jeb214809.

Usui, T., Butchart, S.H.M. and Phillimore, A.B., (2017). Temporal shifts and temperature sensitivity of avian spring migratory phenology: a phylogenetic meta-analysis, *J. Anim. Ecol.*, 86, 250-261.

van Boekel, W., (2019). Food, caching and out-of-season copulations of Kestrels *Falco tinnunculus*, *De Takkeling*, 27, 172-177.

Vergara, P. and Fargallo, J.A., (2008). Sex, melanic coloration, and sibling competition during the postfledging dependence period, *Behav. Ecol.*, 19, 847-853.

Vergara, P., Fargallo, J.A. and Martínez-Padilla, J., (2010). Reaching independence: food supply, parent quality, and offspring phenotypic characters in Kestrels, *Behav. Ecol.*, 21, 507-512.

Vickery, J.A., Tallowin, J.R., Feber, R.E., Asteraki, E.J., Atkinson, P.W., Fuller, R.J. and Brown, V.K., (2001). The management of lowland neutral grasslands in Britain: effects of agricultural practices on birds and their food resources, *J. Appl. Ecol.*, 38, 647-664.

Vickery, J.A., Carter, N. and Fuller, R.J., (2002). The potential value of managed cereal field margins as foraging habitats for farmland birds in the UK, *Agr. Ecosyst. Environ.*, 89, 41-52.

Vickery, J.A., Bradbury, R.B., Henderson, I.G. Eaton, M.A. and Grice, P.V., (2004). The role of agri-environment schemes and farm management practices in reversing the decline of farmland birds in England, *Biological Conservation*, 119, 19-39.

Vickery, J.A., Feber, R.E. and Fuller, R.J., (2009). Arable field margins managed for biodiversity conservation: A review of food resource provision for farmland birds, *Agr. Ecosyst. Environ.*, 133, 1-13.

Vidal. A., Baldomà, L., Molina-López, R.A., Martin, M. and Darwich, L., (2017). Microbiological diagnosis and antimicrobial sensitivity profiles in diseased free-living raptors, *Avian Pathology*, 46, 442-450.

Videler, J.J., (2005). *Avian Flight*, Oxford University Press, Oxford.

Videler, J.J. and Groenewold, A., (1991). Field measurements of hanging flight aerodynamics in the Kestrel (*Falco tinnunculus*), *J. Exp. Biol.*, 155, 519-530.

Videler, J.J., Weihs, D. and Daan, S., (1983). Intermittent gliding in the hunting flight of the Kestrel *Falco tinnunculus*, *J. Exp. Biol.*, 102, 1-12.

Videler, J.J., Vossebelt, G., Gnodde, M. and Groenewegen, A., (1988a). Indoor flight experiments with trained Kestrels: I. Flight strategies in still air with and without weights, *J. Exp. Biol.* 134, 173-183.

Videler, J.J., Groenewegen, A., Gnodde, M. and Vossebelt, G., (1988b). Indoor flight experiments with trained Kestrels: II. The effect of added weigh on flapping flight kinematics, *J. Exp. Biol.* 134, 185-199.

Viitala, J., Korpimäki, E., Palokangas, P. and Koivula, M., (1995). Attraction of Kestrels to vole scent marks visible in ultraviolet light, N*ature*, 373, 425-427.

Village, A., (1980). The ecology of the Kestrel (*Falco tinnunculus*) in relation to vole abundance at Eskdalemuir, south Scotland, PhD thesis, University of Edinburgh.

Village, A., (1982). The home range and density of Kestrels in relation to vole abundance, *J. Anim. Ecol.*, 51, 413-428.

Village, A., The role of nest-site availability and territorial behaviour in limiting the breeding density of Kestrels, *J. Anim. Ecol.*, 52, 635-645, 1983b.

Village, A., (1983a)Seasonal changes in the hunting behaviour of Kestrels, *Ardea*, 71, 117-124

Village, A., (1985). Spring arrival times and assortative mating of Kestrels in south Scotland, *J. Anim. Ecol.*, 54, 857-868.

Village, A., (1990). *The Kestrel*, T&AD Poyser, London.

Waba, V. and Grüll, A., (2009). (Common Kestrels *Falco tinnunculus* adopt Barn Owls *Tyto alba* breeding in the wild), *Vogelwelt*, 130, 201-204.

Walker, I., (2012). More observations of an association between falcons and steam trains, *Brit. Birds*, 105, 626.

Walker, L.A., Chaplow, J.S., Llewellyn, N.R., Pereira, M.G., Potter, E.D., Sainsbury, A.W. and Shore, R.F., (2013). *Anticoagulant rodenticides* in Predatory Birds 2011: a Predatory Bird Monitoring Scheme (PMBS) report and Addendum, Centre for Ecology and Hydrology, Lancaster Environment Centre, Lancaster.

Wallin, K., Wallin, M.L., Jaras, T., and Strandvik, P., (1987). Leap-frog migration in the Swedish Kestrel *Falco tinnunculus* population. *Proceedings of the Fifth Nordic Ornithological Congress, 1985*, 213-22.

Wallis, R. J., (2017). 'As the Falcon her Bells' at Sutton Hoo? Falconry in Early Anglo-Saxon England, *Arch. J.*, 174, 409-436.

Wallis, R.J. (2020). The origins of falconry in England, *The Falconer*, 65-78.

Walls, s. and Kenward, R., (2020). *The Common Buzzard*, T&AD Poyser, London.

Walpole-Bond, J., (1938). *A History of Sussex Birds*, Witherby, London.

Warkentin, I.G., James, P.C. and Oliphant, L.W., (1992). Assortative mating in urban-breeding Merlins, *Condor*, 94, 418-426.

Wassman, R., (1993). (The Kestrel (*Falco tinnunculus*) as a ground nesting bird of open fields), *Egretta*, 36, 40-41.

Watson, R.T., Kolar, P.S., Ferrer, M., Nygård, T., Johnston, N., Hunt, W.G., Smit-Robinson, H.A., Farmer, C.J., Huso, M. and Katzner, T.E., (2013). Raptor interactions with wind energy: case studies from around the world, *J. Raptor Res.*, 52, 1-18.

Welsh Bird Report 2017, (2018). *Birds in Wales*, 15 (3).

Wernham, C., Toms, M., Marchant, J., Clark, J., Siriwardena, G. and Baillie, S. (eds), (2002). *The Migration Atlas: Movements of the birds of Britain and Ireland*, T&AD Poyser, London (reprinted 2008).

Wiebe, K.L., Korpimäki, E. and Wiehn, J., (1998a). Hatching asynchrony in Eurasian Kestrels in relation to the abundance and predictability of cyclic prey, *J. Anim. Ecol.*, 67, 908-917.

Wiebe, K.L., Wiehn, J., and Korpimäki, E., (1998b). The onset of incubation in birds: can females control hatching patterns?, *Anim. Behav.*, 55, 1043-1052.

Wiebe, K.L, Jönsson, K.I., Wiehn, J. and Hakkarainen, H., (2000). Behaviour of female Eurasian Kestrels during laying: are there time constraints on incubation?, *Ornis Fennica*, 77, 1-9.

Wiehn, J. and Korpimäki, E., (1997). Food limitation on brood size: Experimental evidence in the Eurasian Kestrel, *Ecology*, 78, 243-2050.

Wiehn, J., Ilmonen, P., Korpimäki, E., Pahkala, M. and Wiebe, K.L., (2000). Hatching asynchrony in the Eurasian Kestrel *Falco tinnunculus*: an experimental test of the brood reduction hypothesis, *J. Anim. Ecol.*, 69, 85-95.

Wiklund, C.G., (1995). Nest predation and Life-span: Components of variance in LRS among Merlin females, *Ecology*, 76, 1994-1996.

Wiklund, C.G. and Village, A., (1992). Sexual and seasonal variation in territorial behaviour of Kestrels (*Falco tinnunculus*), *Anim. Behav.* 43, 823-830.

Wildman, L., O'Toole, L. and Summers, R.W., (1998). The diet and foraging behaviour of the Red Kite in Scotland, *Scottish Birds*, 19, 134-140.

Wink, M. and Sauer-Gürth, H., (2000). Advances in the molecular systematics of African raptors, in Chancellor, R.D. and Meyburg, B-U. (eds), *Raptors at Risk*, World Working Group on Birds of Prey and Owls/Hancock House, 135-147.

Wink, M., Sauer-Gürth, H., El-Sayed, A.A and Gonzalez, J., (2007). (A look through the magnifying glass of genetics: birds of prey from the DNA perspective), in *Greifvögel und Falknerei 2005/2006* (Yearbook of the German Falcon Order), J, Neumann-Neudamm Verlag, Melsungen.

Won, I.J., Park, M.C., Park, H.D. and Cho, S.R., (2016). (A study on selection of Common Kestrels (*Falco tinunculus*) hunting areas in the breeding season), *J. Wetlands Res.* 18, 350-356.

Wright, A.A., (1972). The influence of ultraviolet radiation on the pigeons' color discrimination, *J. Exp. Analysis Behav.*, I7, 325-337.

Wright, T.F., Schirtzinger, E.E., Matsumoto, T., Eberhard, J.R., Graves, G.R., Sanchez, J.J., Capelli, S., Müller, H., Scharpegge, G.K., Chambers, G.K. and Fleischer, R.C., (2008). A multi-locus molecular phylogeny of the parrots (Psittaciformes): support for a Gondwanan origin during the Cretaceous, *Molecular Biology and Evolution*, 25, 2141-2156.

Yalden, D.W.,(1980). Notes on the diet of urban Kestrels, *Bird Study*, 27, 235-238.

Zampiga, E., Gaibani, G., Csermely, D., Frey, H. and Hoi, H., (2006). Innate and learned aspects of urine UV-reflectance use in hunting behaviour of the Common Kestrel *Falco tinnunculus*, *J. Avian Biol.*, 37, 318-322.

Zampiga, E., Gaibani, G. and Csermely, D., (2008). Ultraviolet reflectance and female mating preferences in the Common Kestrel (*Falco tinnunculus*), *Can. J. Zool.*, 86, 479-483.

Zellweger-Fischer, J., Schaub, M., Müller, C., Rudin, M., Spiess, M. and Jenni, L., (2011). Reproductive success in Common Kestrels *Falco tinnunculus*: results from integrated population monitoring over five years, *Ornithol. Beob.*, 108, 37-54.

Zuberogoitia, I., Martínez, J.A., González-Oreja, J.A., Calvo, J.E. and Zabala, J., (2013). The relationship between brood size and prey selection in a Peregrine Falcon population located in a strategic region on the Western European Flyway, *J. Orn.*, 154, 73-82.

Index

Barn Owl, 15, 48, 68, 189, 196, 197, 234, 235, 268, 269, 287-288, 293, 311, 329, 332, 357-358
***Beast from the East* (2018),** 288-289
Beech Marten, 203
***Book of St Albans*,** 27-28
Boreal Owl, 49, 188, 259-260
Brown Hare, 57
Brown Rat, 306, 316, 357-358
Budgerigar, 63-64

Canary, 63
Carrion (in Kestrel diet), 67
Carrion Crow, 194
Common Buzzard, 20, 194, 333, 334, 359-360
Common Frog, 306
Common or Eurasian Kestrel
   age at first breeding, 166-169
   amphibian prey of, 60-61
   attacks on other species 333-334
   avian prey of, 61-67, 72
   breeding density, 183-192
      dependence on nest site availability, 190-191
      dependence on prey density, 185-186, 187-189
      variation with local temperature, 185, 187, 189-190
   breeding failures,
      reasons for (egg/chick losses), 254-257
   breeding success, 254-258, 262
      effect on site fidelity, 263
   cannibalism, 317
   causes of death, 317-318, 345
      chemical contamination (including pesticides/rodenticides), 354
      collisions, 348-349
      wind farms, 348-351
      disease, 349, 351-352
      parasites, 352
      poisoning, 348-349
      shooting, 348-349
      starvation, 317-318, 345, 348, 351
   chick growth, 235-253
      development of feathers and behaviour, 246-248
      fledging and dispersal, 250-253, 320-324
      graphs of growth, 236, 248
      pre-hatch chick communication, 238
      prey deliveries to chicks, 237-239, 240-241
   chick rearing, 277, 298, 315-318, 320
   clutch size, 214-217
      variation with laying date of first egg, 223
      effect of clutch size on adult survival, 225
      second clutches, 227, 229
      colonial breeding 190-192

   copulation, 183, 204-205, 292
   courtship, 177-179, 183, 287
   derivation of name, 25-26
   diet, 54-80
      seasonal variation of, 74-75
      dietary breadth across the range, 80-81
      dietary difference between males and females, 78
      dietary difference between adults and juveniles, 78-79
      dietary difference urban and country Kestrels, 78-80
   dimensions, 45
   displays, 176-177
   egg hatching, 229-234
   egg laying, 206-215, 218n 288-289
      period/interval, 218-219, 288, 293, 311-312
      timing of, 206-215
      influence of prey availability on, 207-210
      influence of daylight hours on, 210-212
      influence of weather on, 213, 215
   eggs, 219-221
      weight loss during incubation, 222
      egg dumping, 222
   energy balance, 88-89, 150-159
   energy consumption, variation through year, 153, 157
   feather weights, 43
   fidelity (nest, territory, mate), 258-264
   flicker-fusion frequency (FFF) of, 114
   flight
      characteristics, 90-112
      cruising speed, 104-108
      effect of attaching IMUs on, 98-99
      flapping flight, 95-112
      hovering' see flight-hunting,
      migration, 110-111
      position of bill, tail and wings during, 108-109
      soaring and gliding, 109-110
      take-off, 102-104
      wing beat frequency, 95-102
      wing loading, 92-93
   flight-hunting
      body attitude during, 127-128
      eye/head position during, 131-136
      flight and, 122
      local topography and, 127
      prey deliveries to chicks, (see above)
      vision and, 113-119
      wind speed and, 124-126
   food caching, 83-88
   gastroliths, 55
   habitat, 52
   home range see territory/home range

hunting,
    effect of environmental factors
    effect of habitat, 147-148
    effect of weather conditions, 146
    energy cost of, 140-142
    timing of, 81-83
    strategies, 81-83, 137
    strategy variation with season, 140-141
    success of, 142-144
incest, 176, 259
incubation, 226-227. 229-235, 271-275, 289-290, 293-294, 312-314
    female brood patch, 230
    male brood patch, 229
insect prey of, 59, 72-73
life expectancy, 342-345
mammal prey of, 57-58, 60, 67, 70-73
mate selection, 172-176
migration, 324-327
    differential migration by sex and age, 326-327
    effect of climate change on, 329
moult,
    sequence and timing, 10-11, 40-45
    energy requirements of, 43
nest predation, 254-255
nest sites, 192-204
    choice of, 192-193
    ground nesting, 198
non-migratory populations, 324-326
pair formation, 179-183
parasites of, 41, 173, 234-235, 352-354
pellets, 54-55
plumage,
    adults, 35-37, 174
    juvenile, 38-40, 174, 250-251
    nestling, 38
polygyny, 179-183
population, 355-373
    effect of potential predators on, 359-363
    effect of changes in agriculture on, 367-373
post breeding dispersal, 320-324
prey deliveries to nest, 239, 271, 277, 294-295, 302-303, 314, 318
    effect of weather on, 280, 304-305
    fractions by prey type, 281-282, 306
    timing of, 277, 303, 315
    weight of, 282
    conversion to fledgling Kestrels, 286, 306-307
prey preparation, 54
relationship with Curlews, 330
replacement clutches, 227-229
reptile prey of, 59, 60 -61, 307
Reverse Sexual Size Dimorphism of, 46-51
second clutches, 227, 229

sex ratio of chicks, 239-240
survival, 225, 340-354
sub-species, 19-20
Tasty Chick Hypothesis, 234-235
territory/home range, 170-172
    variation with season, 172, 185, 186
    variation with prey density, 185-186, 187-189
time energy budget 163-165
use of traditional nest sites, 194-196, 201, 263-264
UV vision of, 115-119, 174
voice, 52-53
weight, 43, 45-46
    of body parts, 46
    of feathers, 43
    variation with time of day, 88
    variation with time of year, 158-159
wintering grounds, 324
winter population density, 328-329
Courtship feeding, 158, 178
Curlew, 330

Dame Juliana Barnes, 27

Eagle Owl, 259-260, 334
Earthworms (as Kestrel prey), 74

Falcon, derivation of name, 26
Falconidae, origins of, 10-12
Falconry
    origins of, 26-27
    literature on, 27-28
Falcons,
    relationship to hawks, 10
    method of killing, 32-33n 59
    'True Falcons', 12-19
    general characteristics, 29-33
    use of live prey, 33
Food caching, 83-88

Gastroliths, 55
Geraint Thomas, 156
Golden Eagle, 334, 364-367
Green Woodpecker, 196
Griffon Vulture, 350
Grey Squirrel, 194
Goshawk, 334, 359, 362-4

Harleian Manuscripts, 28
Hierofalcons, 12
Hobbies, 15
Hopkins, Gerard Manley, 20

Inertial Measuring Units (IMUs), 97-99, 131-132

Jackdaw, 63, 196, 275-277, 281, 286, 298, 306, 317

Kleptoparasitism, 67-69

Life expectancy, 342-345
Lifetime reproductive success (LRS), 167-168

Kestrels
how to define, 19
position among 'True Falcons', 15-17

Magpie, 68, 194, 333
Mice, 73, 74, 282
Mole, 58

Neonicotinoids, 348

Organochlorine contamination, 348, 354, 357

Parrots, 10-11
Pellets, 54-55
Peregrines, 12-15, 334
Peregrine Falcon, 22, 91, 330, 331-332, 359, 364
Population, 355-373
effect of potential predators on, 359-363
effect of changes in agriculture on, 367-373

Rabbit, 57, 282, 306
Red Kite, 334-335, 357-358, 364
Reverse Sexual Size Dimorphism (RSD), 46-51

Short-eared Owl, 68, 122, 187, 188, 362-363, 364
Slow Worm, 60, 304-305, 306, 317
Snowy Owl, 187, 188, 330
Stock Dove, 196, 311, 318
Sparrowhawk, 61, 68, 91, 194, 334, 362
Starling, 61, 153, 190, 335-339

Tawny Owl, 20, 196, 236, 256, 293, 334, 362
Tengmalm's Owl, see Boreal Owl
Third birds at nest, 168, 169, 291
Tomia, 32
'True Falcons', 12-19

Ural Owl, 201, 259-260
Usain Bolt, 156

Voles, 70-71, 74, 153, 198, 282, 358-359
time taken to consume, 151

Weasels, 33, 58, 72, 282
Windhover, The, 24
Wood Pigeons, 194, 269, 277, 293, 295-297

FALCO (CERCHNEIS) TINNUNCULUS LIN. $\tfrac{1}{10}$
Tornfalk ♀